CONTEMPORARY MATHEMATICS

Titles in this Series

Titles in this Series

Volume

28 **Fluids and plasmas: geometry and dynamics,** Jerrold E. Marsden, Editor

29 **Automated theorem proving,** W. W. Bledsoe and Donald Loveland, Editors

30 **Mathematical applications of category theory,** J. W. Gray, Editor

31 **Axiomatic set theory,** James E. Baumgartner, Donald A. Martin and Saharon Shelah, Editors

32 **Proceedings of the conference on Banach algebras and several complex variables,** F. Greenleaf and D. Gulick, Editors

33 **Contributions to group theory,** Kenneth I. Appel, John G. Ratcliffe and Paul E. Schupp, Editors

34 **Combinatorics and algebra,** Curtis Greene, Editor

35 **Four-manifold theory,** Cameron Gordon and Robion Kirby, Editors

36 **Group actions on manifolds,** Reinhard Schultz, Editor

37 **Conference on algebraic topology in honor of Peter Hilton,** Renzo Piccinini and Denis Sjerve, Editors

38 **Topics in complex analysis,** Dorothy Brown Shaffer, Editor

39 **Errett Bishop: Reflections on him and his research,** Murray Rosenblatt, Editor

40 **Integral bases for affine Lie algebras and their universal enveloping algebras,** David Mitzman

41 **Particle systems, random media and large deviations,** Richard Durrett, Editor

Particle Systems, Random Media and Large Deviations

CONTEMPORARY MATHEMATICS

Volume 41

Particle Systems, Random Media and Large Deviations

Richard Durrett, Editor

AMERICAN MATHEMATICAL SOCIETY

Providence · Rhode Island

PROCEEDINGS OF THE AMS-IMS-SIAM JOINT SUMMER RESEARCH CONFERENCE IN THE MATHEMATICAL SCIENCES ON MATHEMATICS OF PHASE TRANSITIONS

HELD AT BOWDOIN COLLEGE
JUNE 24–30, 1984

These proceedings were prepared by the American Mathematical Society with partial support from the National Science Foundation Grant DMS-8218075.

1980 *Mathematics Subject Classification.* Primary 60K35.

Library of Congress Cataloging in Publication Data

AMS-IMS-SIAM Joint Summer Research Conference in the Mathematical Sciences on Mathematics of Phase Transitions (1984: Bowdoin College)
Particle systems, random media, and large deviations.

(Contemporary mathematics, ISSN 0271-4232; v. 41)
"Proceedings of the AMS-IMS-SIAM Joint Summer Research Conference in the Mathematical Sciences on Mathematics of Phase Transitions, held at Bowdoin College, June 24–30, 1984"–T.p. verso.
"Prepared by the American Mathematical Society"–T.p. verso.
Bibliography: p.
1. Stochastic processes–Congresses. 2. Dynamics of particle–Congresses. 3. Phase transformations (Statistical physics)–Congresses. I. Durrett, Richard, 1951– II. American Mathematical Society. III. Institute of Mathematical Statistics. IV. Society for Industrial and Applied Mathematics. V. Title. VI. Series: Contemporary mathematics (American Mathematical Society; v. 41)
QA274.A1A47 1984 519.2 85-6181
ISBN 0-8218-5042-3 (alk. paper)

TABLE OF CONTENTS

INTRODUCTION

This volume contains the proceedings (in a generalized sense) of the AMS Summer Research Conference on "The Mathematics of Phase Transitions" held at Bowdoin College June 24-30, 1984. The papers appearing here are based on talks which were (or should have been) given at that meeting and to quote the instructions given to the authors, they are intended to "be a friendly introduction to the subject under consideration which makes the main results and open problems accessible to nonexperts." A look at the table of contents should reveal that the twenty some odd papers cover a number of different aspects of "Particle Systems, Random Media, and Large Deviations", three topics which I consider to be among the most exciting areas in probability theory today. I hope that the reader will find the papers as pleasant and enjoyable as I found the conference to be, and I especially hope that some of you will become interested in these topics and solve some of the many open problems described below. Finally, I would like to thank the people who made all this possible: the National Science Foundation for its financial support, the American Mathematical Society for allowing me to organize the conference and publish the proceedings, and above all Carole Kohanski for the superb job she did from beginning to end keeping things organized and running smoothly.

Rick Durrett

Contemporary Mathematics
Volume 41, 1985

INEQUALITIES FOR CRITICAL PROBABILITIES IN PERCOLATION

Massimo Campanino

ABSTRACT. We examine methods of comparison between critical percolation probabilities in two cases : a) planar graphs related by inclusion ; b) graphs related by identification of sites.

1. INTRODUCTION. Percolation theory deals with the study of geometrical properties of random systems and in particular with the distribution of clusters. Since the paper of T. E. Harris ([2]) appeared, there has been remarkable progress in the understanding of two-dimensional systems, whereas most of the important problems are still open in higher dimensions.

One of the interesting objects in percolation theory are critical probabilities of specific models, i.e. the probabilities for the occupation of single sites or bonds above which infinite clusters start to arise. In some planar graphs with particular symmetry properties the critical probabilities can be exactly computed. This is in general not possible and a natural problem is to find methods of comparing critical probabilities in different graphs.

It is quite intuitive that the critical probability of a given model strictly decreases if we add, in a periodic way, new connections to a periodic graph. A first result in this direction was obtained by Y. Higushi ([3]) who proved that the critical probability for site percolation on the square lattice with connection is strictly less than that on the square lattice with nearest neighbour connection. The conjecture has then been proved by H. Kesten ([5]) under wide conditions for planar graphs.

1980 Mathematics Subject Classification. Primary 60K35, secondary 82A67.

Another relation between graphs is the following : a graph can be obtained from another one by identification of some of its sites. In the paper [1] it is proved that a situation of this kind leads under additional conditions to a weak inequality between the critical probabilities of the two graphs. This result was used in the same paper to prove that the critical probability for site percolation on the three-dimensional cubic graph is strictly less than 1/2.

In Section 2 we give the essential definitions. In Section 3 we expose the method of Kesten with some simplification in a particular case. In section 4 we prove a general inequality essentially contained in [1] between critical probabilities of graphs related by identification of sites.

2. DEFINITIONS. Given a graph $\mathcal{G}$, the set of its sites is denoted by $S(\mathcal{G})$, the fact that two sites $x,y \in S(\mathcal{G})$ are adjacent in $\mathcal{G}$, i. e. joined by a bond of $\mathcal{G}$ is denoted by

$$x \overset{\mathcal{G}}{\longleftrightarrow} y \qquad . \tag{2.1}$$

A chain of $\mathcal{G}$ is a sequence $(x_1,\dots,x_n)$ of points in $S(\mathcal{G})$ such that

$$x_i \overset{\mathcal{G}}{\longleftrightarrow} x_{i+1} \tag{2.2}$$

for $1 \le i \le n-1$. Supp(c), the support of the chain $c=(x_1,\dots,x_n)$, is the set $\{x_1,\dots,x_n\}$. The chain c is called self-avoiding if $x_i \neq x_j$ for $i \neq j$.

The set of configurations on the sites of the graph $\mathcal{G}$ is $\Omega_{S(\mathcal{G})} = \{-1,1\}^{S(\mathcal{G})}$. Given a configuration $\omega \in \Omega_{S(\mathcal{G})}$, a chain c is a +chain (−chain) in ω if $\omega_x = 1$ (-1) for every $x \in \mathrm{Supp}(c)$.

Given a configuration $\omega \in \Omega_{S(\mathcal{G})}$ and $x \in S(\mathcal{G})$, the +cluster $W_+^{\mathcal{G}}(x)$ is the random set defined by

$$W_+^{\mathcal{G}}(x) = \bigcup_{\substack{c \text{ +chain} \\ x \in \mathrm{Supp}(c)}} \mathrm{Supp}(c) \tag{2.3}$$

The −cluster $W_-^{\mathcal{G}}(x)$ is defined in an analogous way. It is easy to see that the union in (2.3) can be restricted to the self-avoiding

+chains starting from x.

Let $x \in S(\mathcal{G})$ and define

$$p_c^{\mathcal{G}}(x) = \inf \left| p \,/\, 0 \le p \le 1,\ \mu_p^{S(\mathcal{G})}(\,|W_+^{\mathcal{G}}(x)| = \infty) > 0 \right| , \qquad (2.4)$$

where $\mu_p^{S(\ )}$ is the Bernoulli measure with parameter p on $\Omega_{S(\ )}$. We shall consider graphs on which $p_c^{\mathcal{G}}(x)$ does not depend on x. The constant value of $p_c^{\mathcal{G}}(x)$ is then denoted by $p_c^{\mathcal{G}}$ and is called the critical probability for percolation on $\mathcal{G}$ (see[5]).

Given a configuration $\omega \in \Omega_{S(\mathcal{G})}$ and an event A, x is called a pivotal site for A in ω if

$$\chi_A(\omega) \neq \chi_A(\tau_x \omega),$$

where $\tau_x\omega$ is the configuration obtained from ω by changing the value in x and χ_A is the indicator of A.

3. STRICT INEQUALITIES FOR PLANAR GRAPHS. In order to illustrate the method of Kesten we consider an example containing the essential features of the general case : the comparison of the critical probabilities for percolation and * percolation on the square lattice.

Let $\mathcal{G}_1$ and $\mathcal{G}_2$ be the graphs defined by

$$S(\mathcal{G}_1) = S(\mathcal{G}_2) = Z^2$$

$$x \overset{\mathcal{G}_1}{\longleftrightarrow} y \quad \text{if } |x_1-y_1| + |x_2-y_2| = 1$$

$$x \overset{\mathcal{G}_2}{\longleftrightarrow} y \quad \text{if } \max(|x_1-y_1|, |x_2-y_2|) = 1$$

It is useful to translate the problem of percolation on the graphs $\mathcal{G}_1$ and $\mathcal{G}_2$ into the problem of percolation on a single graph $\mathcal{Z}$ with two different probability measures. The graph is defined by

$$S(\mathcal{Z}) = Z^2 \cup ((\tfrac{1}{2}, \tfrac{1}{2}) + Z^2)$$

$$x \overset{\mathcal{Z}}{\longleftrightarrow} y \text{ if } |x_1-y_1| = |x_2-y_2| = \tfrac{1}{2} \text{ or}$$

$$x, y \in Z^2 \text{ and } |x_1-y_1| + |x_2-y_2| = 1.$$

We introduce on $\Omega_{S(\mathcal{Z})}$ a two parameter family of probability measures $\mu_{p,\alpha}$, $0 \le p \le 1$, $0 \le \alpha \le 1$. $\mu_{p,\alpha}$ is a Bernoulli measure with parameter p on the sites of Z^2 and with parameter α on the of $Z^2+(\frac{1}{2},\frac{1}{2})$. It is easy to see that percolation on the graphs $\mathcal{G}_1$ and $\mathcal{G}_2$ with the measure $\mu_p^{Z^2}$ is equivalent to percolation on $\mathcal{T}$ with measures $\mu_{p,0}$ and $\mu_{p,1}$ respectively.

For L a positive integer let

$$R_L = [0,L]\times[0,3L] .$$

Let A_L be the event on $\Omega_{S(\mathcal{T})}$ defined by

A_L=(there exists a +chain of $\mathcal{T}$ with support contained in R_L connecting the two shorter sides of R_L).

It is easy to check that A_L^c, the complementary event of A_L, can be expressed as

A_L^c=(there exists a −chain of $\mathcal{T}$ with support contained in R_L connecting the two longer sides of R_L).

One can exploit the following results of percolation theory that are valid for planar graphs satisfying suitable conditions and are true in particular for the graph $\mathcal{T}$:

a) In order to show that

$$\mu_{p,\alpha}\,(\,|W_+^{\mathcal{T}}(x)| = \infty) > 0 \tag{3.1}$$

for some p and α it is enough to prove that for some L

$$\mu_{p,\alpha}\,(A_L) \geq c, \tag{3.2}$$

where c is a constant, $0<c<1$, independent of L ([6], see Th. 5.4 of [5]).

b) From a formula of L. Russo ([6] , see Prop. 4.2 of [5]) we have

$$\frac{d}{d\alpha}\,\mu_{p,\alpha}\,(A_L) = -\frac{d}{d\alpha}\,\mu_{p,\alpha}\,(A_L^c) = E_{\mu_{p,\alpha}}\,(\tilde{N}_L) \geq \tag{3.3}$$

$$\geq E_{\mu_{p,\alpha}}\,(\tilde{N}_L \mid A_L^c)\quad \mu_{p,\alpha}\,(A_L^c),$$

where E_μ denotes the expectation with respect to the measure μ and $\tilde{N}_L$ denotes the number of pivotal sites for the event A_L (or equivalently for A_L^c) that belong to $(\frac{1}{2},\frac{1}{2})+Z^2$.

c) For $p \geq p_c^{\mathcal{G}_1}$, uniformly in α , $0 \leq \alpha \leq 1$

$$\lim_{L\to\infty} E_{\mu_{p,\alpha}}\,(N_L \mid A_L^c) = \infty \quad . \tag{3.4}$$

If δ_x^L is the event that the site x is pivotal for A_L, we have

$$E_{\mu_{p,\alpha}}(N_L \mid A_L^c) = \sum_{x\in R_L} \mu_{p,\alpha}(\delta_x^L \mid A_L^c) \tag{3.5}$$

$$E_{\mu_{p,\alpha}}(\tilde{N}_L \mid A_L^c) = \sum_{x\in R_L\cap((\frac{1}{2},\frac{1}{2})+Z^2)} \mu_{p,\alpha}(\delta_x^L \mid A_L^c) \tag{3.6}$$

One can prove that for each α with $0<\alpha<1$ there is a positive constant K_α such that for every $x\in Z^2\cap R_L$ there is y, with $y\in R_L\cap((\frac{1}{2},\frac{1}{2})+Z^2)$ and $|x-y|=1$, verifying

$$K_\alpha\, \mu_{p,\alpha}(\delta_y^L \mid A_L^c) \ge \mu_{p,\alpha}(\delta_x^L \mid A_L^c) \tag{3.7}$$

The constant K_α is independent of L and can be bounded uniformly for α in any interval $[\alpha_1,\alpha_2]\subset(0,1)$. In order to prove (3.7) we take $x\in Z^2\cap R_L$ pivotal site for the event A_L^c in the configuration ω. Let Q be a square with side 6 and center in x. Assume for the moment that Q is contained in the interior of R_L. Then it follows from the definition of pivotal site that in the configuration ω there must be two −chains contained in $R_L\cap Q$ joining ∂Q with the two longer sides of R_L and between them two +chains also contained in $R_L\cap Q$ joining ∂Q with the two shorter sides of R_L. Then it is not difficult to construct a configuration ω' equal to ω in $R_L\cap Q$ and such that a site $y\in(\frac{1}{2},\frac{1}{2})+Z^2$ and adjacent to x is pivotal for A_L^c in ω'. The inequality (3.7) follows since we need only to change ω in the sites of Q to obtain ω' and, for $0<\alpha<1$, the change in probability can be bounded by a finite multiplicative constant. The case in which Q is not contained in the interior of R_L can be dealt with in a similar way.

If we take $\bar{p}=p_c{}^1$, then

$$\mu_{\bar{p},1}(A_L^c) \le \mu_{\bar{p},0}(A_L^c)\exp\left(\int_{\alpha_0}^{\alpha_1} \frac{1}{\mu_{\bar{p},\alpha}(A_L^c)} \frac{d}{d\alpha}\mu_{\bar{p},\alpha}(A_L^c)\,d\alpha\right) \le \tag{3.8}$$

$$\le \exp\left(-(\alpha_1-\alpha_0)\inf_{\alpha_0\le\alpha\le\alpha_1} E_{\mu_{\bar{p},\alpha}}(N_L \mid A_L^c)\right) \le \exp(-(\alpha_1-\alpha_0)K_L)$$

with $K_L\to\infty$ as $L\to\infty$, where we have used (3.4), (3.5), (3.6), (3.7). By taking L sufficiently large we can therefore obtain that

$$\mu_{\bar{p},1}(A_L)=1-\mu_{\bar{p},1}(A_L^c)\geq c+(1-c)/2 \ . \tag{3.9}$$

Since A_L is an event that depends only on the values of the configuration on a finite number of sites, its probability depends continuously on p and we can find a $\tilde{p}<p$ such that

$$\mu_{\tilde{p},1}(A_L)\geqslant c \ . \tag{3.10}$$

Therefore, using a),.we obtain

$$p_c^{\mathcal{G}_2} < p_c^{\mathcal{G}_1} \ . \tag{3.11}$$

The proof above can be extended to planar periodic graphs under wide conditions since we have not essentially used any particular property of the graph $\mathcal{T}$. The requirement for the graph to be planar cannot be loosened, since the argument relies on the validity of a), b) and c).

4. IDENTIFICATION OF SITES. Given a graph $\mathcal{G}$ and a partition $\mathcal{J}$ of $S(\mathcal{G})$, we can construct a new graph denoted by $\mathcal{G}/\mathcal{J}$ by identifying sites that belong to the same element of $\mathcal{J}$. Formally

$S(\mathcal{G}/\mathcal{J})=\mathcal{J}$ and for I, $I'\in\mathcal{J}$ $I \overset{\mathcal{G}/\mathcal{J}}{\leftrightarrow} I'$ if and only if there are $x\in I$ and $x'\in I'$ such that $x \overset{\mathcal{G}}{\leftrightarrow} x'$.

Let π denote the natural projection of $S(\mathcal{G})$ onto $\mathcal{J}$. We say that the graph $\mathcal{G}$ and the partition $\mathcal{J}$ satisfy condition C if the following is true :

For every self-avoiding chain $\bar{c}=(y^{(1)},\dots,y^{(n)})$ in $\mathcal{G}/\mathcal{J}$ and for $x\in S(\mathcal{G})$ such that $\pi(x)=y^{(1)}$, there exists a self-avoiding chain $c=(x^{(1)},\dots,x^{(n)})$ in $\mathcal{G}$ such that $x=x^{(1)}$ and $\pi(x^{(i)})=y^{(i)}$ for $i=1,\dots,n$.

Example :

Let $\mathcal{G}$ be the graph defined by

$S(\mathcal{G})=Z^3$ and $x=(x_1,x_2,x_3) \overset{\mathcal{G}}{\leftrightarrow} y=(y_1,y_2,y_3)$ if and only if $|x_1-y_1|+|x_2-y_2|+|x_3-y_3|=1$.

Let $\mathcal{J}$ be the partition of Z^3 in sets of the form

$$I_{(x_1,x_2)}=\left\{ y\in Z^3 |\ y_1+y_3=x_1,\ y_2+y_3=x_2 \right\}$$

for $(x_1,x_2)\in Z^2$. Then $\mathcal{G}/\mathcal{J}$ is isomorphic to the triangular graph $\mathcal{G}'$ defined by

$S(\mathcal{G}')=Z^2$ and $x \overset{\mathcal{G}'}{\leftrightarrow} y$ if and only if $|x_1-y_1|+|x_2-y_2|=1$ or

$x_1-y_1=x_2-y_2=\pm 1$. It is easy to verify that the graph $\mathcal{G}$ and the partition $\mathcal{J}$ satisfy condition C. It is also easy to construct examples of graphs and partitions that do not satisfy condition C.

We have the following result :

Theorem :

Let the graph $\mathcal{G}$ and the partition $\mathcal{J}$ satisfy condition C. Then the critical probability for percolation on the graph $\mathcal{G}$ is less or equal than that on $\mathcal{G}/\mathcal{J}$.

Proof :

Let $\tilde{x} \in S(\mathcal{G}/\mathcal{J})$ and $\tilde{\mathcal{C}}_n$ be the family of all self-avoiding chains of length n starting from x. For every $\tilde{c} \in \tilde{\mathcal{C}}_n$ let c be the chain in $\mathcal{G}$ starting from $\tilde{x}$ corresponding to $\tilde{c}$ whose existence is insured by condition C and let $\bar{\mathcal{C}}_n$ be the family of such chains. Then we have

$$\mu_p^{S(\mathcal{G}/\mathcal{J})}\Big(\bigcup_{\tilde{c} \in \tilde{\mathcal{C}}_n} \{\omega \mid \tilde{c} \text{ is a +chain}\} \Big)=$$

$$=\mu_{p,\mathcal{J}}^{S(\mathcal{G})}\Big(\bigcup_{c \in \bar{\mathcal{C}}_n} \{\omega \mid c \text{ is a + chain}\} \Big), \tag{4.1}$$

where $\mu_{p,\mathcal{J}}^{S(\mathcal{G})}$ is the probability measure on $\Omega_{S(\mathcal{G})}$ for which the configuration is constant on each $I \in \mathcal{J}$, the values on each I are 1 with probability p and the values on different elements of $\mathcal{J}$ are independent.

At this stage we can assume that $S(\mathcal{G})$ is finite since the events in (4.1) involve the values of the configuration in a finite number of sites. We can therefore consider a sequence of partitions $\mathcal{J}_1,\dots,\mathcal{J}_k$ such that $\mathcal{J}_1=\mathcal{J}$, $\mathcal{J}_k$ is the total partition of $S(\mathcal{G})$ and $\mathcal{J}_{j+1}$ is obtained from $\mathcal{J}_j$ by splitting in two parts one of its elements. Let I be the element of $\mathcal{J}_j$ that is split as $I_1 \cup I_2$ and let

$$\mu_{p,j}=\mu_{p,\mathcal{J}_j}^{S(\mathcal{G})}, \quad E=\bigcup_{c\in\bar{\mathcal{C}}_n} E_+(c), \quad E_0=\bigcup_{\substack{c\in\bar{\mathcal{C}}_n \\ \mathrm{Supp}(c)\cap I=\phi}} E_+(c), \quad E_i=\bigcup_{\substack{c\in\bar{\mathcal{C}}_n \\ \mathrm{Supp}(c)\cap I_i\neq\phi}} E_+(c)$$

for i=1,2, where $E_+(c)$ is the event that c is a +chain.

We have

$$\mu_{p,j+1}(E)=\mu_{p,j+1}(E_0)+\mu_{p,j+1}(E_1)+\mu_{p,j+1}(E_2)-\mu_{p,j+1}(E_0\cap E_1)-$$

$$\mu_{p,j+1}(E_0 \cap E_2) - \mu_{p,j+1}((E_1 \cap E_2) \setminus E_0) . \tag{4.2}$$

We have

$$\mu_{p,j+1}(E_i) = \mu_{p,j}(E_i) \text{ for } i=0,1,2, \tag{4.3}$$

because these events do not involve together I_1 and I_2. For the same reason

$$\begin{aligned} \mu_{p,j+1}(E_0 \cap E_1) &= \mu_{p,j}(E_0 \cap E_1) \\ \mu_{p,j+1}(E_0 \cap E_2) &= \mu_{p,j}(E_0 \cap E_2) \end{aligned} \tag{4.4}$$

On the other side

$$\begin{aligned} \mu_{p,j}((E_1 \cap E_2) \setminus E_0) &= p\, \mu_{p,j}((E_1 \cap E_2) \setminus E_0 \mid \omega_x = 1, \quad x \in I) = \\ &= p\, \mu_{p,j+1}((E_1 \cap E_2) \setminus E_0 \mid \omega_x = 1, \quad x \in I) \geq \\ &\geq p^2 \mu_{p,j+1}((E_1 \cap E_2) \setminus E_0 \mid \omega_x = 1, \quad x \in I) = \mu_{p,j+1}((E_1 \cap E_2) \setminus E_0) \end{aligned} \tag{4.5}$$

Using formulae from (4.2) to (4.5) we get

$$\mu_{p,j}(E) \leq \mu_{p,j+1}(E) \tag{4.6}$$

and

$$\begin{aligned} \mu_p^{S(\mathcal{G}/\mathcal{J})}\Big(\bigcup_{\tilde{c} \in \tilde{\mathcal{C}}_n} E_+(\tilde{c})\Big) &= \mu_{p,1}(E) \leq \mu_{p,k}(E) = \\ &= \mu_p^{S(\mathcal{G})}\Big(\bigcup_{c \in \bar{\mathcal{C}}_n} E_+(c)\Big) \leq \mu_p^{S(\mathcal{G})}\Big(\bigcup_{c \in \mathcal{C}_n} E_+(c)\Big), \end{aligned} \tag{4.7}$$

where $\mathcal{C}_n$ is the family of all the self-avoiding chains of length n starting from x. This completes the proof since

$$\mu_p^{S(\mathcal{G}/\mathcal{J})}(|W_+^{\mathcal{G}/\mathcal{J}}(\tilde{x})| = \infty) = \lim_{n \to \infty} \mu_p^{S(\mathcal{G}/\mathcal{J})}\Big(\bigcup_{\tilde{c} \in \tilde{\mathcal{C}}_n} E_+(\tilde{c})\Big) \tag{4.8}$$

and

$$\mu_p^{S(\mathcal{G})}(|W_+^{\mathcal{G}}(x)| = \infty) = \lim_{n \to \infty} \mu_p^{S(\mathcal{G})}\Big(\bigcup_{c \in \mathcal{C}_n} E_+(c)\Big) \tag{4.9}$$

We note that a particular case of this theorem was used in [1]in combination with the methods of Section 3 for the proof that the critical probability for site percolation on the cubic lattice is strictly less than 1/2. In this case the comparison was between the cubic lattice and the triangular lattice whose critical probability is known to be equal to 1/2 (see[5], Chap.3). In Ref.[7]inequalities related to those discussed in this Section are applied to other problems in percolation and reliability theory. As many questions of percolation theory on non-planar graphs it is an open problem if one can get strict inequalities in general situations like that discussed in this Section.

BIBLIOGRAPHY

1. Campanino M. and Russo L.,"An upper bound on the critical percolation probability for the three-dimensional cubic lattice", to appear in Annals of Probability.

2. Harris T.E., "A lower bound for the critical probability in a certain percolation process", Proc. Cambr. Phil. Soc. 56 (1960),13-20.

3. Higushi Y., "Coexistence of the infinite (✱)clusters- A remark on the square lattice site percolation", Z. Wahrsch. verw. Geb. 61 (1982), 75-81.

4. Kesten H., "The critical probability of bond percolation on the square lattice equals 1/2", Comm. Math. Phys. 74 (1980), 41-59.

5. Kesten H., Percolation Theory for Mathematicians, Birkhäuser, Boston-Basel-Stuttgart, 1982.

6. Russo L., "On the critical percolation probabilities", Z. Wahrsch. verw. Geb. 43 (1981), 229-237.

7. Van den Berg J. and Kesten H., "Inequalities with applications to percolation and reliability", preprint.

DIPARTIMENTO DI MATEMATICA
UNIVERSITA DI ROMA "LA SAPIENZA"
PIAZZALE A. MORO 5
00185 ROMA (ITALY)

Contemporary Mathematics
Volume 41, 1985

RANDOM TUBES AS A MODEL OF PAIR CORRELATIONS

J. T. Chayes[1] and L. Chayes[1]

ABSTRACT. A random surface system, related to the correlation functions of certain models in statistical physics, is studied. Using nonperturbative techniques based on an analysis of the surface entropy, the noncritical behavior is characterized for the range of parameter where the model exists.

1. INTRODUCTION. The following is a summary of some recent work [1,2,3], done in collaboration with D. B. Abraham and reported on by one of us (L.C.) at the A.M.S. conference: The Mathematics of Phase Transitions.

The central topic concerns the statistical behavior of a class of tubular surfaces, the study of which is motivated by the analysis of pair correlations occurring in certain physical systems, as described below. In the course of our investigations, several techniques were developed which may be of wider interest because of a close analogy with some existing probabilistic methods, including renewal theory and the theory of large deviations.

1.1. SURFACE EXPANSIONS: AN OVERVIEW. It is perhaps worthwhile to begin with a brief summary of the successes and failures of random surface theory in the study of phase transitions. The principal physical motivation for analyzing

1980 Mathematics Subject Classification. 60K35, 82A05, 82A25.
[1]J.T.C. and L.C. hold National Science Foundation Postdoctoral Research Fellowships. Work supported in part by the National Science Foundation under Grant No. PHY-82-03669.

random surfaces is that the correlation functions of

1) lattice gauge theories in the strong coupling regime, and

2) three-dimensional spin systems in the low temperature regime

have natural expansions as weighted sums over appropriate classes of surfaces. In other words, generic correlation functions can be expressed in the form

$$G(\beta) = \sum_{\mathbf{s}\in\mathbf{S}} w(\mathbf{s};\beta) , \qquad (1.1)$$

where $\beta\in(0,\infty)$ is a parameter (e.g., the inverse coupling or inverse temperature), $\mathbf{S}$ denotes a particular class of surfaces on the lattice $\mathbb{Z}^d$, and $w(\mathbf{s};\beta)$ is the weight assigned to the surface $\mathbf{s}\in\mathbf{S}$. In general, both $\mathbf{S}$ and $w(\mathbf{s};\beta)$ depend on the theory under consideration, and $\mathbf{S}$ depends also on the particular correlation function under study. For example, the *Wilson loop*, $G_\gamma(\beta)$, is given by summing over all surfaces of the appropriate class which span a fixed loop γ, while the *glueball propagator* (or *truncated two-point function*), $G_L(\beta)$, may be expressed as a sum over surfaces which span two elementary squares (i.e., plaquettes) separated by a distance L. The higher order correlation functions are defined analogously.

It was originally pointed out by Symanzik [4] that the n-point functions of certain mesonic theories (e.g., $\varphi^4{}_d$) have expressions of the form (1.1) with $\mathbf{S}$ replaced by a class of random *paths*. This insight has been the starting point for a number of remarkable results in the analysis of φ^4 theory [5,6]. Unfortunately, similar progress has not yet been made in the random surface approach to the study of lattice gauge theories. The difficulties with this approach are two-fold:

1) The issue of immediate concern is that, even on a formal level, the existing expansions for most theories of interest are intractable. There are problems with both the size of the class of surfaces $\mathbf{S}$, and with the

expressions for the associated weights. As an example of the first point, it turns out that if there is no self-avoiding restriction and no restriction on topology -- as is the case in certain lattice gauge theories -- then the number of surfaces of area A containing a fixed plaquette grows factorially in A [7]. The weights $w(s;\beta)$ are also unmanageable; indeed, in lattice gauge theories with nonabelian continuous symmetry groups, they are not of a definite sign.

Obviously, the above two problems are related and should in principle compensate each other, in the sense that there should exist clever resummations yielding sums over smaller classes of surfaces. Although there has been some progress in this direction [8,9], this approach has not yet produced convergent expansions for the correlation functions of most theories of interest.

2) Ignoring for the moment the above difficulties, there is a more fundamental problem. At the present time, our understanding of random surface phenomena is severely limited. This is in marked contrast to random paths which have been studied for a number of years, and for which there are explicitly solvable examples that form the basis of much of our "physical intuition."

Most of the (admittedly scarce) rigorous analysis of random surfaces has addressed the second issue. However, there is some hope that progress on this will also lead to more manageable expansions. In any case, the work that has been done can be roughly divided into two classes:

1) First there has been an attempt to analyze random surfaces in the context of relatively simple theories, such as plaquette percolation. In this case, expressions such as (1.1) are replaced by the probabilities of certain surface events, with respect to Bernoulli measure on the plaquettes. In a joint endeavor with M. Aizenman, J. Fröhlich and L. Russo [10], we considered the analogue of the Wilson loop for the three-dimensional Bernoulli plaquette system and proved that there is a (presumably sharp) transition between phases

in which the Wilson loop has asymptotic decay of the "perimeter law" and "area law" form.

A related model for the study of random surface phenomena is the first covering problem, which is the natural generalization of first passage percolation to plaquette systems. We have recently been informed that H. Kesten has made substantial progress in understanding the properties of this system for low densities of "zero tension" plaquettes [11]. The authors have some related, but weaker results for all densities below the surface condensation threshold [12].

2) The second approach which has been followed is somewhat more ad hoc. Here one takes a relatively complex theory and "mutilates" the surface expansion in some physically motivated fashion, with the hope that the resulting theory approximates (in some sense) a resummation of the original model for a renormalized value of the parameter. To be specific, one considers a representative subset of all surfaces contributing to the correlations of the actual theory, and then chooses a simple weight $w(\mathbf{s};\beta)$. The correlations to be studied are then *defined* by expressions of the form (1.1). The question of whether such expressions actually do approximate those of the original theory and, if so, in what sense, must be addressed separately; in any case, the correlations so defined can be studied in their own right. This is the approach followed below. Before describing our system, it should be remarked that this second approach has also been followed in the analysis of planar random surfaces [13,14,15]. Such surfaces are purported to correspond to the "$N \to \infty$" limit of SU(N) lattice gauge theories.

Here we consider a model of the form (1.1) for a glueball propagator or truncated two-point function. The surfaces S, to be defined precisely in Sect. 1.2, are a subclass of self-avoiding surfaces with an additional constraint along a particular lattice axis. The weight we choose is of the form

$$w(s;\beta) = e^{-\beta|s|} \tag{1.2}$$

where $|s|$ denotes the area (i.e., number of plaquettes) of the surface s. The definition of this model is motivated by the random surface expansions of Ising-type gauge theories and three-dimensional uniaxial ferromagnets. Indeed, the surfaces contributing to the correlations of these theories are self-avoiding, and the corresponding weights tend to $e^{-\beta|s|}$ as $\beta \uparrow \infty$.

1.2. THE MODEL. For simplicity in our description, here and in the remainder of these notes, we will restrict attention to three dimensions. (This is not only essential for visualization; it also helps with the graphical presentation.) See, however, Ref. 2 for a definition of the model in higher dimensions and for the d-dimensional statements of the theorems.

We must specify the class of surfaces which define the model. For peculiar reasons, we take these surfaces to be composed of plaquettes on the *dual* of the lattice $\mathbb{Z}^3$. Let p_0 denote the plaquette centered at $(\frac{1}{2},0,0)$ and let $p_{L;a,b}$ denote its translation by the lattice vector (L,a,b). The class of surfaces, $S_{L;a,b}$ we consider are those self-avoiding surfaces s with boundary $\partial s = \partial p_0 \cup \partial p_{L;a,b}$ which satisfy the additional constraint that the intersection of s with each of the planes $x = k$, $1 \le k \le L$, is a single (closed) ring, r_k. (See Fig. 1.)

Remarks. 1) The self-avoiding nature of the surfaces forbids: i) multiple occupation of any given plaquette, and ii) junction of more than two plaquettes at any given edge. Thus all edges of the dual lattice, save those in $\partial p_0 \cup \partial p_{L;a,b}$, have zero or two plaquettes in their coboundary.

2) Due to the additional constraint, the surfaces in $S_{L;a,b}$ have no overhangs in the x-direction. By analogy with other surface models in statistical mechanics which forbid overhangs, we have called these surfaces solid-on-solid (S.O.S.) tubes. Note, however, that in contrast to the standard

S.O.S. surface model, our S.O.S. tubes may have overhangs in the y and z directions (i.e., the rings r_k need not be convex).

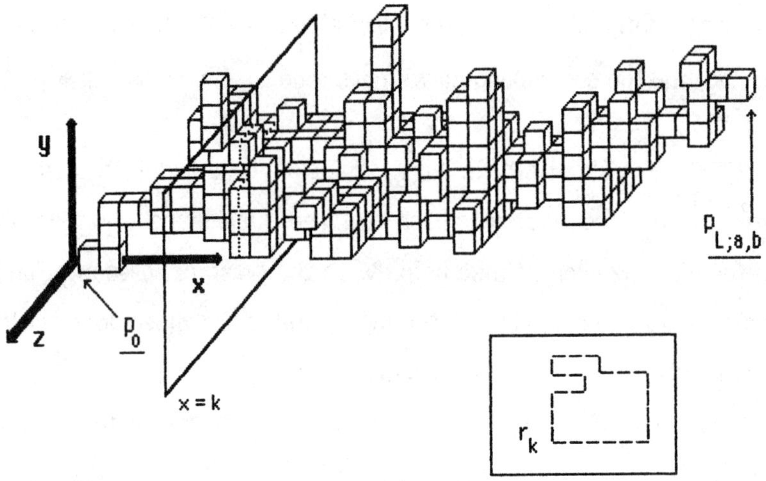

Fig. 1. An S.O.S. tube

We now examine the properties of the random surface correlation function, or generating function, defined as the exponentially weighted sum over S.O.S. tubes:

$$Q_{L;a,b}(\beta) = \sum_{s \in S_{L;a,b}} e^{-\beta|s|} . \tag{1.3}$$

It turns out that many of the expressions we derive are cleaner if we consider the (trivially modified) generating function defined by summing over S.O.S.tubes which begin in a "node" consisting of four plaquettes spanning ∂p_0. (This only modifies Q_L by a multiplicative factor of $e^{-4\beta}$ and, of course, changes the definition of "L".) Hereafter, we take $S_{L;a,b}$ to be defined with this boundary condition.

Remark. Observe that in three dimensions, each surface is determined uniquely by the sequence of rings $(r_1, r_2, \ldots, r_L)$. This allows an equivalent formulation of the problem in terms of a transfer matrix, with transition elements between adjacent rings, r^a and r^b, given by $\exp(-\frac{1}{2}\beta[\mathrm{Per}(r^a) + \mathrm{Per}(r^b) + 2\,\mathrm{Area}(\mathrm{int}[r^a] \,\Delta\, \mathrm{int}[r^b])])$. Here $\mathrm{int}[r]$ denotes the interior of the ring r, and Δ indicates the symmetric difference between interiors of overlapping rings. An implicit transfer matrix approach can also be given in $d \geq 4$ by considering the *equivalence class* of surfaces sharing a common set of rings. (It is not difficult to show that the weights factor.) Although the transfer matrix approach has some intuitive -- as well as historical -- appeal, it also has some limitations. First, the S.O.S. constraint is essential for its implementation. As will become apparent, the approach we follow uses the S.O.S. constraint only to facilitate certain subadditive estimates, so that with some (albeit nontrivial) modifications our methods should extend to a larger class of aggregate models, including self-avoiding walks and polymers. The second limitation is that, without additional input, it appears difficult to use the transfer matrix to obtain results for nonperturbative values of β, in contrast to the methods we shall use. It is nonetheless gratifying that the *existence* of a transfer matrix implies that the model under consideration actually represents an underlying (Euclidean) quantum field theory.¥

2. SUMMARY OF RESULTS

2.1. SOME BASIC PROPERTIES. Before reviewing our principal results, it is necessary to establish several basic features of the model. It turns out that, for β large, the Q's decay exponentially. Below we demonstrate the existence of the decay rate and prove some elementary properties which will allow us to characterize the phase structure.

Proposition 2.1. *Let* $Q_L \equiv Q_{L;0,0}$ *and* $\mathbb{Q}_L \equiv \Sigma_{a,b}\, Q_{L;a,b}$. *Then the limits*

¥ We would like to thank M. O'Carroll for pointing out to us this consequence of the existence of a transfer matrix.

$$M(\beta) \equiv \lim_{L\to\infty} [\,-L^{-1}\log Q_L\,]$$

and

$$M(\beta) = \lim_{L\to\infty} [\,-L^{-1}\log \mathbb{Q}_L\,]$$

exist (*in* $\mathbb{R}^*$). *Furthermore*:

a) $\forall L,\ \mathbb{Q}_L \le e^{-M(\beta)L}$.

b) $M(\beta)$ *is a concave, nondecreasing function of* β.

c) $M(\beta)\uparrow 4\beta$ *as* $\beta\uparrow\infty$.

d) *For* β *sufficiently small,* $M(\beta) = -\infty$.

Remark. The use of the notation $M(\beta)$ for the decay rate of both Q_L and $\mathbb{Q}_L$ will be justified by the analysis of Sect. 3.

Proof. Observe that a surface in $\mathbf{S}_{L_1}$, when pasted to (a translate of) a surface in $\mathbf{S}_{L_2}$, produces a surface in $\mathbf{S}_{L_1+L_2}$. Thus we obtain the subbadditive bound

$$Q_{L_1+L_2} \ge Q_{L_1} Q_{L_2}, \tag{2.1}$$

with a similar argument pertaining to the $\mathbb{Q}$'s. The existence of the limits follows in a straightforward fashion. Furthermore, for finite L and for any integer n, it is seen that

$$Q_{nL} \ge [Q_L]^n . \tag{2.2}$$

Taking logs and dividing by n, result (a) follows from the existence of the limit. Next, an application of the Hölder inequality yields

$$Q_L(\lambda\beta_1 + (1-\lambda)\beta_2) \le [Q_L(\beta_1)]^{\lambda}\, [Q_L(\beta_2)]^{(1-\lambda)}, \tag{2.3}$$

from which one establishes result (b). Finally, it is easy to verify that

$$(\#\text{ of S.O.S. tubes of area } n) \sim e^{\mu n}, \tag{2.4}$$

with $\mu < \infty$. Since the minimal tube contributing to $\mathbb{Q}_L$ has area $4L$, we obtain

$$M(\beta) \ge 4(\beta-\mu), \tag{2.5}$$

provided $\beta > \mu$. It is also clear that $M(\beta) \leq 4\beta$, which implies (c). Similar minded estimates show that for β too small, the finite L correlation functions diverge. These considerations establish (d). ▨

Remark. Since $M(\beta)$ is concave, it is also continuous save, perhaps, at the "endpoint". Thus, if $M(\beta)$ has a jump discontinuity (which we suspect), thereafter $M(\beta) = -\infty$. At any rate, results (c) and (d) imply the existence of two a priori different critical points:

$$\beta_c = \inf\{\beta \in (0,\infty) | M(\beta) > 0\} \tag{2.6a}$$

$$\beta^* = \inf\{\beta \in (0,\infty) | M(\beta) > -\infty\} . \tag{2.6b}$$

As of yet, we have no proof that $\beta_c = \beta^*$. We will, however, be able to identify β_c as the growth, or Peierls constant counting the number of S.O.S. tubes.

2.2. PRINCIPAL RESULTS. Having established the "existence" of this model -- in the sense of nontrivial critical point(s) -- it is of interest to investigate its features in detail. The questions we address in these notes, and the results we obtain fall into three categories, as described below. The condition under which these results are proved is simply $\beta > \beta^*$ (i.e., the results hold for *all* noncritical temperatures).

A) First, we examine the large L asymptotics of $Q_{L;a,b}$ and continuity properties of the mass $M(\beta)$. The principal result along these lines (Th. 3.3) is that $Q_{L;a,b}$ can be systematically expanded in an asymptotic series, the leading term of which is

$$Q_{L;a,b} \sim \text{const}\, L^{-1} e^{-M(\beta)L} e^{-[a^2+b^2]/\alpha L}[1 + O(L^{-1/2})] . \tag{2.7}$$

This implies that the hitting of endpoints in the plane $x = L + \frac{1}{2}$ is described by a normal distribution. Of particular interest is the leading term evaluated at a

= b = 0. Such behavior is known as Ornstein-Zernike decay, and is a characteristic feature of many systems in QFT and statistical mechanics. Ordinarily, it is only possible to establish these results in the *high* temperature phase.

We also prove (Th. 3.4) that for $\beta > \beta^*$, $M(\beta)$ is real analytic, indicating the absence of (intermediate) phase transitions.

B) Next, we examine the structural properties of typical tubes. We characterize the width of typical tubes by means of a "breathing criterion." It is shown (Th. 4.3) that whenever $\beta > \beta^*$, the tubes do not breathe.

C) Finally, we define and study the behavior of certain conjugate functions (which have the interpretation of surface entropy). Here we examine the S.O.S. tubes from the viewpoint of a different, but in hindsight equivalent ensemble. This is defined by considering the function $\Gamma'_k(L)$ which counts the number of tubes of area kL in the set $\mathbf{S}_L$. Among other results, we show (Th. 5.3) that, for rational k,

$$\Gamma'_k(L) \sim \text{const}\, L^{-3/2}\, e^{\zeta(k)L}\, [1 + O(L^{-1/2})], \tag{2.8}$$

again in the sense of asymptotic expansions. The entropy ensemble not only has a certain æsthetic appeal, but it is essential for our nonperturbative analysis.

The strategy of the proofs of these results involves two steps. First, we state a nonperturbative criterion (easily verified for $\beta \gg 1$) involving various correlation lengths associated with the tubes. Using this as an input hypothesis, the results in (A) - (C) are established. Then, by extending some of the ideas developed in the context of surface thermodynamics, we show that the hypothesis is satisfied whenever $\beta > \beta^*$. The first step is treated in Sects. 3 - 5, the second in Sect. 6. For the purposes of these notes, we omit many of the more technical proofs; the reader is referred to Refs. 2 and 3 for the details.

Of notable absence in this report are any features of the behavior *at* β =

β^*. It is, however, worth (modestly) pointing out that, with the exception of a few exactly solvable models, this is the only example of a system for which the noncritical behavior has been completely characterized.

3. ASYMPOTICS OF THE PAIR FUNCTION. Here we investigate the *detailed* behavior of Q_L and $\mathbb{Q}_L$ at large distances. The first step is to introduce an auxiliary correlation function which is the cornerstone of our later analysis.

Recall that in the (re)definition of $Q_{L;a,b}$ all our surfaces began with a node of four plaquettes. Consider the subset $\mathbf{J}_{L;a,b} \subset \mathbf{S}_{L;a,b}$ for which *each subsequent ring is strictly larger than four* (and thus ≥ 6):

$$\mathbf{J}_{L;a,b} = \{s \in \mathbf{S}_{L;a,b} \mid |r_k| > 4 \quad \forall\, k > 1\} \ . \tag{3.1}$$

We will denote the correlation functions obtained by summing over only the surfaces in $\mathbf{J}_{L;a,b}$ by $C_{L;a,b}$, $\mathbb{C}_L$, etc.

By a slightly enhanced pasting argument (which also involves some "cutting"), one can derive the analogue of Prop. 2.1 for the C functions. In particular, one has

$$-M_C(\beta) \equiv \lim_{L\to\infty} L^{-1} \log \mathbb{C}_L(\beta) \ , \tag{3.2}$$

which satisfies

$$M_C(\beta) \geq M(\beta) \ . \tag{3.3}$$

Now by the same reasoning that was used in Prop. 2.1c, we get

$$\lim_{\beta\to\infty} \beta^{-1} M_C(\beta) = 6 \ , \tag{3.4}$$

which implies that for $\beta \gg 1$, the inequality in (3.3) is strict. Strict separation of these correlation lengths is the necessary ingredient for our proofs of (A) - (C). We start with:

Proposition 3.1. *Whenever* $M_C(\beta) > M(\beta)$, $\mathbb{Q}_L(\beta)$ *has "pure" exponential* *decay in the sense that* $\exists$ *constants* $\chi(\beta) > 0$ *and* $\chi'(\beta) < \infty$ *s.t.*

$$|Q_L(\beta)e^{+M(\beta)L} - \chi'(\beta)| \le e^{-\chi(\beta)L} .$$

Remark. From the field theoretic point of view, $\chi(\beta)$ may be identified with the "upper gap" of the spectrum.

Proof. The contribution to Q_L which comes from surfaces containing no nodes (except the one at the origin) has been expressed as the function C_L. The contribution from surfaces which have no node until the plane $x = N$ is easily seen to be $C_N Q_{L-N}$. Since any surface must have a first node or none at all, we have the identity

$$Q_L = C_L + \sum_N C_N Q_{L-N} . \tag{3.5}$$

Let us define the Laplace transform:

$$Q^{\sim}(z) = \sum_L Q_L z^L , \tag{3.6}$$

with a similar expression for $C^{\sim}(z)$. Now $Q^{\sim}(z)$ is defined a priori only for $|z| < e^M$. On the other hand, since (3.5) is of the form of a convolution, it is easily seen that

$$Q^{\sim}(z) = [1 - C^{\sim}(z)]^{-1} . \tag{3.7}$$

The r.h.s. of (3.7) defines a meromorphic extension of $Q^{\sim}(z)$ which is valid in the larger region $|z| < e^{M_c}$.

We now argue that $Q^{\sim}(z)$ has a simple pole at $z = e^M$, and no other singularities in some larger region $|z| < R$ with $R > e^M$. Since each Q_L and C_L is positive, $Q^{\sim}(z)$ has its first singularity at $z = e^M$. To establish that e^M is a *simple* pole of $Q^{\sim}(z)$, by (3.7) we need only rule out the possibility of a higher order singularity. This is easily verified by noting that each Q_L satisfies $Q_L \le e^{ML}$, so that for x real and positive (and smaller than e^M),

$\mathbb{Q}^{\sim}(x) \le [1 - xe^{-M}]^{-1}$. Evidently, the pole is simple. We are therefore permitted to write

$$\mathbb{Q}^{\sim}(z) = F(z)/[1 - ze^{-M}] , \tag{3.8}$$

with F analytic for $|z| < R$. Noting that $\mathbb{Q}_L$ is simply the coefficient of z^L in the expansion of (3.8), we obtain

$$\mathbb{Q}_L e^{+ML} = [F_0 + F_1 e^M + \ldots + F_L e^{ML}] . \tag{3.9}$$

Finally, recalling the Cauchy bound, $|F_n| \le (\text{const})R^{-n}$, the desired result follows immediately. ▨

To obtain the results advertised in (A), we have to employ transforms that retain more detailed information. This time the goods will not come so cheap. Consider the transform

$$Q^{\sim}(z,\omega_1,\omega_2) = \sum_{L;a,b} Q_{L;a,b}\, z^L e^{i\omega_1 a} e^{i\omega_2 b} , \tag{3.10}$$

with a similar expression for $C^{\sim}$.

Since $Q_{L;a,b}$ and $C_{L;a,b}$ are also related by convolution (which here involves summation over the in-plane coordinates), (3.7) admits the generalization

$$Q^{\sim}(z,\omega_1,\omega_2) = [1 - C^{\sim}(z,\omega_1,\omega_2)]^{-1} . \tag{3.11}$$

Equation (3.7) may be obtained by setting $\omega_1 = \omega_2 = 0$. The coefficients $Q_{L;a,b}$ are recovered from (3.10) via contour integration:

$$Q_{L;a,b} = [2\pi i]^{-1} \oint dz\, z^{-L+1} \iint [1 - \mathbb{C}^{\sim}(z,\omega_1,\omega_2)]^{-1} e^{-i\omega_1 a} e^{-i\omega_2 b}\, d\omega_1 d\omega_2 . \tag{3.12}$$

In order to analyze the above integral, we must establish that $C^{\sim}$ has certain continuity properties. For these purposes, it is useful to regard $e^{i\omega_1}$ and $e^{i\omega_2}$ as complex variables, ξ_1 and ξ_2, restricted to the unit circle. Below we will show that for those values of z at which we know that $\mathbb{C}^{\sim}(z)$ is analytic, $C^{\sim}(z,\xi_1,\xi_2)$ is (separately) regular in ξ_1 and ξ_2 for a neighborhood of

$|\xi_1| = |\xi_2| = 1$.

Lemma 3.2. *Suppose that* $|z| < e^{M_c(\beta)}$ *with* $\beta > \beta^*$. *Then, provided that* $|\xi_1|, |\xi_2|$ *are sufficiently close to* 1, $C^{\sim}(z,\xi_1,\xi_2)$ *is regular in* ξ_1 (ξ_2) *for fixed* ξ_2 (ξ_1).

Proof. If $e^{M_c(\beta)} > |z|$ with $\beta > \beta^*$, then for η sufficiently small, so is $e^{M_c(\beta-\eta)}$. (Since $M_c(\beta)$ is concave nondecreacing and since $\forall\beta$, $M_c(\beta)$ is not exceeded by $M(\beta)$, $M_c(\beta)$ is continuous whenever $M(\beta)$ is.) By the definition of $C_{L;a,b}$, we have

$$C_{L;a,b}(\beta-\eta) = \sum_{\mathbf{s}\in \mathbf{J}_{L;a,b}} e^{-\beta|\mathbf{s}|} e^{+\eta|\mathbf{s}|}$$

$$\geq C_{L;a,b}(\beta) e^{4\eta(|a| + |b|)} , \tag{3.13}$$

since each $\mathbf{s}\in \mathbf{J}_{L;a,b}$ has at least $4(|a| + |b|)$ plaquettes. This implies that $\forall a,b$

$$C_{L;a,b}(\beta-\eta) \geq C_{L;a,b}(\beta) |\xi_1|^a |\xi_2|^b , \tag{3.14}$$

provided that $|\xi_1|, |\xi_2|, |\xi_1|^{-1}, |\xi_2|^{-1} < e^{4\eta}$. Observe also that for each fixed L, we have the subadditive bound

$$(\text{const}) e^{-M_c(\beta-\eta)L} \geq \mathbf{C}_L(\beta-\eta) \equiv \sum_{a,b} C_{L;a,b}(\beta-\eta) . \tag{3.15}$$

The bounds (3.13) - (3.15) imply that the power series for $C^{\sim}(z,\xi_1,\xi_2)$ is absolutely convergent. ▨

Theorem 3.3. *Whenever* $M_c(\beta) > M(\beta)$, $Q_{L;a,b}(\beta)$ *admits the asymptotic expansion*

$$Q_{L;a,b}(\beta) \sim (\chi'/\pi\alpha) L^{-1} e^{-M(\beta)L} e^{-[a^2+b^2]/\alpha L}[1 + O(L^{-1/2})] ,$$

where one may obtain systematic corrections in powers of $L^{-1/2}$. *The above is valid when* $|a| + |b| < cL$ *for a small constant* $c > 0$.

Proof (Sketch). Examine the integrand in (3.12). If $\omega_1 = \omega_2 = 0$, we have good control: $1-\mathbb{C}^{\sim}(z) \equiv 1-C^{\sim}(z,0,0)$ has a simple zero at $z = e^M$ and no other zeros (or poles) in the larger region $|z| < R$. Employing (some of) the continuity provided in Lemma 3.2, a Rouché argument shows that this situation persists if $|\omega_1|$ and $|\omega_2|$ are sufficiently small. The implicit function theorem allows us to "track" this zero, $e^{M(\omega_1,\omega_2)}$, in a neighborhood of $|\omega_1|=|\omega_2|=0$. Finally, a detailed argument shows that the contribution to the integral from *outside* any small neighborhood of $|\omega_1|=|\omega_2|=0$ is (relatively) exponentially small in L as $L\uparrow\infty$. Thus we are left with

$$Q_{L;a,b}(\beta) \sim (\text{const})\left[\int\int d\omega_1 d\omega_2\, e^{-M(\omega_1,\omega_2)L}\, e^{i\omega_1 a}\, e^{i\omega_2 b}\right][1+e^{-O(L)}] \quad (3.16)$$

The analytic inverse function theorem allows the development of an asymptotic series of the above stated form via the usual methods (e.g., Laplace's method).

▨

Remarks. 1) The reader is invited to torture themself by checking Ref. 2 (Theorems 2.7, 4.1 and Appendix B) for a more detailed proof.

2) Observe that the results automatically imply the equality of the mass "M" for the Q_L with the mass "M" for the $\mathbb{Q}_L$.

3) As a corollary, one has a central limit theorem for the hitting distribution of endpoints in the final plane $x = L$. Evidently, the typical tube is found a distance $O(L^{1/2})$ from the x-axis. It is even possible to obtain (implicit) expressions for the large deviations by further elucidating the pole structure of $\mathbb{Q}^{\sim}(z)$.

Observe that one can *define* a C function as the (unique) solution to an equation of the form (3.5), (3.7) or (3.11). This is the approach initiated by Ornstein and Zernike in their study of continuum fluids [16], and is the usual procedure followed in nonrigorous work. However, under such circumstances, one must first prove that a solution exists, which is in itself a serious undertaking involving some delicate functional differentiation [17,18]. Even having done so, one is not guaranteed that the solution has the desired properties (e.g., $M_C > M$). At present, it seems that the only way to demonstrate that a function with the desired properties exists is to produce a concrete (analyzable) object "obeying" the Ornstein-Zernike equation.

We conclude this section with a result on the continuity of $M(\beta)$:

Theorem 3.4. *Whenever* $M_C(\beta) > M(\beta)$, $M(\beta)$ *is real analytic.*

Proof. Due to the considerations of Th. 3.3, we may *define* the mass via the smallest real zero of $[1 - \mathbb{C}^\sim(z)]$. Now by an argument which is similar in spirit (although somewhat different in detail) to that used in the proof of Lemma 3.2, it can be established that $\mathbb{C}^\sim$ is analytic in $g \equiv e^{-\beta}$ in a neighborhood of $g_0 \equiv e^{-\beta_0}$, whenever $|z| < e^{M_C(\beta_0)}$. The result follows from an application of the analytic implicit function theorem. ▨

4. **BREATHING.** In this section, we will elucidate, to some extent, certain *geometric* properties of S.O.S tubes at temperatures $\beta > \beta^*$. In particular, we will propose and discuss a "breathing criterion" which characterizes the width of typical tubes. Our principle result is that for all $\beta > \beta^*$, tubes do not breathe. (This is actually established here under the proviso that $M_C(\beta) > M(\beta)$). Unfortunately, we do not have the space to prove the "key lemmas" (Lemmas 4.1 and 4.2); the reader is referred to Ref. 3 for the details.

4.1. GENERAL CONSIDERATIONS. The issue which concerns us here is the structural properties of typical tubes. In particular, we examine whether the size of rings can exhibit any scaling with the length of the tube. Our criterion is as follows: For a given $s \in S_{2L+1}$, let n_c denote the perimeter of the central ring (i.e., $|r_L|$). The possibility that the central ring gets large is determined by the probability

$$\mathcal{l}_L(h) \equiv \text{Prob}_L(n_c > h) = [Q_{2L+1}]^{-1} \sum_{s \in S_{2L+1}} e^{-\beta|s|}\, \mathbb{1}_{\{n_c > h\}}(s) \ . \tag{4.1}$$

If $\lim_{h\to\infty} \limsup_{L\to\infty} \mathcal{l}_L(h) = 0$, the tubes are said to be *stable*; otherwise, they *breathe*. A breathing point β_b is then the smallest β above which tubes do not breathe. If $\beta_c < \beta_b < \infty$, the system is said to exhibit a *breathing transition* at β_b .

Remark. The above criterion is of course somewhat arbitrary; we focus attention on the *central* ring due to lack of imagination. (We certainly must rivet our attention somewhere because, with probability tending to 1 as $L\uparrow\infty$, an arbitrarily large number of rings will exceed any h.) Nonetheless, the breathing of tubes, by any reasonable criterion, would constitute a fascinating geometrical phenomenon which would not be without physical significance.

Indeed, if tubes did undergo a breathing transition, it is "obvious" (see discussion below) that thereafter $M_C = M$, and that our Ornstein-Zernike (or renewal) analysis would fail. This suggests that in a breathing phase, one would observe a power law correction different than L^{-1} (or in general different than $L^{-(d-1)/2}$). On this note, it is worth remarking that for the analogous questions in two dimensions, one finds that $\forall\beta>\beta^*$: the breathing criterion is satisfied, $M_C(\beta) = M(\beta)$ *and* there is an anomalous correction of the form L^{-2} to exponential decay [18]. It has been known for quite some time [19] that this power law correction is a characteristic of the low-temperature phase in the 2-d Ising magnet.

4.2. ABSENCE OF BREATHING WHEN $M_c(\beta) > M(\beta)$. Heuristically, if $M_c(\beta) > M(\beta)$, absence of breathing can be gleaned from the following argument (which is due to M. Aizennman [20]). Recall that the first -- and perhaps most fundamental -- consequence of $M_c(\beta) > M(\beta)$ is the *pure* exponential decay of $Q_L(\beta)$ (c.f. Prop. 3.1). The probability of observing a node at the midplane is therefore bounded below by $(Q_L)^2e^{-4\beta}/Q_{2L+1} \longrightarrow x'(\beta)e^{M(\beta)-4\beta} > 0$. By a similar calculation, it is equally easy to show that any finite ring has nonzero probability of occurring at the center, uniformly in L.

This is already a strong indication that there is no breathing. The full argument follows by combining the above reasoning with a coarse-graining up to a scale determined by $|M(\beta)-M_c(\beta)|^{-1}$.

Lemma 4.1 . *Let* $M_c(\beta) > M(\beta)$ *and take* $A < L$. *Then, uniformly in* L,

$$\text{Prob}_L\{\exists \text{ nodes within a distance } A \text{ of the midplane, on both sides of the midplane}\} \geq 1 - (\text{const})e^{-(M_c-M)A},$$

with the constant independent of A.

Proof. See Ref. 3 (Lemma 4.2). ▨

Given that these nodes occur, it will be necessary to ensure that their azimuthal separation is not too large. That this is likley to be the case is the subject of our next lemma

Lemma 4.2 *Let* $N(A;x_-,x_+)$ *denote the event that nodes occur within a distance* $A < L$ *to the left and right of the central plane, and that the leftmost has x-coordinate* x_- *while the rightmost has x-coordinate* x_+. *Let* y_-,z_- (y_+,z_+) *be the other coordinates of the leftmost (rightmost) node within* A *of the central plane. If* $M_c(\beta) > M(\beta)$, *then there are constants* $\xi < \infty$, $\xi' > 0$ *such that, uniformly in* L,

$$\mathrm{Prob}_L\{|y_+ - y_-| + |z_+ - z_-| \le \xi A \mid N(A;x-,x+)\} \ge 1 - (\mathrm{const})e^{-\xi' A}$$

Proof. See Ref. 3 (Lemma 4.3). ▨

These are the tools needed for our principal result:

Theorem 4.3 *Whenever* $M_c(\beta) > M(\beta)$, *tubes do not breathe, i.e.*

$$\lim_{h\to\infty} \limsup_{L\to\infty} \mathfrak{l}_L(h) = 0 \quad .$$

Remark. By the results of Sect. 6, Th. 4.3 implies the absence of breathing $\forall\beta>\beta^*$. In fact, for $\beta>\beta_c$, using Corollary (v) of Th. 5.2, one can establish *exponential* decay in h of $\overline{\lim}\,\mathfrak{l}_L(h)$.

Proof. Choose $1 \ll A < L$ and $\eta > \xi+1$. Let $g_{L,A,\eta}$ denote the event that:

i) there are nodes a distance < A both on the left and right of the midplane

ii) the separation between r_- and r_+, the locations of the outermost pair of nodes within the 2A-slab, does not exceed ηA.

Evidently,

$$\mathfrak{l}_L(h) \le \mathrm{Prob}_L\{n_c > h \mid g_{L,A,\eta}\} + \mathrm{Prob}_L\{g_{L,A,\eta}{}^c\} \ . \tag{4.2}$$

The second term is set up for the results of the previous lemmas. The first term is analyzed by further conditioning to all pairs (r_-, r_+):

$$\mathfrak{l}_L(h) \le \mathrm{Prob}_L\{g_{L,A,\eta}{}^c\} + \sum_{(r_-,r_+)} \mathrm{Prob}_L\{n_c > h \mid (r_-,r_+)\}\, \mathrm{Prob}_L\{(r_-,r_+) \mid g_{L,A,\eta}\} \ . \tag{4.3}$$

Observe that the event (r_-, r_+) uncouples the behavior inside the nodes from that of the rest of the tube. Thus $\mathrm{Prob}_L\{n_c > h \mid (r_-,r_+)\}$ is independent of L, and may be computed by looking exclusively at the statistical behavior of tubes in $\mathbf{S}_{r_+ - r_-}$. However these are just "finite" tubes (i.e., their endpoints are separated by finite, L-independent distances). Since each $Q_{r_+ - r_-}$ is a convergent sum, the large n_c "tails" can be made arbitrarily small.

From this and the bounds described in Lemmas 4.1 and 4.2, it is easy to establish that tubes do not breathe. ▨

The result on the absence of breathing, as it stands, sheds little light on what to expect in more realistic systems. On the one hand, one may transmutate these results into the statement that there is no breathing in magnetic systems until the Curie point (i.e., the point where the mass vanishes). On the other hand, it may be that due to the simplicity of the weights we have chosen, the system is not rich enough to exhibit such a transition without ceasing to exist. Assuming for the moment that $\beta_c = \beta^*$, it would therefore be of interest to know whether $M(\beta_c) > 0$. If it turns out that this is the case, there is some reason to hope for a nontrivial "T_b" in three-dimensional uniaxial ferromagnets.

5. **ENTROPY OF RANDOM SURFACES.** In many popular aggregate models, for which estimates of the form (2.4) are readily obtained, the growth constant is often referred to (by people who ought to know better) as the *aggregate entropy*. While the growth constant does count the number of surfaces, and even appears in an exponent, its identification as an entropy seems (to us) to be rather unsubstantiated. This follows from the fact that it is simply a geometric constant, and does not depend on an intrinsic parameter. Our definition of entropy follows a more traditional approach.

Let us rewrite our generating function in the form

$$Q_L(\beta) = \sum_k \Gamma_k(L) e^{-\beta kL}, \tag{5.1}$$

where $\Gamma_k(L)$ is the number of S.O.S. tubes of area kL contributing to Q_L. The $\Gamma_k(L)$ are relatively interesting objects to study in their own right. First observe that although for most of the real k's, $\Gamma_k(L)$ is zero, for each rational $k > 4$, there is a sequence of lengths $\mathbf{N}(k)$ such that $\forall\, L \in \mathbf{N}(k)$, $\Gamma_k(L) > 0$. The following proposition -- which is not at all difficult -- shows that, for these L's, the $\Gamma_k(L)$ have well-defined asymptotic behavior.

Proposition 5.1. *For each rational* $k > 4$, *the limit*

$$\zeta(k) \equiv \lim_{L \to \infty} (L^{-1} \log \Gamma_k(L) | L \in \mathbf{N}(k))$$

exists and is finite. Furthermore, the function $\zeta(k)$ *so defined is concave in* k. *Finally, for every finite* L, $\Gamma_k(L) \le e^{\zeta(k)L}$.

Proof. The reader is referred to Ref. 2 (Props. 5.1 and 5.2). ▨

Remarks. 1) Since $\zeta(k)$ is concave, it is also continuous. Thus it extends to a continuous function on all real $k \ge 4$. It is often convenient to define $\zeta(k) = -\infty$ for $k < 4$. In this regard, it is satisfying to observe that $d\zeta/dk$ diverges (logarithmically) as $k \downarrow 4$.

2) Since $\zeta(k) > 0$ for $k > 4$, concavity implies that it is increasing. Elementary considerations (e.g., $\zeta(k) < \mu k$) show that ζ does not grow dangerously fast.

At this point, it is already tempting to identify $\zeta(k)$ as the surface entropy. Indeed, regarding Q_L as a partition function, the decay rate $M(\beta)$ may be viewed as a free energy and the "covering factor" k as an energy. An encouraging sign is that the "thermodynamic derivative" $dM/d\beta$ is (informally) seen to equal the *expected* covering factor at inverse temperature β. That these variables are all canonically related is a consequence of the following.

Theorem 5.2. $$-M(\beta) = \sup_k [\zeta(k) - \beta k] .$$

Proof (Sketch). For any rational $k > 4$,

$$Q_L(\beta) \geq \Gamma_k(L) e^{-\beta k L} . \qquad (5.2)$$

By continuity, this implies

$$-M(\beta) \geq \sup_k [\zeta(k) - \beta k] . \qquad (5.3)$$

If the r.h.s. of (5.3) is infinite, we are done. If $M(\beta) = -\infty$, *or* if $\beta > \beta^*$, then a bound supplementary to (5.3) is relatively easy to provide. The difficultly comes from the single point $\beta = \beta^*$. There one must utilize a finite-size cutoff, as is provided in Th. 5.3 of Ref. 2. ▨

As corollaries, one can derive:

Corollaries. i) For a.e. $\beta \in (\beta^*, \infty)$, $\exists! \, k(\beta)$ which maximizes $\zeta(k) - \beta k$. At these points, $k(\beta)$ is given by $dM/d\beta$. For those $\beta \in (\beta^*, \infty)$ at which $M'(\beta)$ fails to be continuous (which is, in general, a set of zero measure, and for this model, is empty), the compact set $[dM/d\beta^+, dM/d\beta^-]$ contains the maximizing k's.

ii) $M(\beta)$ is right continuous at $\beta = \beta^*$.

iii) $\beta_c = \sup_k [\zeta(k)/k]$.

iv) $\beta^* = \lim_{k \to \infty} [\zeta(k)/k]$.

v) The growth constant μ, which counts the # of S.O.S. tubes, is equal to β_c.

Proof. The result (i) follows from standard variational arguments and from concavity of $M(\beta)$. Results (ii)-(iv) are proved in Ref. 2 (Sect.5).

To prove (v), let $\mathcal{A}(n)$ denote the number of S.O.S. tubes spanning ∂p_0 that have n plaquettes. By equation (2.4), we may write $\mathcal{A}(n) \sim e^{\mu n}$. Clearly $\forall k$,

$$\Gamma_k(L) \leq \mathcal{A}(n \mid n = kL) , \qquad (5.4)$$

and hence

$$\mu \geq \sup_k [\zeta(k)/k] . \tag{5.5}$$

Next, we use the identity

$$\mathcal{A}(n) = \sum_{k,L\,:\,kL=n} \Gamma_k(L) . \tag{5.6}$$

Now the number of nonzero terms in the sum is easily seen to be smaller than n. We may replace $\Gamma_k(L)$ by its a priori upper bound $e^{\zeta(k)L}$. Then we have

$$\mathcal{A}(n) \leq n[\sup_k(e^{\zeta(k)L} | kL = n)] ; \tag{5.7}$$

hence

$$\mu \leq \sup_k [\zeta(k)/k] . \tag{5.8}$$

▨

Remark. Analogues of the above results hold for a large class of aggregate models.

Thus far, we have only obtained the crudest asymptotic description of the $\Gamma_k(L)$. The following theorem provides additional detail.

Theorem 5.3. *For all* k *for which the corresponding* $\beta(k)$ satisfies $M(\beta) < M_c(\beta)$, $\Gamma_k(L)$ *admits the asymptotic expansion*

$$\Gamma_k(L) \sim (\text{const}) L^{-1/2} e^{\zeta(k)L} [1 + O(L^{-1/2})] .$$

Moreover, if we denote by $\Gamma'_k(L)$ *the analogue of* $\Gamma_k(L)$ *corresponding to the* $Q_L(\beta)$,

$$\Gamma'_k(L) \sim (\text{const}) L^{-3/2} e^{\zeta(k)L} [1 + O(L^{-1/2})] .$$

Proof. Regard $g \equiv e^{-\beta}$ as a complex variable, and write

$$Q_L(g) = \sum_S g^{|S|} . \tag{5.9}$$

The above is well-defined provided that $|g| < e^{-\beta^*}$. The "microcanonical" (entropy) ensemble is recovered via contour integration:

$$\Gamma_k(L) = (1/2\pi i)\oint dg\, g^{-kL+1} Q_L(g) = (-1/2\pi)^2 \oint dg\, g^{-kL+1} \oint dz\, z^{-L+1} Q^{\sim}(g;z)\,. \quad (5.10)$$

First, consider performing the integration at some fixed modulus of g: $g = e^{-\beta_0} e^{i\varphi}$. Observe that by the $g \leftrightarrow -g$ symmetry, we may integrate φ in $[-\pi/2, +\pi/2]$ and double the answer. Next, let us replace $Q^{\sim}$ by its expression in terms of $C^{\sim}$. A detailed look at things shows that outside any small neighborhood $|\varphi| < \delta$, the entire contribution has modulus bounded by

$$|\text{error term}| < (\text{const}) e^{-(M(\beta_0)+\varepsilon)L}\, e^{\beta_0 kL}\,. \quad (5.11)$$

(Of course, ε depends on δ.) This leaves

$$\Gamma_k(L) = (\text{const}) \int_{|\varphi|<\delta} d\varphi\, e^{-ikL\varphi} \oint dz\, z^{-L+1} [1 - C^{\sim}(g;z)]^{-1} e^{\beta_0 Lk} + \text{error term}\,. \quad (5.12)$$

After a minor struggle, making liberal use of the implicit function theorem, the z integration is performed with the result that

$$\Gamma_k(L) = (\text{const}) \int_{|\varphi|<\delta} d\varphi\, e^{-ikL\varphi} e^{M(\beta_0,\varphi)L} e^{\beta_0 Lk} + \text{error term}\,. \quad (5.13)$$

In the above expression, $M(\beta_0,\varphi)$ is the *analytic continuation* of $M(\beta)$ to complex values of temperature. We assume, of course, that β_0 was picked small enough to satisfy $M(\beta_0) < M_c(\beta_0)$. Now when $\varphi \neq 0$, both $e^{M(\beta_0,\varphi)L}$ and $e^{-ikL\varphi}$ oscillate so strongly that, in general, the purported "error term" dominates the "principal contribution." To avoid this difficulty, β_0 must be selected so that the integrand has stationary phase. This is done by choosing the β_0 which satisfies $k = k(\beta_0)$. It is seen that at the point of stationary phase, the coefficient of L in the exponent is exactly $\zeta(k)$. The φ integration is now easily carried out, yielding the above quoted $L^{-1/2}$ correction.

A similar, but titanic struggle can be waged for $\Gamma'_k(L)$ with $Q^{\sim}(z)$ replaced by $Q^{\sim}(z,\omega_1,\omega_2)$, and two additional integrations over the ω variables in (5.10). The end result (the details of which can be found in Ref. 2, Th. 5.6) is an extra power of $L^{-1/2}$ for each of the in-plane integrations that must be performed. ▨

It is clear that the entropy formalism provides yet another ensemble in which one can study aggregate models of this type. Although it has not -- as of yet -- enhanced our knowledge of the critical behavior, the ideas developed in this framework can be used (at least for the S.O.S. model) to gain some understanding of the *entire* noncritical regime. This is done in the final section.

6. NONPERTURBATIVE ANALYSIS. All results derived in previous sections have relied on the criterion $M_c(\beta) > M(\beta)$. Although this, in some sense, makes these results "nonperturbative", the analysis is incomplete unless it can be verified that the condition is satisfied outside a perturbative region.

Let us pause for a moment and examine why $M_c(\beta) > M(\beta)$ if $\beta \gg 1$. The reason is simple: tubes which contribute to $\mathbb{Q}_L$ "want to" have an area essentially as small as possible, while J_L forces the tubes to have area $\geq 6L$. In a different light, according to the analysis of the previous section, this is just the statement that $k(\beta) < 6$ for β large, while the $\mathbb{C}$ function does not permit a covering that small. Now we do not expect to be able to verify $M_c(\beta) >$ $M(\beta)$ at all temperatures by such crude reasoning, but the above does provide the key idea as to what generalization of the $\mathbb{C}$ function will have a bigger mass -- we should consider tubes with area exceeding $k(\beta)L$. Using such objects (which are constructed below) we can then prove, in the end, that the $\mathbb{C}$ mass does exceed M.

There will, however, be a price. Recall from the discussion in Sect. 3 that the choice of $\mathbb{C}$ is unique. Fortunately, it has proved possible, in a variety of circumstances, to actually construct geometric objects which correspond to the $\mathbb{C}$ function [21] (see also [22]). Now in quantum field theory there is a generalization of the Ornstein-Zernike equation for which (practically by definition) this cannot be done. This is known as the Schwinger-Dyson equation, which relates the one-particle irreducible Green function to the ordinary Green function. The important point is that, in these circumstances, neither Green function determines the other due to the existence of nontrivial *vertex terms*. (In the Ornstein-Zernike equation these terms are 1.) It is exactly this freedom that we exploit below to produce -- nonuniquely -- an object (or rather a collection of objects) which is manageable, and yet has a larger mass than M. The equation relating the $\mathbb{Q}$'s to these objects is analagous to (3.5), but contains an additional ("vertex-like") degree of freedom.

Definition. Let $\beta > \beta^*$ and denote by $[k(\beta)]$ the smallest integer which exceeds $k(\beta)$.[¥] Let us enumerate all two-dimensional lattice rings, modulo translations, that have perimeter smaller than $[k(b)]$: $(\mathbf{r}^0, \mathbf{r}^1, \ldots, \mathbf{r}^{n(\beta)})$. For the sake of definiteness, we place the origin of coordinates in the center of the upper leftmost square surrounded by the rings. Our old friends, the $Q_{L;a,b}$, have the following generalization:

Let $Q^{\mu\nu}_{L;a,b}$ denote the correlation function obtained by summing over the class of tubes $S^{\mu\nu}_{L;a,b}$ which obey the S.O.S. constraint, but have boundary

---

¥At this point in the analysis, one cannot rule out the possibility of a nonunique maximum of $\zeta(k) - \beta k$. Nevertheless, the condition $\beta > \beta^*$ implies that the largest maximizer (which is *officially* our definition of $k(\beta)$) must be finite.

$\partial s = \mathbf{r}^{\mu} \cup \mathbf{r}^{\nu}_{L;a,b}$. (We use $\mathbf{r}^{\nu}_{L;a,b}$ as notation for the translation of $\mathbf{r}^{\nu}$). For convenience in our later analysis, we also include the stipulation that the first slice of each $s \in S^{\mu\nu}_{L;a,b}$ actually "is" $\mathbf{r}^{\mu}$ (just as the first slice of $Q_{L;a,b}$ was required to be the node $\mathbf{r}^{0}$.

One can easily establish the following:

Proposition 6.1. *For each pair* μ,ν, *the limit*

$$\lim_{L\to\infty} L^{-1} \log \mathbb{Q}^{\mu\nu}{}_{L}$$

exists and is equal to $M(\beta)$. *Furthermore, there are "ring-dependent" constants,* $d^{\mu\nu}$, *such that*

$$\mathbb{Q}^{\mu\nu}{}_{L} \le d^{\mu\nu} e^{-M(\beta)L} \quad .$$

Proof. See Prop. 2.1. ▨

We now define our generalization of the $\mathbb{C}$ function. Denote by $H^{\mu\nu}_{L;a,b}$ the subset of $S^{\mu\nu}_{L;a,b}$ for which *none* of the rings $(\mathbf{r}^{0},\mathbf{r}^{1}, \ldots ,\mathbf{r}^{n(\beta)})$ appear (save the one nearest the origin):

$$H^{\mu\nu}_{L;a,b} = \{ s \in S^{\mu\nu}_{L;a,b} \mid |r_2|,|r_3|, \ldots , |r_L| > [k(\beta)] \} \quad . \tag{6.1}$$

Denoting the corresponding correlation functions by $\mathbb{B}^{\mu\nu}_{L;a,b}$, etc., we have the following:

Proposition 6.2. *For each pair* μ,ν, *the limit*

$$\lim_{L\to\infty} L^{-1} \log \mathbb{B}^{\mu\nu}{}_{L} \equiv -M_B(\beta)$$

exists and is independent of μ *and* ν. *Furthermore,* $M_B(\beta) > M(\beta)$.

Proof. The existence of M_B independent of boundary conditions is straightforward. Let us concentrate on $\mathbb{B}^{00}{}_L$. Observe that

$$\mathbb{B}^{00}{}_L \le \sum_{k > k(\beta)} \Gamma_k(L) e^{-\beta k L} . \tag{6.2}$$

We split the r.h.s. of (6.2) into a "light piece" ($k \ge 2[k(\beta)]$) and a "heavy piece" ($[k(\beta)] < k < 2[k(\beta)]$). There are at most $[k(\beta)]L$ nonzero terms in the "heavy piece", and it is straightforward to show that the "light piece" is relativley exponentially small. Using the a priori bound of Prop. 6.1 and taking logs, we have

$$-M_B(\beta) \le \sup_{k > [k(\beta)]} \{\zeta(k) - \beta k\} < -M(\beta) , \tag{6.3}$$

the strict inequality following from the fact that k has been bounded strictly away from the maximizer. ▨

We are now prepared to derive our "Schwinger-Dyson" equation. The key result is the extention of Prop. 3.1 to all noncritical temperatures.

Lemma 6.3. $\forall\, \beta > \beta^*$, $\mathbb{Q}^\sim(z)$ *has a simple pole at* $z = e^{M(\beta)}$ *and is otherwise analytic in a disk of radius* $R > e^{M(\beta)}$.

Proof. From those surfaces s in which none of the rings $(\mathbf{r}^0, \mathbf{r}^1, \ldots, \mathbf{r}^{n(\beta)})$ occur, we get a contribution $\mathbb{B}^{00}{}_L$. As for the rest of the surfaces, let N denote the place where the earliest infraction occurs. If the ring causing the disturbance is $\mathbf{r}^\mu$, we obtain the contribution $\mathbb{B}^{0\mu}{}_N \mathbb{Q}^{\mu 0}{}_{L-N}$. Summing over all intermediate rings and planes, we have the identity

$$\mathbb{Q}_L = \sum_\mu \sum_{N=1}^{L} \mathbb{B}^{0\mu}{}_N \mathbb{Q}^{\mu 0}{}_{L-N} , \tag{6.4}$$

with the definition $\mathbb{Q}^{\mu 0}{}_0 \equiv \delta^{\mu 0}$.

More generally, we have

$$\mathbb{Q}^{\mu\nu}{}_L = \sum_{\alpha} \sum_{N=1}^{L} \mathbb{B}^{\mu\alpha}{}_N \mathbb{Q}^{\alpha\nu}{}_{L-N} , \tag{6.5}$$

again with some special definitions if $L = 0$. Let us define the matrix $\mathbf{Q}_L$ by the matrix elements $\mathbb{Q}^{\mu\nu}{}_L$, and similarly for $\mathbf{B}_L$. The above, after a Laplace transform, reads

$$\mathbf{Q}^{\sim}(z) = \mathbb{1} + \mathbf{Q}^{\sim}(z)\mathbf{B}^{\sim}(z) . \tag{6.6}$$

Inverting (6.6) for $|z| \ll 1$, one can obtain an explicit expression for $\mathbb{Q}^{\sim}(z)$ in terms of rational functions of the $\mathbb{B}^{\sim}(z)$'s. Using Prop. 6.1 and the argument of Prop. 3.1 which demonstrated a *simple* pole at $z = e^{M(\beta)}$, we obtain the desired result. ▨

Remark. At this point, it is possible to establish all the results of previous sections based on the pole structure of $\mathbb{Q}^{\sim}(z)$ and the fact that $M_B > M$. However, this is technically a bit painful because there are certain features of the $\mathbb{C}$ functions that are not shared by the $\mathbb{B}^{\mu\nu}$. It is instead easier to make direct use of the previous theorems by proving the single result below.

Theorem 6.4. $\forall \beta > \beta^*$, $M_C(\beta) > M(\beta)$.

Proof. If $M_C(\beta) = M_B(\beta)$, we are through. In the opposite case, let us demonstrate that the transform $\mathbb{C}^{\sim}(z)$ has a simple pole of its own at $z = e^{M_C(\beta)}$. This is done by following mutatis mutandis the arguments of Lemma 6.3 with $\mathbb{C}$'s taking the role of $\mathbb{Q}$'s. Now observe that regardless of whether or not $M_C(\beta) > M(\beta)$, $\mathbb{C}^{\sim}$ and $\mathbb{Q}^{\sim}$ are still related by (3.7). Restricting this to $z = x \in \mathbb{R}^+$, we find that as $x \uparrow e^{M(\beta)}$, $\mathbb{C}(x) \uparrow 1$. There is no surer indication than this that $e^{M(\beta)}$ is *not* a pole of $\mathbb{C}^{\sim}(z)$. Evidently, we have $M_C(\beta) > M(\beta)$. ▨

REFERENCES

1. Abraham,D.B., Chayes,J.T. and Chayes,L., "Statistical mechanics of lattice tubes", Phys. Rev. D**30**, 841 (1984).

2. Abraham,D.B., Chayes,J.T. and Chayes,L., "Random surface correlation functions", Commun. Math. Phys. **96**, 439 (1984).

3. Abraham,D.B., Chayes,J.T. and Chayes,L., "Nonperturbative analysis of a model of random surfaces", Nucl. Phys. B [FS], in press.

4. Symanzik,K. "Euclidean quantum field theory" in *Local Quantum Theory*, (R. Jost, ed.), Academic Press, New York, 1969.

5. Aizenman,M., "Geometric analysis of φ^4 fields and Ising models. Parts I and II", Commun. Math. Phys. **86**, 1 (1982).

6. Fröhlich,J., "On the triviality of $\lambda\varphi^4{}_d$ theories and the approach to the crtitical point in $d > 4$ dimensions", Nucl. Phys. B**200** [FS4], 281 (1982).

7. Eguchi,T. and Kawai,H., "The number of random surfaces on the lattice and the large-N gauge theory", Phys. Lett. **110**B, 143 (1982).

8. Kazakov,V.A., "U(∞) lattice gauge theory as a free string theory", Phys. Lett. **128**B, 316 (1983).

9. Kostov,I., seminar presented at the summer school on Critical Phenomena, Random Systems and Gauge Theories (Les Houches, 1984).

10. Aizenman,M., Chayes, J.T., Chayes,L., Fröhlich,J. and Russo,L., "On a sharp transition from area law to perimeter law in a system of random surfaces", Commun. Math. Phys. **92**, 19 (1983).

11. Kesten,H., private communication.

12. Chayes,J.T. and Chayes,L., in preparation.

13. Durhuus,B., Fröhlich,J. and Jónsson,T., "Self-avoiding and planar random surfaces on the lattice", Nucl. Phys. B**225** [FS9], 185 (1983).

14. Durhuus,B., Fröhlich,J. and Jónsson,T., "Critical properties of a model of planar random surfaces", Phys. Lett. **137**B, 93 (1984).

15. Durhuus,B., Fröhlich,J. and Jónsson,T., "Critical behaviour in a model of planar random surfaces", Nucl. Phys. B**240** [FS12], 453 (1984).

16. Ornstein,L.S. and Zernike,F., Proc. Acad. Sci. Amst. **17**, 793 (1914).

17. Chayes,J.T., Chayes,L. and Lieb,E.H., "The inverse problem in classical statistical mechanics", Commun. Math. Phys. **93**, 57 (1984).

18. Chayes,J.T. and Chayes,L., "On the validity of the inverse conjecture in classical density functional theory", J. Stat. Phys. **36**, 471 (1984).

18. Abraham,D.B., "Two-point functions and bubbles", Phys. Rev. Lett. **50**, 291 (1983).

19. Wu,T.T., "Theory of Toeplitz determinants and the spin correlations of the two-dimensional Ising model", Phys. Rev. **149**, 380 (1966).

20. Aizenman,M., private communication.

21. Abraham,D.B. and Kunz,H., "Ornstein-Zernike theory of classical fluids at low density", Phys. Rev. Lett. **39**, 1011 (1977).

22. Simon,B., "The rate of falloff of the Ising model correlations at large temperature", J. Stat. Phys. **26**, 53 (1981).

DEPARTMENTS OF PHYSICS AND MATHEMATICS
HARVARD UNIVERSITY
CAMBRIDGE, MA. 02138

Contemporary Mathematics
Volume 41, 1985

LARGE DEVIATIONS FOR SOME INFINITE PARTICLE SYSTEM OCCUPATION TIMES

J. Theodore Cox[1] — Syracuse University

David Griffeath[2,3] — University of Wisconsin

ABSTRACT. We survey large deviation theorems for occupation times of several infinite particle systems. In low dimensions the large deviation tails are "fat."

1. INTRODUCTION. We are interested in large deviation probabilities for the occupation time process of several infinite particle systems ξ_t : independent random walks on $\mathbb{Z}^d$, critical binary branching Brownian motion on $\mathbb{R}^d$, simple exclusion on $\mathbb{Z}^d$, and the voter model on $\mathbb{Z}^d$. Except for the branching Brownian motion, each process ξ_t is $\{0,1\}^{\mathbb{Z}^d}$-valued, with

$$\xi_t(x) = 1 \text{ if } x \text{ is "occupied"},$$
$$= 0 \text{ otherwise.}$$

The occupation time process is

$$T_t(A) = \sum_{x\in A} \int_0^t \xi_s(x)\, ds \quad , \text{ finite } A \subset \mathbb{Z}^d$$

AMS 1980 Subject Classification: Primary 60K35
[1]Partially supported by N.S.F. grant MCS84-01317
[2]Alfred P. Sloan Research Fellow
[3]Partially supported by N.S.F. grant DMS-830549

for the three lattice processes, and

$$= \int_A \int_0^t \xi_s(dx)\, ds \quad , \text{ bounded Borel } A \subset \mathbb{R}^d$$

for the branching Brownian motion.

In all the cases we consider here ξ_t is <u>density preserving</u>, which implies that for some constant θ, for all $t \geq 0$, $E[T_t(A)] = \theta\, |A|\, t$ ($|A|$ = cardinality of A in the lattice case, = Lebesgue measure of A in the branching Brownian case). Usually one also has

$$(1.1) \qquad \frac{T_t(A)}{|A|t} \to \theta \quad \text{a.s.} \quad \text{as } t \to \infty ,$$

and so it is natural to consider the question of large deviations for $T_t(A)$. We seek theorems of the form

$$(1.2) \qquad P\left(\frac{T_t(A)}{|A|t} > \alpha \right) \sim \exp[-I(\alpha)\, a_t] \quad \text{as } t \to \infty ,$$

for $\alpha > \theta$ and suitable $a_t \to \infty$. By (1.2) we mean

$$\lim_{t\to\infty} a_t^{-1} \log P\left(\frac{T_t(A)}{|A|t} > \alpha \right) = -I(\alpha) \quad ,$$

insisting that $I(\alpha) \in (0,\infty)$. For simplicity, the corresponding results below θ will be omitted.

Typically in such problems $a_t = t$. However we are particularly interested in systems with <u>strong dependence</u>, in which case it is natural to expect

$$a_t = o(t) \quad \text{as } t \to \infty .$$

For the models we consider here this occurs in "low" dimensions, whereas $a_t = t$ is the rule in "high" dimensions. Section 2 describes our basic technique for getting (1.2). Then in Sections 3 and 4 we present a summary of results from [3], [4], [1], and our current research, starting with the independent models and finishing with the interacting ones. Proofs of the results announced for the voter model in Section 4 will appear in a future publication.

2. BASIC STRATEGY. A useful tool for obtaining results like (1.2) is the following lemma, in the spirit of Sievers [8], Plachky and Steinebach [7], and Ellis [5]. The hypotheses can be weakened somewhat; cf. [5].

LARGE DEVIATIONS CONVERGENCE LEMMA. Given real random variables Y_t $(t \geq 0)$ and $a_t \to \infty$ as $t \to \infty$, let

$$\Psi_t(\lambda) = a_t^{-1} \log E[e^{\lambda Y_t}] .$$

Suppose that for some open interval (λ_-, λ_+) containing 0,

$$\lim_{t\to\infty} \Psi_t(\lambda) = \Psi(\lambda) < \infty \qquad \text{on} \quad (\lambda_-, \lambda_+)$$

and that Ψ is strictly convex and C^2 on (λ_-, λ_+). Then, with $\mu = \Psi'(0)$, $\alpha_+ = \lim_{\lambda\uparrow\lambda_+} \Psi'(\lambda)$, for all $\alpha \in (\mu, \alpha_+)$,

$$(2.1) \qquad \lim_{t\to\infty} a_t^{-1} \log P(Y_t/a_t > \alpha) = -I(\alpha) \in (-\infty, 0) ,$$

where $I(\alpha) = \alpha\lambda_\alpha - \Psi(\lambda_\alpha)$, and λ_α is the unique solution of $\Psi'(\lambda_\alpha) = \alpha$.

Similar results hold below the mean. There is also a law of large numbers and, under minimal additional hypotheses, a central limit theorem. See [5] for a nice account of these matters.

To apply the lemma to our occupation time problems, consider the cumulant generating function

$$\Phi_t(\lambda) = \log E[\exp\{\lambda \frac{T_t(A)}{|A|t}\}]$$

$$= \sum_{n=1}^{\infty} \frac{\varphi_n(t)}{n!} \lambda^n ,$$

$\varphi_n(t)$ = the nth cumulant of $\frac{T_t(A)}{|A|t}$. Now suppose that for some constants c_n,

(2.2) $$\varphi_n(t) \sim c_n/a_t^{n-1} \quad \text{as} \quad t \to \infty .$$

Then choosing $Y_t = a_t \frac{T_t(A)}{|A|t}$ we have

$$\Psi_t(\lambda) = a_t^{-1} \log E[e^{\lambda Y_t}]$$

$$= a_t^{-1} \Phi_t(a_t \lambda)$$

$$\sim a_t^{-1} \sum_{n=1}^{\infty} \frac{c_n}{n!} a_t^{1-n} (a_t \lambda)^n \quad \text{as} \quad t \to \infty \quad \text{by (2.2)}$$

$$= \sum_{n=1}^{\infty} \frac{c_n}{n!} \lambda^n \stackrel{def}{=} \Psi(\lambda) .$$

Applying the Large Deviations Lemma we obtain

$$P(a_t^{-1} a_t \frac{T_t(A)}{|A|t} > \alpha) \sim \exp\{-I(\alpha) a_t\} \quad \text{as} \quad t \to \infty,$$

as desired. So the goal is to verify (2.2), and then justify the interchange of limits performed above.

Of course the first step is to determine a_t, but this is easy. Since $\varphi_2(t) = \mathrm{Var}(T_t(A)/|A|t)$ and (2.2) requires that $\varphi_2(t) \sim c_2/a_t$, a_t must grow like $t^2/\mathrm{Var}(T_t(A))$ as $t \to \infty$.

3. INDEPENDENT MODELS. Our program has been carried out very successfully for the independent random walk system ζ_t (see [3]) and somewhat less completely for critical binary branching Brownian motion β_t (see [4]). In both cases we start the process from a Poisson field with positive intensity θ. The particles of ζ_t execute independent rate one simple random walks on $\mathbb{Z}^d$. The particles in β_t undergo independent Brownian motions on $\mathbb{R}^d$ subject to death (at rate one) and binary branching (at rate one). The random walk system ζ_t is stationary in all dimensions; the branching Brownian motion clusters in dimensions one and two, but is stable in higher dimensions. To summarize our results we will need the following notation:

X_t = rate 1 simple random walk $\qquad$ W_t = Brownian motion

$p_t(x,y) = P_x(X_t = y)$ $\qquad$ $u_t(x,y)\,dy = P_x(W_t \in dy)$

$G_t(x,y) = \int_0^t p_s(x,y)\,ds$ $\qquad$ $\Gamma_t(x,y) = \int_0^t s\,u_s(x,y)\,ds$

INDEPENDENT RANDOM WALKS. Our results for ζ_t are particularly explicit. The asymptotics (2.2) can be derived exactly, with the scalings

$$
\text{(3.1)} \qquad
\begin{aligned}
a_t &= \sqrt{t} && d = 1, \\
&= t/\log t && d = 2, \\
&= t && d \geq 3.
\end{aligned}
$$

The limiting cumulant generating functions are

$$\Psi(\lambda) = 2\theta\left[\frac{e^{\lambda^2/2}-1}{\lambda} + \sqrt{2/\pi}\ \frac{e^{\lambda^2/2}\int_0^\lambda e^{-s^2/2}ds - \lambda}{\lambda}\right] \quad (\lambda\in\mathbb{R}) \quad d=1$$

$$= \frac{\pi\theta\lambda}{\pi-\lambda} \quad (-\infty < \lambda < \pi) \quad d=2$$

$$= \theta \sum_{x,y\in A}\left[\frac{|A|}{\lambda}\,\underline{1} - G_A\right]^{-1}(x,y) \quad (-\infty < \lambda < e_A^{-1}) \quad d\geq 3.$$

Here $(G_A) = (G_\infty(x,y))$ restricted to A, e_A is the largest eigenvalue of G_A and $\underline{1}$ is the identity matrix. It is even possible to compute $I(\alpha)$ in some special cases. For instance,

$$I(\alpha) = \pi\,(\sqrt{\alpha} - \sqrt{\theta})^2 \quad d=2,$$

$$= [G_\infty(0,0)]^{-1}\,(\sqrt{\alpha} - \sqrt{\theta})^2 \quad d\geq 3, \quad |A|=1.$$

A key step in computing the cumulant asymptotics (2.2) is the "Campbell's formula"

$$\text{(3.2)} \qquad \log E[e^{\lambda T_t(A)}] = \theta \sum_{x\in\mathbb{Z}^d}\left(E[\exp\{\lambda T_t^x(A)\}] - 1\right),$$

where $T^x(A)$ is the occupation time of A due to a single particle starting at x. Thus the *cumulants* of the infinite system reduce to sums of *moments* of one particle systems. See [3] for the details.

CRITICAL BINARY BRANCHING BROWNIAN MOTION. The results for β_t are fairly complete, but not as explicit. Since Fleischman and Sawyer [6] have shown that $T_t(A) \to T_\infty(A) < \infty$ a.s. for $d = 1$, we will consider only $d \geq 2$. The proper scalings are

$$
\begin{aligned}
a_t &= 1 & d &= 2\ , \\
&= \sqrt{t} & d &= 3\ , \\
&= t/\log t & d &= 4\ , \\
&= t & d &\geq 5\ .
\end{aligned}
$$

With these scalings one can check (2.2) and then prove that the limiting cumulant generating function $\Psi(\lambda)$ exists for λ sufficiently close to 0, at least when $d \neq 4$. In addition there is an inductive recipe (in terms of u_t and Γ_t) for the c_n. So (1.2) again holds for α close to θ, and presumably for all $\alpha > \theta$. The most interesting case is $d = 2$, where we prove that

$$\frac{T_t(A)}{|A|t} \Rightarrow Y \qquad \text{as} \quad t \to \infty\ ,$$

with Y a nondegenerate (non-Gaussian) random variable.

To derive these results one first applies the continuous state space analogue of (3.2) to reduce the problem of cumulants to one of moments. A second key step is the one-sided version of (2.2):

$$\text{(3.3)} \qquad \varphi_n(t) \leq \theta\, n!\, (K/a_t)^{n-1} \qquad t \geq 1\ ,$$

with K independent of n and t.

For $d = 4$ we have only succeeded in proving a statement weaker than (1.2): namely, there are positive constants c and C such that for large enough t and α sufficiently close to θ,

$$\text{(3.4)} \qquad \exp\left[-Ct/\log t\right] \leq P\left(\frac{T_t(A)}{|A|t} > \alpha\right) \leq \exp\left[-ct/\log t\right]\ .$$

Our method of deriving (3.4) may be of use in other complex large deviation problems. The idea is that if one cannot prove the actual convergence of Ψ_t required by the Convergence Lemma, then reasonable upper and lower bounds on Ψ_t should at least provide reasonable order estimates for the large deviation probabilities. We have formalized this strategy as follows.

LARGE DEVIATIONS COMPARISON LEMMA. Given real random variables Y_t $(t \geq 0)$ and $a_t \to \infty$ as $t \to \infty$, let $\Psi_t(\lambda) = a_t^{-1} \log E[e^{\lambda Y_t}]$. Let $\underline{\Psi}$ and $\overline{\Psi}$ be functions of λ such that for some $t_0 > 0$ and $0 < \lambda_0 < \infty$,

(i) $\underline{\Psi}$ and $\overline{\Psi}$ are strictly convex on $[0,\lambda_0)$, and $\overline{\Psi}$ is differentiable on $[0,\lambda_0)$,

(ii) $\underline{\Psi}(0) = \overline{\Psi}(0) = 0$, $\underline{\Psi}'(0) = \overline{\Psi}'(0) = \mu$, and

(iii) $\underline{\Psi} \leq \Psi_t \leq \overline{\Psi}$ on $[0,\lambda_0)$ for all $t \geq t_0$.

Then there exists $\overline{\alpha} > \mu$ such that for all $\alpha \in (\mu,\overline{\alpha})$,

$$-\infty < \liminf_{t \to \infty} a_t^{-1} \log P(a_t^{-1} Y_t > \alpha)$$

$$\leq \limsup_{t \to \infty} a_t^{-1} \log P(a_t^{-1} Y_t > \alpha) < 0 .$$

An analogous result holds below μ.

To check that this lemma proves (3.4), we use (3.2), (3.3) and the fact that $\varphi_2(t) \sim c_2/a_t$; (iii) follows for $d=4$ with

$$\underline{\Psi}(\lambda) = \theta\lambda + k\lambda^2 , \qquad \overline{\Psi}(\lambda) = \theta\lambda + \frac{JK\lambda^2}{1-K\lambda}$$

for suitable $0 < k,J,K < \infty$, $\lambda < \lambda_0$. See [4] for the details.

4. INTERACTING MODELS. Life now becomes more difficult. Start both the simple exclusion process η_t and the voter model ξ_t in Bernoulli product measure with density θ, $0<\theta<1$. The particles of η_t execute independent rate one simple random walks on $\mathbb{Z}^d$, except that jumps to occupied sites are suppressed. The "voters" in ξ_t change opinions according to the rule:

$$\xi_t(x) \to 1-\xi_t(x) \quad \text{at rate } (2d)^{-1}|\{y: \xi_t(y)\neq\xi_t(x),\ |y-x|=1\}|.$$

Like the independent random walks, the exclusion process is stationary; like the branching Brownian motions, the voter model clusters for $d = 1,2$ but is stable for $d \geq 3$. So it is natural to expect parallel large deviations behavior.

Unfortunately the Campbell trick is unavailable in the presence of interaction, and the cumulants $\varphi_n(t)$ seem intractable. So what *can* be done? To simplify matters in the discussion which follows let us set $A = \{0\}$ and write T_t for $T_t(A)$. First, the variance calculations are not difficult. For η_t, the proper choices of a_t agree exactly with the independent random walk scalings (3.1). For ξ_t the appropriate scalings are given by

$$\begin{aligned} a_t &= 1 && d = 1, \\ &= \log t && d = 2, \\ &= \sqrt{t} && d = 3, \\ &= t/\log t && d = 4, \\ &= t && d \geq 5. \end{aligned} \tag{4.1}$$

Note that for $d \geq 3$ these agree with the scalings of β_t. However the one dimensional case here most resembles the two dimensional case for branching Brownian motion, and the two dimensional case here is qualitatively new. Using an invariance principle due to Arratia we show in [2] that for the one dimensional voter model, $T_t/t \Rightarrow Z$ as $t \to \infty$, Z a nondegenerate random variable (with support in [0,1]). The intriguing scenario suggested by our techniques for the voter model on $\mathbb{Z}^2$ is:

$$P(T_t/t > \alpha) \sim t^{-I(\alpha)} \qquad \text{as} \quad t \to \infty .$$

One would like to find a "soft" proof that (1.2) holds for some $I(\alpha) \in (0,\infty)$. For instance, it can be shown that η_t satisfies

$$\begin{aligned} P(T_{s+t} > \alpha(s+t)) &\geq P(T_s > \alpha s\,,\ T_{s+t} - T_s > \alpha t) \\ &\geq P(T_s > \alpha s)\ P(T_{s+t} - T_s > \alpha t) \\ &= P(T_s > \alpha s)\ P(T_t > \alpha t) \quad . \end{aligned}$$

By subadditivity,

$$\lim_{t\to\infty} \frac{1}{t} \log P(T_t/t > \alpha) = -\bar{I}(\alpha)$$

exists. The trouble is that $\bar{I}(\alpha) \equiv 0$ for $d = 1,2$, which indicates that the large deviation tails are "fat". For $d \geq 3$ this result is satisfactory, although we get little information about the limits $\Psi(\lambda)$ and $I(\alpha)$. Similar arguments hold for the voter model, but now the results are nontrivial only when $d \geq 5$. We have no existence proofs for $\Psi(\lambda)$ and $I(\alpha)$ in low dimensions.

In the face of complex models with strong interaction it is perhaps inevitable to settle for estimates like (3.4), which capture at least the correct order of decay in the large deviation probabilities. We would like to report some progress along these lines for the two models under discussion.

SIMPLE EXCLUSION. Arratia [2] has obtained some partial results. He has shown that there are constants $0 < c < C < \infty$ such that

$$(4.2) \qquad \exp\{-Ca_t\} \le P(T_t=t) \le \exp\{-ca_t\} ,$$

with a_t given by (3.1). The constants c and C relate to the range of simple random walk. The left inequality clearly provides a lower bound on $P(T_t/t > \alpha)$ for all $\alpha \in (\theta,1)$. We do not yet have a corresponding upper bound for the exclusion process.

VOTER MODEL. We are able to prove good upper *and* lower bounds for the voter model occupation time large deviation probabilities, with a_t given by (4.1). To get the upper bound we use the Comparison Lemma, producing a $\overline{\Psi}_t$ of the form

$$\overline{\Psi}_t(\lambda) = \frac{\theta\lambda}{1-K\lambda} , \quad t \ge 0 , \quad \lambda < K^{-1} ,$$

for a suitable positive constant K . The method is to estimate $E[\exp\{\lambda T_t\}]$ directly, using duality, a correlation inequality, and an improvement of inequality (4.3) of [2]. Corresponding lower bounds, obtained by estimating $P(T_t = t)$ as in (4.1) , are currently being worked out in collaboration with Maury Bramson. This technique doesn't quite work in dimension two, where yet another approach is needed.

BIBLIOGRAPHY

1. Arratia, R., "Symmetric exclusion processes: a comparison inequality and a large deviation result", Ann. Probability, 13 (1985), to appear.

2. Cox, J.T. and Griffeath. D., "Occupation time limit theorems for the voter model", Ann. Probability, 11 (1983), 876-893.

3. Cox, J.T. and Griffeath, D., "Large deviations for Poisson systems of independent random walks", Z. Wahr. verw. Gebiete, 66 (1984), 543-558.

4. Cox, J.T. and Griffeath, D., "Occupation times for critical branching Brownian motions", Ann. Probability, 13 (1985), to appear.

5. Ellis, R.S., "Large deviations and other limit theorems for dependent random variables with applications to the Ising model in statistical mechanics", (1983), preprint.

6. Fleischman, J. and Sawyer, S., "Maximum geographic range of a mutant allele considered as a subtype of a Brownian branching field", Proc. Natl. Acad. Sci. U.S.A., 76 (1978), 872-875.

7. Plachky, D. and Steinebach, J., "A theorem about probabilities of large deviations with an application to queueing theory", Period. Math. Hungar., 6 (1975), 343-345.

8. Sievers, G.L., "On the probability of large deviations and exact slopes", Ann. Math. Statist., 40 (1969), 1908-1921.

DEPARTMENT OF MATHEMATICS
SYRACUSE UNIVERSITY
SYRACUSE, NEW YORK 13210

DEPARTMENT OF MATHEMATICS
UNIVERSITY OF WISCONSIN
MADISON, WISCONSIN 53706

Contemporary Mathematics
Volume 41, 1985

FUNCTION-VALUED DUALS FOR MEASURE-VALUED PROCESSES AND APPLICATIONS

Donald A. Dawson[1] and Kenneth J. Hochberg[2]

ABSTRACT. A class of function-valued stochastic processes is derived, and it is shown that these processes are dual to certain measure-valued diffusion processes. The duality relationships that are obtained are then applied to the studies of scaling limits, ergodic behavior, and support properties of measure-valued processes. Specific applications of these techniques are presented.

1. INTRODUCTION. Important insight into the behavior of stochastic processes in finite dimensions has been obtained from in-depth studies of such processes as the Wiener process, the Poisson process, Gaussian and stable processes, and branching processes, in which one can explicitly compute the finite-dimensional distributions. It is reasonable to expect that a similar role may be played in the developing theory of infinite-dimensional stochastic systems by special classes of processes whose finite-dimensional distributions can be explicitly determined. One of the main techniques in the computation of the distribution of elementary stochastic processes is the use of characteristic functions or Laplace transforms. However, in the context of systems of infinitely many interacting particles, a new method involving the use of "dual processes" was introduced by Spitzer (1970) and was further developed by Holley and Liggett (1975) and by Holley and Stroock (1979). Dual processes have also been applied to the study of a class of finite-dimensional diffusion processes in Holley, Stroock and Williams (1977), and to measure-valued stochastic processes in Shiga (1981) and in Dawson and Hochberg (1982). Dawson and Kurtz (1982) give sufficient conditions under which existence of a process that is dual to a measure-valued diffusion implies uniqueness of the solution to an associated

1980 Mathematics Subject Classification. 60J60, 60J70, 60J25, 60G57, 60K35.

[1]Research supported in part by the Natural Sciences and Engineering Research Council of Canada.

[2]Research supported in part by the U. S. National Science Foundation under grant DMS83-11972, while the author was a visitor of the Hebrew University of Jerusalem and of Bar-Ilan University, Israel. The hospitality extended by these institutions is gratefully acknowledged.

0271-4132/85 $1.00 + $.25 per page

martingale problem.

In this paper, we describe a class of function-valued dual processes which are used in the study of measure-valued diffusion processes, and discuss how one can apply the dual process to the studies of scaling limits, ergodic behavior, and support properties of measure-valued processes.

The model used to illustrate these techniques is the probability-measure-valued process introduced by Fleming and Viot (1978, 1979). However, these methods are also applicable to the voter model (Holley and Liggett (1975)), measure-valued branching processes (Dawson and Hochberg (1982)), the random Schrödinger equation (Dawson and Papanicolaou (1983)), and several models in genetics (Ethier and Kurtz (1985), Shiga (1981)).

2. MEASURE-VALUED STOCHASTIC PROCESSES

The general setting for measure-valued processes and their duals is given in Dawson and Kurtz (1982). We begin with a locally compact complete separable metric space S and the space $\mathcal{M}(S)$ of bounded Radon measures on S. $\mathcal{M}(S)$ can be furnished with an appropriate metric so as to be a complete, separable metric space with a topology equivalent to that of weak convergence of measures. For a closed subset E of $\mathcal{M}(S)$, let $\mathcal{B}(E)$ denote the σ-algebra of Borel subsets of E, $C_b(E)$ denote the space of continuous bounded functions on E, and $L_\infty(E)$ denote the space of bounded measurable functions on E. Let $\Omega_C^E = C([0,\infty),E)$ be the space of continuous functions mapping $[0,\infty)$ into E, and let $D_C^E = D([0,\infty),E)$ be the space of functions mapping $[0,\infty)$ into E that are right continuous with limits from the left. A measure-valued (or E-valued) stochastic process $\{X(t):t \geq 0\}$ is then given by a mapping $X:[0,\infty) \times \Omega_C^E$ (or Ω_D^E) $\to E$ defined by the canonical process $X(t,\omega) \equiv \omega(t)$ for $t \geq 0$, $\omega \in \Omega_C^E$ (or Ω_D^E). The distribution of a measure-valued (or E-valued) stochastic process $\{X(t):t \geq 0\}$ is determined by a measurable mapping $\mu \to P_\mu$ from E into $P(\Omega_C^E)$ or $P(\Omega_D^E)$, the spaces of probability measures on Ω_C^E and Ω_D^E, respectively, furnished with the topology of weak convergence.

For a subset A of $C_b(E) \times L_\infty(E)$, a family of probability distributions $\{P_\mu:\mu \in E\}$ is a $P(\Omega_C^E)$- (or $P(\Omega_D^E)$-) solution to the initial-value martingale problem for A if for every pair $(f,g) \in A$, the following hold:

(i) $P_\mu \in P(\Omega_C^E)$ (or $P(\Omega_D^E)$) for each $\mu \in E$,

(ii) $P_\mu\{\omega:X(0,\omega) = \mu\} = 1$ for each $\mu \in E$,

(iii) For each P_μ, the canonical process $\{X(t):t \geq 0\}$ is a solution to the martingale problem for A in the sense that for every $(f,g) \in A$,

$$Z(t) \equiv f(X(t)) - \int_0^t g(X(t))\, ds \tag{2.1}$$

is a P_μ-martingale.

The following theorem gives conditions for asserting uniqueness of the solution to the initial-value martingale problem for A and thereby will serve as motivation for the selection of an appropriate class of functions on which to define the dual process.

(2.2) THEOREM (Stroock and Varadhan (1979)). Let Φ be any subset of $L_\infty(E)$ that generates $L_\infty(E)$ under bounded pointwise convergence. If for each $t \geq 0$ and $F \in \Phi$, there is a function $G_{F,t} \in L_\infty(E)$ such that for $\mu \in E$,

$$E_\mu\{F(X(t))\} = G_{F,t}(\mu) \tag{2.3}$$

for any $P(\Omega_C^E)$- (or $P(\Omega_D^E)$-) solution $\{P_\mu : \mu \in E\}$ of the initial value martingale problem for A, then there exists at most one such solution. If such a solution does exist, then it is measurable and satisfies the strong Markov property.

In the next section, we construct a dual process whose state space is contained in $C(E)$ and is sufficiently rich so as to generate $L_\infty(E)$ under bounded pointwise convergence.

3. FUNCTION-VALUED DUAL PROCESSES

Let S^* denote the one-point compactification $S \cup \{\infty\}$ of S, and let S^{*N} denote the N-fold Cartesian product of S^*. Let $\mathcal{C} = \cup_{N=0}^\infty C(S^{*N})$, furnished with the topology given by the inductive limit of the sup norm topologies, and let $C(S^{*0}) = R^1$. For $f \in \mathcal{C}$, define the function $N(f) = N$ if $f \in C(S^{*N})$. Similarly, let $\mathcal{D} = \cup_{N=0}^\infty D(S^{*N})$ where, for each N, $D(S^{*N})$ is a dense subset of $C(S^{*N})$.

A <u>monomial</u> on $\mathcal{M}(S^*)$ is a function of the form

$$\begin{aligned} F_f(\mu) &= \int_S \cdots \int_S f(x_1,x_2,\dots,x_{N(f)})\mu^{(N(f))}(dx) \\ &\equiv \int_S \cdots \int_S f(x_1,x_2,\dots,x_{N(f)})\mu(dx_1) \cdots \mu(dx_{N(f)}) \end{aligned} \tag{3.1}$$

for $f \in \mathcal{C}$. For a compact subset E of $\mathcal{M}(S^*)$, let $\Pi_1(E)$ denote the set of monomials restricted to E, and let the <u>algebra of polynomials</u> $\Pi(E)$ denote the smallest algebra of functions on E which contains $\Pi_1(E)$. Similarly, define $\Pi_1^{\mathcal{D}}(E)$ and $\Pi^{\mathcal{D}}(E)$ as the monomials and polynomials on E with coefficients $f \in \mathcal{D}$. Since $\Pi^{\mathcal{D}}(E)$ separates points of E and is dense in E, it follows from the Stone-Weierstrass theorem that $\Pi^{\mathcal{D}}(E)$ is dense in $C(E)$. But $C(E)$ generates $L_\infty(E)$ under bounded pointwise convergence, so $\Pi^{\mathcal{D}}(E)$ does so as well.

Let Q be the linear space of functions on $\mathcal{D}$ which contains functions of the form

(3.2) $$F(\mu,f) = F_\mu(f) \equiv F_f(\mu) \qquad \text{for } \mu \in E,\ f \in \mathcal{D}\ .$$

Consider a pregenerator L with domain $\mathcal{D}(L) \subset \Pi^{\mathcal{D}}(E)$ such that for $F_f \in \mathcal{D}(L)$, L can be related to another pregenerator $L^{\#}$ via the relationship

(3.3) $$LF_f(\mu) = L^{\#}F_\mu(f) + V(N(f))F_\mu(f)$$

where V is a function defined on $\{0,1,2,\ldots\}$.

If the martingale problem associated with $\{(F_f,LF_f):f \in \mathcal{D}(L)\}$ has an $\mathcal{M}(S^*)$-valued solution $\{X(t):t \geq 0\}$ and the martingale problem associated with $\{(F_\mu,L^{\#}F_\mu):\mu \in E\}$ has a $\Pi^{\mathcal{D}}(E)$-valued solution $\{Y(t):t \geq 0\}$, then $X(t)$ and $Y(t)$ are said to form a pair of <u>dual processes</u>.

Under various sets of hypotheses, Dawson and Kurtz (1982) proved that the existence of solutions to both the L and $L^{\#}$ martingale problems implies the uniqueness, measurability and strong Markov property of a $P\{\Omega_C^{\mathcal{M}(S^*)}\}$-valued solution $\{P_\mu:\mu \in \mathcal{M}(S^*)\}$ of the initial-value martingale problem associated with the pair (F_f,LF_f) for $f \in \mathcal{D}$. The result is demonstrated by producing the function

(3.4) $$G_{F,t}(\mu) = E_\mu\{F_f(X(t))\}\ ,$$

required in the theorem of Stroock and Varadhan, from the duality relationship (3.3) and the Feynman-Kac formula:

(3.5) $$E_\mu\{F_f(X(t))\} = E\{F_\mu(Y(t)) \exp(\int_0^t V(Y(u))du) \mid Y(0) = f\}\ .$$

Many measure-valued diffusion processes can be characterized as unique solutions of martingale problems associated with second order differential operators of the form

(3.6) $$LF(\mu) \equiv A(\mu)\,\frac{\delta F(\mu)}{\delta\mu(x)} + \tfrac{1}{2}\,B(\mu)\,\frac{\delta^2 F(\mu)}{\delta\mu(x)\delta\mu(y)}\ , \qquad F \in \mathcal{D}(L)$$

where

$$\frac{\delta F(\mu)}{\delta\mu(x)} = \left.\frac{\partial F(\mu + \varepsilon\,\delta_x)}{\partial\varepsilon}\right|_{\varepsilon = 0}$$

and where

$A(\mu)$ is a linear functional on $\mathcal{D}(A) \subset C(S)$:

$B(\mu)$ is a bilinear functional on $\mathcal{D}(B) \subset C(S \times S)$;

for $F \in \mathcal{D}(L)$, $\frac{\delta F(\mu)}{\delta\mu(\cdot)} \in \mathcal{D}(A)$, $\frac{\delta^2 F(\mu)}{\delta\mu(\cdot)\delta\mu(\cdot)} \in \mathcal{D}(B)$; and

$\delta_x \in \mathcal{M}(S)$ consists of a unit atom at x .

The computations $\frac{\delta F_f(\mu)}{\delta\mu(x)}$ and $\frac{\delta^2 F_f(\mu)}{\delta\mu(x)\delta\mu(y)}$ are now straightforward:

$$(3.7)\quad \frac{\delta F_f(\mu)}{\delta\mu(x)} = \sum_{j=1}^{N(f)} \int_{S^*} \cdots \int_{S^*} f(x_1,\ldots,x_{j-1},x,x_{j+1},\ldots,x_{N(f)}) \prod_{\substack{\ell=1\\ \ell\neq j}}^{N(f)} \mu(dx_\ell) ,$$

and

$$(3.8)\quad \frac{\delta^2 F_f(\mu)}{\delta\mu(x)\delta\mu(y)} =$$

$$\sum_{\substack{j=1\\ j\neq k}}^{N(f)} \sum_{k=1}^{N(f)} \int_{S^*}\cdots\int_{S^*} f(x_1,\ldots,x_{j-1},x,x_{j+1},\ldots,x_{k-1},y,x_{k+1},\ldots,x_{N(f)})$$

$$\cdot \prod_{\substack{\ell=1\\ \ell\neq j,k}}^{N(f)} \mu(dx_\ell) .$$

4. AN APPLICATION OF DUALITY TO A MEASURE-VALUED PROCESS

An example of a measure-valued stochastic process which is characterized as a solution to a martingale problem with differential operator L of the form (3.6) is provided by the Fleming-Viot process, which arises as a diffusion approximation to the continuous-time Ohta-Kimura process of population genetics. According to this model, two evolutionary forces act on some natural population of fixed size N : random genetic drift, which results in multinomial sampling from the empirical distribution of genetic types in the host population at a rate γ, and mutation to neighboring genetic states at a rate D. If $p(t,k)$ denotes the relative frequency of genetic type $k = (k_1,\ldots, k_d) \in Z^d$, then the Markov process $\{p(t,\cdot):t \geq 0\}$ has generator, for $p \in \mathcal{M}_1(Z^d)$, the space of probability measures on Z^d, and $f \in C_b(\mathcal{M}_1(Z^d))$, given by

$$(4.1)\quad L_N f(p) = \sum_{i\neq j} [\gamma p(i)p(j)+Dp(i)\sum_k p(k)\theta_{kj}][f(p^{ij})-f(p)] ,$$

where

$$(4.2)\qquad p^{ij}(k) = \begin{cases} p(k) - \frac{1}{N} & \text{if } k = i \\ p(k) + \frac{1}{N} & \text{if } k = j \\ p(k) & \text{if } k \neq i,j \end{cases}$$

and

$$(4.3)\qquad \theta_{ij} = \begin{cases} 1 & \text{if } |i-j| \equiv \sum_{k=1}^{d} |i_k - j_k| = 1 \\ -2d & \text{if } i = j \\ 0 & \text{otherwise}. \end{cases}$$

If we rescale in space and time by letting

$$(4.4)\qquad Y_N(t,A) \equiv \sum_{\frac{j}{\sqrt{N}} \in A} p(N^2 t, j),$$

the resulting process $Y_N(\cdot)$ is a Markov process with state space $\mathcal{M}_1^N(R^d)$, the space of probability measures concentrated on the lattice $N^{-\frac{1}{2}}Z^d$ and consisting of the superposition of N atoms of mass $1/N$. The generator is given by

$$(4.5)\qquad G_N F(\mu) = DN^2 \int \sum_{\substack{k \\ |k-j|=1}} [F(\mu - \frac{\delta_j}{N} + \frac{\delta_k}{N}) - F(\mu)]\mu(dj)$$

$$+ \gamma N^2 \iint [F(\mu - \frac{\delta_j}{N} + \frac{\delta_k}{N}) - F(\mu)]\mu(dj)\mu(dk)$$

$$= \int [A_N \frac{\partial F(\mu)}{\partial \mu(x)}]\,\mu(dx)$$

$$+ \frac{\gamma}{2} \iint [\frac{\partial^2 F(\mu)}{\partial \mu(x)^2} + \frac{\partial^2 F(\mu)}{\partial \mu(y)^2} - 2\,\frac{\partial^2 F(\mu)}{\partial \mu(x)\partial \mu(y)}]\mu(dx)\mu(dy)$$

$$+ R(F)/N^{\frac{1}{2}},$$

where

$$A_N g(x) = D \sum_{\substack{y \in N^{-\frac{1}{2}}Z^d \\ |x-y| = \frac{1}{N^{\frac{1}{2}}}}} \frac{g(y) - g(x)}{N} \quad \text{for } x \in N^{-1/2}Z^d$$

provided that F has bounded third derivatives. Hence in the notation of (3.6),

$$(4.6)\qquad G_N F(\mu) = LF(\mu) + R(F)/N^{\frac{1}{2}}$$

where

$$(4.7)\quad LF(\mu) = \int_{S^*} [A\,(\frac{\delta F(\mu)}{\delta\mu(x)})]\mu(dx) + \int_{S^*}\int_{S^*} \frac{\delta^2 F(\mu)}{\delta\mu(x)\delta\mu(y)}\, Q(\mu{:}dx \times dy) + R(F)/N^{1/2}\,.$$

Here,

(4.8) $S^* = R^d \cup \{\infty\}$, $A = D \cdot \Delta$ where Δ denotes the Laplace operator, $R(F)$ is a bounded remainder term provided that F has bounded third derivatives, and $Q(\mu{:}dx \times dy) = \gamma[\mu(dx) + \delta_x(dy) - \mu(dx)\mu(dy)]$.

Fleming and Viot (1979) derived the existence and uniqueness of a solution to the initial-value martingale problem associated with the pair $(L,\mathcal{D}(L))$ as a probability-measure-valued diffusion process.

A dual for this process was introduced in Dawson and Hochberg (1982). Let $E = \mathcal{M}_1(S^*)$, the set of probability measures on the one-point compactification of R^d, and form the algebra of polynomials $\Pi(\mathcal{M}_1(S^*))$ as before. For $f \in C^\infty(S^{*N})$,

$$(4.9)\qquad LF_f(\mu) = \int\cdots\int Kf(x_1, \ldots, x_N)\,\mu(dx_1)\ldots\mu(dx_\mu),$$

where

$$(4.10)\quad Kf(x_1,\ldots,x_N) = D\sum_{j=1}^{N} \Delta_j f(x_1, \ldots, x_N) + \gamma \sum_{j=1}^{N} \sum_{\substack{k=1\\k\neq j}}^{N} [\Phi_{jk}(f(x_1,\ldots,x_N)) - f(x_1, \ldots, x_N)]$$

where Δ_j denotes the d-dimensional Laplacian acting on the variable x_j, and

$$(4.11)\qquad \Phi_{jk} : C(S^N) \to C(S^{N-1})$$

such that

$$\Phi_{jk}(f(x_1,\ldots,x_N)) = (\Phi_{jk}f)(y_1, \ldots, y_{N-1})$$

where for $j \neq k$,

$$\begin{aligned} x_i &= y_i && \text{for } i = 1, \ldots, k-1,\\ x_k &= y_j && \text{if } j < k\\ x_k &= y_{j-1} && \text{if } j > k\\ x_i &= y_{i-1} && \text{for } i = k+1, \ldots, N\,. \end{aligned}$$

Equivalently, we can write

$$(4.12)\quad LF_f(\mu) = F_{\Delta_{N(f)}f}(\mu) + \gamma \sum_{j=1}^{N(f)} \sum_{\substack{k=1 \\ j\neq k}}^{N(f)} [F_{\Phi_{jk}f}(\mu) - F_f(\mu)]$$

where Δ_N is the Laplacian operating on functions $f \in C(S^{*N})$.
Thus, we have the relationship (3.3) in the form

$$LF_f(\mu) = L^{\#}F_{\mu}(f),$$

where $L^{\#}$ is a linear operator defined on $Q(C)$. $L^{\#}$ is the infinitesimal generator of a Markov process $\{Y(t)\}$ on C evolving according to the following mechanisms:

(i) $Y(t)$ jumps from $C(S^N)$ to $C(S^{N-1})$ for $N \geq 2$;

(ii) At the time of a jump, f is replaced by $\Phi_{jk}f$;

(iii) Between jumps, $Y(t)$ is deterministic on $C(S^N)$ and evolves according to the Brownian motion semigroup H_t^N on $(R^d)^N$, i.e.,

$$(4.13)\quad H_t^N f(x) = \frac{1}{(2\pi t)^{Nd/2}} \int e^{-|x-y|^2/2t} f(y)dy \ ;$$

(iv) Once $Y(t)$ is in $C(S)$, i.e., it is a continuous function of only one variable, no further jumps occur.

To make explicit the process $Y(t)$, let $T_2, T_3, \ldots, T_N$ be a sequence of independent, identically distributed exponential random variables with means

$$E(T_i) = \frac{1}{\gamma i(i-1)}, \ i = 2, 3, \ldots, N,$$

and let

$$(4.14)\quad \begin{aligned} S_N &= 0 \\ S_{N-1} &= T_N \\ S_{N-2} &= T_N + T_{N-1} \\ &\vdots \\ S_1 &= T_2 + T_3 + \ldots + T_N, \end{aligned}$$

so that

$$(4.15)\qquad T_{N-j} = S_{N-j-1} - S_{N-j} \ .$$

Finally, let $u_2, u_3, \ldots, u_n$ be independent random variables such that u_i has a discrete uniform distribution on the $i(i-1)$ ordered pairs from the set $\{1, 2, \ldots, i\}$.

Assume that $Y(0) = F \in C(S^N)$. Then,

$$
\begin{aligned}
(4.16) \qquad Y(t) &= H_t^N Y(0) \text{ for } 0 = S_N \leq t < S_{N-1} \\
&= H_{t-S_{N-1}}^{N-1} \Phi_{u_N}(H_{S_{N-1}}^N Y(0)), \; S_{N-1} \leq t < S_{N-2} \\
&= H_{t-S_{N-2}}^{N-2} \Phi_{u_{N-1}}(Y(S_{N-2}-)), S_{N-2} \leq t < S_{N-3} \\
&\;\;\vdots \\
&= H_{t-S_2}^{2} \Phi_{u_3}(Y(S_2 -)), \; S_2 \leq t < S_1 \\
&= H_{t-S_1}^{1} \Phi_{u_2}(Y(S_1 -)), \; t \geq S_1 \;.
\end{aligned}
$$

The duality relationship can be expressed as follows:

$$(4.17) \qquad E_\mu \{F_f(X(t))\} = E_f \{F_\mu(Y(t))\}$$

where

$$
\begin{aligned}
(4.18) \qquad E_\mu \{F_f(X(t))\} &= T_t F_f(\mu) \\
&= E_\mu \left\{ \int \ldots \int f(x_1,\ldots,x_N) X(t,dx_1) \ldots X(t,dx_N) \right\}
\end{aligned}
$$

and

$$
\begin{aligned}
(4.19) \qquad E_f \{F_\mu(Y(t))\} &= F_\mu(E_f\{Y(t)\}) \\
&= \int \ldots \int E_f\{Y_t(x_1, \ldots, x_{N(t)})\}\mu(dx_1) \ldots \mu(dx_{N(t)}).
\end{aligned}
$$

5. ERGODIC BEHAVIOR VIA DUALITY

To see how one may utilize the duality to obtain results on the limiting behavior of the Fleming-Viot process, we now examine the behavior of the empirical centered moments of $X(t)$. Let the <u>empirical centered moments</u> of a probability measure-valued process $\{X(t) : t \geq 0\}$ be defined by

$$(5.1) \qquad R_{k_1,\ldots,k_d}(t) \equiv \int_{R^d} \prod_{i=1}^{d} (x_i - x_i(t))^{k_i} X(t,dx),$$

where the empirical mean process $x(t) \equiv (x_1(t), \ldots, x_d(t))$ is given by

$$x_i(t) = \int_{R^d} x_i X(t,dx) \quad , \ i = 1, \ldots, d .$$

Then,

$$(5.2) \quad R_{k_1,\ldots,k_d}(t) = \int \ldots \int g(x,x^{(1)},\ldots,x^{(N_0)})X(t,dx)X(t,dx^{(1)})\ldots X(t,dx)^{(N_0)})$$

where

$$(5.3) \quad g(x,x^{(1)},\ldots,x^{(N_0)}) = \prod_{\ell=1}^{d} \prod_{j=1}^{k_\ell} (x_\ell - x_\ell^{(j)}), \ N_0 = \sum_{\ell=1}^{d} k_\ell .$$

We then have the following result:

(5.4) THEOREM (Dawson and Hochberg (1982)). If $\int |x|^{N_0} \mu(dx) < \infty$, then

(i) $E_\mu(R_{k_1,\ldots,k_d}(t)) < \infty$ for $t \geq 0$,

and

(ii) $\lim_{t\to\infty} E_\mu(R_{k_1,\ldots,k_d}(t)) \equiv r_{k_1,\ldots,k_d}$ exists and is finite.

Proof: By duality,

$$(5.5) \quad E_\mu(R_{k_1,\ldots,k_d}(t)) = E_g\Big\{\int \ldots \int Y_t(x_1,\ldots,x_{N(t)})\mu(dx_1) \ldots \mu\, (dx_{N(t)}) \cdot \exp\Big(\int_0^t V(Y(u))du\Big)\Big\} < \infty,$$

since the dual preserves an initial growth condition. Moreover, $P_f\{\inf\{t : Y(t) \in C(S)\} < \infty\} = 1$ by elementary properties of pure death processes, so, by the translation invariance of g,

$$(5.6) \quad \lim_{t\to\infty} E_\mu\{R_{k_1,\ldots k_d}(t)\} = E_g\{\int Y_\infty(x)\mu(dx)\} = E_g\{Y_\infty\} < \infty ,$$

since the exponential distribution has finite moments of all orders. Here, $Y_\infty(x) = \lim_{t\to\infty} Y_t(x_1,\ldots,x_N)$ because $\lim_{t\to\infty} N(f) = 1$.

From the dual process we can also obtain the ergodic behavior of the random cluster

$$(5.7)\quad X^*(t,dx) \equiv X(t,\theta_{x(t)}dx) \equiv X(t,\{a-x(t) : a\in dx\})$$

centered at the empirical mean X(t). X^* is a stationary process, and the Birkhoff ergodic theorem implies that

$$(5.8)\qquad I(f) = \lim_{T\to\infty} \frac{1}{T}\int_0^T \int_{R^d} f(x)\, X^*(t,dx)dt$$

exists a.s. as a linear functional on C(S) and is independent of the initial measure μ . Thus, we have for any $\mu \in \mathcal{M}_1(S^*)$,

$$(5.9)\qquad \lim_{t\to\infty} E_\mu\left\{\int f(x)\, X^*(t,dx)\right\} = I(f).$$

For example, we have the non-Gaussian limit

$$(5.10)\quad \lim_{T\to\infty} \frac{1}{T}\int_0^T X^*(t)dt = E\{X^*(0)\}$$

$$= \gamma \in \mathcal{M}_1(R^d) \quad \text{a.s. ;}$$

that is,

$$(5.11)\qquad \lim_{T\to\infty} \frac{1}{T}\int_0^T X^*(t,A)dt = \nu(A),$$

where $\nu(A)$ is the expected value of the equilibrium random measure on the set A . The collection

$$(5.12)\qquad \{r_{k_1,\ldots,k_d}\} = \{\lim_{t\to\infty} E_\mu(R_{k,\ldots k_d}(t))\}$$

forms the joint moment system of the expected steady-state distribution of the random cluster centered at the empirical mean.

6. SCALING LIMIT VIA DUALITY

In this section, we illustrate an application of duality to obtain the scaling limit of the Fleming-Viot process. Let

$$(6.1)\qquad X_\varepsilon(t,A) \equiv X(t/\varepsilon^2\ ;\ A_\varepsilon)$$

where $\{X(t) : t \geq 0\}$ denotes the Fleming-Viot process and

$$A_\varepsilon \equiv \{x : \varepsilon x\in A\}\ .$$

It can be verified that the rescaled process X_ε has generator

(6.2) $$L_\varepsilon F_f(\mu) = L_\varepsilon^{\#} F_\mu(f),$$

where

(6.3) $$L_\varepsilon^{\#} F_\mu(f) = F_\mu(\Delta_{N(f)} f) + \frac{\gamma}{\varepsilon^2} \sum_{\substack{j=1 \\ j\neq k}}^{N(f)} \sum_{k=1}^{N(f)} [F_\mu(\Phi_{jk} f) - F_\mu(f)] .$$

From the explicit form (4.16)-(4.17) of the duality relation, it follows that for $f \in C(S^N)$,

(6.4) $$\lim_{\varepsilon\to 0} E_{\delta_x}\{F_f(X_\varepsilon(t))\} = \lim_{\varepsilon\to 0} E_f\{F_{\delta_x}(Y_\varepsilon(t))\} = H_t^1(\Psi^1 f)(x),$$

where for $f \in C(S^N)$,

(6.5) $$\Psi^1 f(x) \equiv f(x, x, \ldots, x).$$

But if $\{b(t) : t \geq 0\}$ denotes the standard Wiener process in R^d, then we have

(6.6) $$E(F_f(\delta_{b(t)})) = H_t^1(\Psi^1 f)(x) .$$

In other words, we have shown that the finite-dimensional distributions of the rescaled Fleming-Viot process converge to those of the probability-measure-valued process which consists of a single atom in R^d moving according to a standard Wiener process.

7. SUPPORT PROPERTIES VIA DUALITY

In this section, we illustrate how the dual process constructed above can be used to obtain information on the distribution of the random closed set which supports the random probability measure $X(t)$.

Let $\{Y_t : t \geq 0\}$ denote the dual process constructed earlier. The mapping $f \to \Psi(f)$ given by

(7.1) $$\Psi(f) \equiv E_f \left\{ \int \cdots \int (Y_t(x_1, \ldots, x_{N(t)}))\mu(dx_1)\ldots\mu(dx_{N(t)}) \right\}$$

defines a positive linear functional on $C(S^N)$ with $\Psi(1) = 1$. It follows from the Riesz representation theorem that this mapping defines a probability measure P_t^N on S^N . From the duality relation (4.17), it follows that

P_t^N is the Nth moment measure of the random probability measure $X(t)$. Furthermore, the $\{P_t^N : N \geq 1\}$ form a consistent family of measures and therefore, by the Kolmogorov extension theorem, they extend to a probability measure P_t^∞ on S^∞.

For $\omega \in S$, let $Z_k(\omega) = \omega(k)$, the coordinate process. Then, from the construction of P_t, it is easy to verify that $\{Z_k : k = 1, 2, 3, \ldots\}$ is a sequence of S-valued exchangeable random variables. The random variables $Z_k(k = 1, 2, 3, \ldots)$ may also be viewed as the locations of a countable collection of particles in S.

Now let $\mathcal{A}$ be a countable algebra that generates $B(S)$. The de Finetti theorem on exchangeable random variables implies that for any $A \in \mathcal{A}$ and $\omega \in S^\infty$, the limit

$$(7.2) \qquad Y^*(\omega,A) \equiv \lim_{N\to\infty} \frac{1}{N} \sum_{j=1}^{N} I_A (Z_j(\omega))$$

exists P_t^∞ - a.s., where I_A denotes the indicator function of the set A. Furthermore, de Finetti's theorem asserts that Y^* is $\mathcal{E}$-measurable, where $\mathcal{E}$ denotes the σ-algebra of exchangeable events.

It is proved in Dawson and Hochberg (1982) that Y^* can be extended to a random probability measure on S that is a version of $X(t)$. Furthermore, for each ω, the support of $Y^*(\omega)$ is contained in the closure of the set $\{Z_k(\omega) : k = 1, 2, 3, \ldots\}$.

In the case of the Fleming-Viot process, we can identify the locations of the $\{Z_k : k = 1, 2, 3, \ldots\}$ which yield the random probability measure X_t with the locations of the particles in a branching Brownian motion. The appropriate branching Brownian motion is one in which the branching rate grows linearly with the total number of particles and, consequently, is explosive. This branching Brownian motion is obtained by replacing the Brownian motion semigroups in the description of the dual process (4.16) by expectation with respect to a system of Brownian motions.

To give an explicit description, we begin with $k \leq N$ particles located at $x_1, \ldots x_k$ in R^d and consider times $\tau_{k+1}, \ldots, \tau_N$. Consider the branching Brownian motion such that at each of the times $\tau_{k+1}, \ldots, \tau_N$, one of the existing particles is chosen at random, the chosen particle undergoes binary fission at that moment, and then all the particles continue to perform independent Brownian motions. Let

$$\{Z_k^{\tau_{k+1},\ldots,\tau_N}(t) : t \geq 0\}$$

denote this process with state space $\bigcup_{j=k}^{N} S^j$. Then we obtain the following representation of the dual process $Y(t)$:

$$(7.3)\quad Y(t) = \sum_{j=1}^{N} I_{\{S_{N-j}\le t<S_{N-j-1}\}} E\{f(Z_{N-j}^{t-S_{N-j},T_{N-j+1},\ldots,T_N})(x_1,\ldots,x_{N-j})\},$$

where

$$T_{N-j} = S_{N-j-1} - S_{N-j} ,$$

and

$$(7.4)\quad E_f\{Y(t)\}$$
$$= \sum_{j=1}^{N} E\left\{I_{S_{N-j}\le t<S_{N-j-1}} E\{f(Z_{N-j}^{t-S_{N-j},T_{N-j+1},\ldots,T_N})(x_1\ldots,x_{N-j})\}\right\} .$$

From this we conclude that the joint distribution of $\{Z_1, \ldots, Z_N\}$ is given by the distribution of the location of the N particles in the branching Brownian motion at a time t such that $S_{N-j} \le t < S_{N-j-1}$, given that there were $(N-j-1)$ particles initially at time zero which are independent and identically distributed with distribution μ .

The joint distribution of the $\{Z_k : k \ge 1\}$ then corresponds to the positions of the countable set of particles in the branching Brownian motion at time t starting with $(k-1)$ particles at time zero, given that

$$\sum_{j=k}^{\infty} T_j \le t < \sum_{j=k-1}^{\infty} T_j .$$

This characterization of the collection $\{Z_k : k \ge 1\}$ in the de Finetti representation can be used to obtain results concerning the support of the random probability measure $X(t)$. For example, in Dawson and Hochberg (1982) it is proved that the support of $X(t)$, $t > 0$, is compact and has Hausdorff-Besicovitch dimension not larger than 2 .

BIBLIOGRAPHY

1. D.A. Dawson and K.J. Hochberg, "Wandering Random Measures in the Fleming-Viot Model", Annals of Probability, 10 (1982), 554-580.

2. D.A. Dawson and T.G. Kurtz, "Application of Duality to Measure-Valued Diffusion Processes", in Advances in Filtering and Optimal Stochastic Control, Lecture Notes in Control and Information Sciences, Springer-Verlag, Berlin-Heidelberg-New York, Vol. 42, 1982, pp. 91-105.

3. D.A. Dawson and G.C. Papanicolaou, "A Random Wave Process", Laboratory for Research in Statistics and Probability, Carleton University and University of Ottawa, Technical Report 18, 1983.

4. S.N. Ethier and T.G. Kurtz, "The Infinitely-Many-Alleles Model with Selection as a Measure-Valued Diffusion", to appear (1985).

5. W.H. Fleming and M. Viot, "Some Measure-Valued Population Processes", in Proc. International Conference on Stochastic Analysis, Northwestern University, Academic Press, New York, 1978, pp. 97-108.

6. W.H. Fleming and M. Viot, "Some Measure-Valued Markov Processes in Population Genetics Theory", Indiana Univ. Math. J., 28 (1979), 817-843.

7. R. Holley and T. Liggett, "Ergodic Theorems for Weakly Interacting Systems and the Voter Model", Annals of Probability, 3 (1975), 643-663.

8. R. Holley and D.W. Stroock, "Dual Processes and Their Applications to Infinite Interacting Systems", Advances in Mathematics, 32 (1979), 149-174.

9. R. Holley, D.W. Stroock, and D. Williams. "Applications of Dual Processes to Diffusion Theory", in Proc. Symp. Pure Math., American Mathematical Society, Providence, Vol. 31, 1977, pp. 23-26.

10. T. Shiga, "Diffusion Processes in Population Genetics", J. Math. Kyoto Univ., 21 (1981), 131-151.

11. F. Spitzer, "Interaction of Markov Processes", Advances in Mathematics, 5 (1970), 246-290.

12. D. W. Stroock and S.R.S. Varadhan, Multidimensional Diffusion Processes, Grundlehren der Mathematischen Wissenschaften, Springer-Verlag, Berlin-Heidelberg-New York, Vol. 233, 1979.

DEPARTMENT OF MATHEMATICS AND STATISTICS
CARLETON UNIVERSITY
OTTAWA, ONTARIO, CANADA
K1S 5B6

DEPARTMENT OF MATHEMATICS AND STATISTICS
CASE WESTERN RESERVE UNIVERSITY
CLEVELAND, OHIO 44106

Contemporary Mathematics
Volume 41, 1985

INVARIANCE PRINCIPLE FOR REVERSIBLE MARKOV PROCESSES WITH APPLICATION TO DIFFUSION IN THE PERCOLATION REGIME

A. De Masi[1], P.A. Ferrari[2], S. Goldstein[3] and W.D. Wick[4]

ABSTRACT. We present an invariance principle for antisymmetric functionals of a reversible Markov process which immediately implies convergence to Brownian motion for a wide class of random motion in random environments. We here apply it to establish convergence to Brownian motion for a walker moving in the infinite cluster of the two dimensional bond percolation model.

1. INTRODUCTION.

Recently there has been considerable interest, stimulated by physical applications, in the study of the asymptotic motion of a tagged particle in a random (static or dynamic) environment. In symmetric cases one expects the motion to be asymptotically Brownian, with a finite diffusion matrix, which one would like to compute as explicitly as possible. In this note we discuss an invariance principle applicable when the process of the environment as seen from the tagged particle is reversible in time. The main theorems - which are useful in other contexts as well - are parsimonious in hypotheses; in particular, there are no mixing properties or spectral conditions assumed on the variables which might be difficult to check in practice. As a result we are able to improve upon several results in the literature, and give short proofs of many other results. Details will appear in [2].

Before discussing the abstract theorems we mention in passing a typical example to which our methods apply (to be discussed more fully below). On the integer lattice $\mathbb{Z}^d$ let there be given rates (or conductivities) $a_e(x) > 0$ for the bond $(x,x+e)$, $x \in \mathbb{Z}^d$, $e \in \mathbb{Z}^d$, $|e| = 1$. Let X_t be the position of a

1. Partially supported by CNR.
2. Partially supported by CNP_q Grant # 201682-83.
3. Partially supported by NSF Grant # PHY 8201708.
4. Partially supported by NSF PHY 83 42570

particle started at $t = 0$ at $x = 0$ and performing a random walk with these rates. Assume in addition that the $a_e(x)$ are chosen from a translation-invariant, ergodic distribution (random bond conductivities). Then, with only moment conditions on the rates, we prove that $\epsilon X_{t\epsilon^{-2}}$ tends, in distribution as $\epsilon \to 0$, to a Brownian motion with a finite, non-degenerate diffusion matrix. (Compare [1], [7], [8], in which $0 < b_1 < a_e(x) < b_2 < \infty$ is assumed). We can even treat the case when $P[a_e(x) > 0] = p < 1$, assuming that $p > p_c$ (percolation threshold) and conditioning on the percolation cluster containing the origin. In other words, we prove an invariance principle for a random walk on an infinite percolation cluster.

2. ABSTRACT THEOREMS. Invariance principles for anti-symmetric functionals of reversible Markov processes.

Let ξ_t, $t \in \mathbb{Z}$ on $\mathbb{R}$ (discrete or continuous time) be a time-reversible Markov chain with state space X. We are interested in proving an invariance principle for a functional $X_{[0,t]}$ of paths of the process (depending on ξ_τ, $0 \leq \tau \leq t$), which has a certain symmetry property. Let $R_{t/2}$ be reflection in the midpoint of the interval, defined by $(R\xi)_\tau = \xi_{t-\tau}$. We call $X_{[0,t]}$ anti-symmetric if

$$X_{[0,t]} \circ R_{t/2} = -X_{[0,t]} , \tag{2.1}$$

with a similar definition for a general closed interval $I = [a,b]$.

Anti-symmetric variables arise naturally in many contexts. For instance, let $X_{[0,t]}$ be the increment made, in the interval $[0,t]$, by one component of a multi-component process. Alternatively, let ξ_t be a jump process on X, let x and y be points in X and let $X_{[0,t]}$ be the "current" from x to y, i.e. the number of jumps from x to y minus the number from y to x. In an infinite particle system $X_{[0,t]}$ might be the current across a bond.

We first present the discrete time version of our theorem whose only assumptions concern symmetry and square integrability and in which the antisymmetry appears in simplest form.

THEOREM 2.1. Let ξ_i, $i = 0,1,2,\ldots$ be an ergodic reversible Markov process with state space X and initial distribution μ, $Y : X \times X \to \mathbb{R}$ be antisymmetric : $Y(\xi,\eta) = -Y(\eta,\xi)$, let $Y_i \equiv Y(\xi_{i-1},\xi_i)$ be square integrable.

Define

$$X_n = \sum_{i=1}^{n} Y_i \tag{2.2}$$

$$X_t^\varepsilon = \varepsilon X_{[\varepsilon^{-2}t]}$$

Let $\mathbb{E}_D$ be the expectation with respect to the Brownian motion W_D starting at the origin with variance Dt and let $\mathbb{E}_\xi$ be the expectation with respect to the Markov process ξ_i starting at ξ. Then for all bounded continuous function F on $D(\mathbb{R},\mathbb{R})$

$$\mathbb{E}_\xi(F(X_\cdot^\varepsilon)) \underset{\varepsilon\to 0}{\longrightarrow} \mathbb{E}_D(F) \tag{2.3}$$

in μ-measure, where D is given by

$$D = \mathbb{E}(X_1^2) - 2\langle\varphi(1-T)^{-1}\varphi\rangle \tag{2.4}$$

with $\varphi(\xi) = \mathbb{E}_\xi(X_1)$ and where T is the (time one) transition operator.

In particular, as $\varepsilon \to 0$ the process $X_\cdot^\varepsilon$ converges (in distribution) weakly to $W_D(\cdot)$. □

(2.5) We call the convergence given in (2.3) convergence in distribution in μ probability.

The continuous time version of our invariance principle applies to a family X_I, indexed by closed intervals $I = [a,b]$ and taking values in $\mathbb{R}^d$, which in addition to being antisymmetric (i.e. each X_I is antisymmetric) satisfies certain additional natural properties (except for A.1, automatically satisfied in practice) which we list below.

(A.1) X_I is square-integrable for each I

(A.2) The family X_I is covariant

$$X_{[a,b]} \circ \theta_\tau = X_{[a+\tau,b+\tau]}$$

where θ_τ denotes the time-shift operator, defined by $(\theta_\tau\xi)_t = \xi_{t+\tau}$

(A.3) The family X_I is additive :

$$X_I + X_{I'} = X_{I\cup I'}$$

a.s. if $\mathring{I} \cap \mathring{I}' = \emptyset$. ($\mathring{I}$ is the interior of I).

We also require that $X_t \equiv X_{[0,t]}$ has a version with paths in $D(\mathbb{R},\mathbb{R}^d)$.

THEOREM 2.2. Let ξ_t, $t \in \mathbb{R}$, be an ergodic reversible Markov process with initial measure μ. Let X_I be an anti-symmetric family of $\mathbb{R}^d$-valued random variables satisfying (A.1)-(A.3). Assume in addition that the mean forward drift velocity exists and is square-integrable, more precisely, that

$$\lim_{\delta\to 0^+} \delta^{-1}\, \mathbb{E}_\mu[X_{[0,\delta]}/F_o] = \varphi \tag{2.6}$$

exists weakly in $L^1(\mu)$ and $\varphi \in L^2(\mu)$ (φ is vector-valued). Then the following hold :

(i) The velocity auto-correlation function is integrable, that is, for each i,j

$$\int_o^\infty |\langle \varphi_i T_t \varphi_j\rangle|\, dt < \infty$$

where T_t is the Markov semigroup determined by the process ξ_t, and $\langle\cdot\rangle$ is the expectation w.r.t. μ, where $\varphi_i = e_i.\varphi$, with e_i, $i = 1,\dots,d$ an orthonormal basis on $\mathbb{R}^d$.

(ii) Let $X_t^\varepsilon \equiv \varepsilon X_{[0,\varepsilon^{-2}t]}$. Then as $\varepsilon \to 0$

$$X_\cdot^\varepsilon \to W_{\mathbb{D}}(\cdot) \tag{2.7}$$

in distribution in μ-probability, where $W_{\mathbb{D}}(\cdot)$ is a d-dimensional Brownian motion starting at zero with covariance matrix $\mathbb{D}t$ and where

(iii) $\mathbb{D}$ is given by

$$D_{ij} = C_{ij} - 2\int_o^\infty \langle \varphi_i T_t \varphi_j\rangle\, dt \tag{2.8a}$$

$$D_{ij} = C_{ij} - 2\langle \varphi_i L^{-1} \varphi_j\rangle \tag{2.8b}$$

where $\mathbb{C} = (C_{ij})$ is determined by

$$e^T.\mathbb{C}.e = \lim_{\delta\to 0^+} \delta^{-1}\, \mathbb{E}_\mu (e.X_{[0,\delta]})^2 \tag{2.9}$$

and $-L$ is the generator of the semigroup $T_t : L^2(\mu) \to L^2(\mu)$.

$\mathbb{C}$ is positive definite and finite.

(iv) Let $X_t = X_{[0,t]}$. Then

$$\lim_{t\to 0} \frac{\mathbb{E}(X_{t,i}^2)}{t} = D_{ii} \qquad \square$$

REMARKS. (1) Ergodicity is required only for part (ii) of the theorem. (2) $\mathbb{D}$ degenerate is not excluded. We call the limiting Brownian motion nondegenerate if all the eigenvalues of $\mathbb{D}$ are positive. (3) By averaging with respect to μ it is immediate from (2.7) (see eq. (2.3)) that

$X^{\varepsilon}_{\cdot} \underset{\varepsilon\to 0}{\to} W_{\mathbb{D}}(\cdot)$ in distribution (under the law $\mathbb{P}_{\mu}$ of the process ξ_t starting from the stationary distribution μ).

The proof of the theorem (2.1) builds on earlier work [6], [3]. The gist of the proof is to write $\varepsilon X_{t\varepsilon^{-2}}$ as the sum of a square-integrable martingale and a small remainder, and then to apply the martingale Central limit theorem [4]. The first step in the proof is, roughly speaking, to establish (2.8), simultaneously proving the integrability of the velocity-autocorrelation function, which is (equivalent to) the main assumption in earlier work. (Familiarity with Spohn's Green-Kubo formulas for interacting particle systems [12] lead us to this formula).

More precisely we first derive the equation obtained from (2.8) by $\lim_{t\to\infty} \mathbb{E}_{\mu}(X_{t,i}X_{t,j})$, which we later identify with D_{ij} . Assume, for notational convenience, that $X_t \in \mathbb{R}$. The key observation is that from the easily obtained decompostion $X_t = S_t+M_t$ where $S_t = \int_0^t \varphi(\xi_s)ds$ and M_t is a square integrable martingale with stationary increments, it follows that $\mathbb{E}_{\mu}(M_t^2) = \mathbb{E}_{\mu}(X_t^2)+\mathbb{E}_{\mu}(S_t^2)$, since the cross term $2\,\mathbb{E}_{\mu}(X_tS_t)$ vanishes by symmetry. Here we use the fact that the measure $\mathbb{P}_{\mu}$ is symmetric under time reversal $R_{t/2}$ (i.e. μ is reversible) while S_t is manifestly symmetric, and by assumption X_t is anti-symmetric.

We turn next to the problem of obtaining bounds on the diffusion matrix (in particular a positive lower bound). Note that from eq. (2.8.b) and writing D for D_{ii} , φ for φ_i and C for C_{ii} , an upper bound is immediate : since L is non-negative, $D \leq C$. For a lower bound one can often exploit the following situation. Suppose $L = L_1+L_2$, with $-L_i$, $i = 1,2$ generators of reversible Markov processes. Then operator monotonicity of the inverse function suggests the lower bound :

$$D \geq C - 2(\varphi L_i^{-1}\varphi) \tag{2.10}$$

which is often useful in practice and frequently has an interesting probabilistic meaning. We mention several : (i) For reversible infinite particle systems e.g. interacting Brownian particles [11], [10], [13] (2.10) can be used to prove that the diffusion constant for a particle moving in a random, moving environment is greater than that for the same particle in a "frozen" environment. (ii) When $-L$ generates a multi-dimensional process and $-L_1$ generates a process in fewer dimensions, (2.10) implies that D is an increasing function of dimension.

3. RANDOM WALK IN THE PERCOLATION CLUSTER.

In this section we use the theorem 2.2 to establish the diffusive behavior of a random walker on an inhomogeneous lattice, i.e. the invariance principle with non degenerate $\mathbb{D}$. The model is the following. In the two dimensional lattice $\mathbb{Z}^d$ with each bond $(x,x+e)$, $x,e \in \mathbb{Z}^2$, $|e| = 1$ associate random rates $a_e(x) \geq 0$ (sometimes we write $a_i(x)$ instead of $a_{e_i}(x)$) . A configuration of rates is called an environment and the space of environments we denote by $\mathcal{X} = \{a : \mathbb{Z}^2 \times \{e_1,e_2\} \to [0,\infty)\}$.

Let μ a measure on $\mathcal{X}$ satisfying

C.1. $a_i(x)$ are identically distributed and mutually independent.

Given a configuration $a \in \mathcal{X}$, put a particle at the origin and let it move to its nearest neighbours with rates given by a . That is, define the jump process $X(a,t)$, $t \in \mathbb{R}$ as the Markov process with state space $\mathbb{Z}^2$ and transition probabilities determined by

$$P(X(a,t+h) = x + e \,/\, X(a,t) = x) = a_e(x)h + o(h) \tag{3.1}$$

$$P(X(a,t+h) = x \,/\, X(a,t) = x) = 1 - \sum_{|e|=1} a_e(x)h + o(h)$$

We are interested in establishing the convergence to Brownian Motion of the rescaled process $\varepsilon X(a,\varepsilon^{-2}t)$ as $\varepsilon \to 0$.

Given $a \in \mathcal{X}$, let

$$W(x,a) = \text{connected cluster of } x \text{ in } W(a) = :\{\text{bonds } (x,x+e) \text{ with } a_e(x) > 0\} \tag{3.2}$$

It is clear that for configurations $a \in \mathcal{X}$ for which $|W(0,a)| < \infty$ ($|A|$ is the cardinality of the set A), the position $X(a,t)$ of the particle will be bounded for all t . So $|W(0,a)| = \infty$ is the minimal condition for convergence to Brownian Motion.

It is well known [5] that if

$$\text{C.2.} \quad \mu\{a_1(0) > 0\} = p > \frac{1}{2} \tag{3.3}$$

then $\mu\{|W(0,a)| = \infty\} = \theta_p > 0$. Define

$$\hat{\mu} = :\mu\{\cdot / |W(0,a)| = \infty\} . \tag{3.4}$$

Consider also the following conditions on μ , which will be discussed later :

C.3. $\mu\{a_1(0) > b\} = 0$ for some $b > 0$.

C.4. $\mu(a_1(0)^{-1}/a_1(0) > 0) > 0$.

The main result of this section is the following

THEOREM 3.1. Let μ a measure on the space of environments X satisfying conditions C.1-4 above, and let $X(a,t)$ be the jump Markov process with state space $\mathbb{Z}^2$, $X(a,0) \equiv 0$ and transition probabilities given by eq. (3.1) above.

Then

(A) as $\varepsilon \to 0$, $\varepsilon X(a,\varepsilon^{-2}.) \to W_{\hat{\mathbb{D}}}(\cdot)$ in distribution in $\hat{\mu}$-probability (see Def. 2.5), where $\hat{\mu} = \mu(\cdot/|W(0,.)| = \infty)$ and $W_{\hat{\mathbb{D}}}(t)$ is two dimensional Brownian motion with covariance matrix $\hat{\mathbb{D}}$, where

(B) $\hat{\mathbb{D}} = \hat{D}I$, where I is the identity matrix, and

$$\theta_p \hat{D} = D =: 2\mu(a_1(0)) - 2\int_0^\infty dt\ \mu[(a_1(0) - a_1(-e_1)) \cdot \mathbb{E}_a(a_1(X(a,t)) - a_1(X(a,t) - e_1))] \tag{3.5}$$

(C) $\hat{D} > 0$ □

REMARKS.

1) Convergence to (non degenerate) Brownian Motion of the process $X(a,t)$ was established by [1], [7], [8] for any ergodic translation invariant μ such that

$$0 < b_1 < a_i(x) < b_2 < \infty \quad \text{for all} \quad x \in \mathbb{Z}^2 , \quad i = 1,2 .$$

In [2] the result is extended up to $b_1 = 0$ and $b_2 = \infty$ with only the conditions that $\mu(a_i(0)) < \infty$, and $\mu(a_i(0)^{-1}) < \infty$. This is done by introducing a related discrete time process and invoking Theorem 2.1, but we prefer not to introduce this approach here and so we require that $b_2 < \infty$ and apply Thm. 2.2 directly. We also need $b_2 < \infty$ as a technical requirement for proving that $\hat{D} > 0$ (see Lemma 3.1 below).

2) The condition that μ is a product measure is needed in two parts : first to prove that the induced process $a(t)$ (the environment as seen from $X(a,t)$) is ergodic with respect to $\hat{\mu}$. Second to establish the positivity of $\hat{D}$. Here we use strong results of percolation theory, namely we relate $\hat{D}$ to the electric current flowing through the environment a when a potential gradient is applied.

The condition $\mu(a_1(0)^{-1} / a_1(0) > 0) < \infty$ is needed only to establish the positivity of $\hat{D}$, since it appears as a divisor in the Kesten's lower bound for the lim inf of the current.

3) For the case $a_i(x) > 0$, the result of convergence to Brownian Motion is independent of the dimension if $0 < a_i(x)$ and $\mu[a_i(x)] < \infty$. In [2] the inequalities $\mu(a_1(0))^{-1} \leq D \leq \mu(a_1(0))$ (known in special case) are derived generally for rotation (by $\pi/2$) invariant μ, while if the measure is not rotation invariant similar equalities are proven for each of the coefficients. It is also proven there that the diffusion coefficient(s) D (D_i) is a non decreasing function of the dimension.

Proof. (A) We place the problem within the framework of Theorem 2.2. The role of the Markov process ξ_t is played by the process $a(t) \in X$ defined by

$$\begin{aligned} a(0) &\equiv a \\ a(t) &= \tau_{X(a,t)} a \end{aligned} \tag{3.6}$$

where τ_x is the translation by x on X. $a(t)$ is clearly reversible for both μ and $\hat{\mu}$.

Moreover, $X(a,t)$ is an adapted (to $a(t)$) antisymmetric process. Since the rates are bounded, the function $\varphi_i = a_i(0) - a_i(-e_i)$ is well defined and X_t is square integrable. So, the only hypothesis of Theorem 2.2 left to check is the ergodicity of $\hat{\mu}$ with respect to the process $a(t)$. This follows from the fact that in two dimensions there exists only one infinite cluster μ a.s., and the ergodicity of the random environment under translations (see [2]) □

(B) The fact that $\mathbb{D}$ is a multiple of the identity is an easy consequence of the rotation invariance of μ.

The equation $\theta_p \hat{D} = D$ can be obtained as follows : in view of (2.8), by decomposing μ into its ergodic components (for the process $a(t)$) we obtain from the definition, eq. (3.5), of D, its ergodic decomposition $D = \int \nu(d\beta) D_\beta$. Since the only ergodic component of μ with positive diffusion constant is $\hat{\mu}$, the equation follows (see [2] for details). □

The idea of the proof of (C) is the following. We consider periodic configurations with period $2N$ and call D_N the diffusion coefficient (i.e. $\lim \frac{E(X_t^2)}{t}$) for the corresponding random walk. Since with positive probability all the rates are strictly positive one expects that $D_N > 0$, which can, in fact, be easily shown. The problem left is to see if (a) as N increases D_N

remains bounded away from 0 and b) D_N and D are suitably related.

The relationship between D_N and D is easily established by looking at the "explicit" formula for D in eq. (3.5). In that formula the integrand of the second term is positive and involves E (X(t)) . For each t fixed the expectation $E_N(\varphi(X(t))$ related to the periodic configurations converges to $E(\varphi(X(t))$. So we can use Fatou's lemma to obtain the inequality.

$$D \geq \overline{\lim}\, D_N \quad .$$

The positivity of $\overline{\lim}\, D_N$ is not so clear : here deep percolation problems enter. We solve this problem without really solving it ! In fact we prove that

$$\frac{1}{2} D_N = \sigma_N \tag{3.7}$$

where σ_N is the current flowing in a box of size N when a unit potential difference is established, which is better thought of as the effective conductivity. Then we observe that Kesten (and also Grimmet and Kesten) solved our problem : they proved that $\underline{\lim}\, \sigma_N > 0$. [5], [9].

Thus to prove (C) we define the periodic random walk, its diffusion coefficient D_N ; we establish that $D \geq \overline{\lim}\, D_N$ and then define the conductivity σ_N and determine its relationship with D_N .

DEFINITION. The periodic random walk.

Let $B_N \subset \mathbb{Z}^2$ be the box

$$B_N = \{x = (x_1,x_2) \in \mathbb{Z}^2 \ : -N \leq x_1 \leq N \quad , \quad -N \leq x_2 \leq N-1\}$$

Given a configuration $a \in X$ of rates, let $\tilde{a}$ be the periodic rate configuration, with period 2N , (on bonds linking nearest neighbours points) in the strip

$$\Sigma_N = \{(x_1,x_2) \in \mathbb{Z}^2 \ : -N \leq x_2 \leq N-1\}$$

which agrees with a on the box B_N , except that the vertical bonds on $x_1 = \pm N$ are given infinite rates (this is required to obtain a simple relationship between D_N and σ_N).

Since bonds leading out of Σ_N on $x_2 = N-1$ and on $x_2 = -N$ do not exist for $\tilde{a}$, Σ_N has "reflecting boundary conditions" on top and bottom. In order to construct a random walk on Σ_N with rate $\tilde{a}$, it is necessary to identify the points of $x_1 = -N$, and similarly for the points of $x_1 = N, 3N, -3N, \ldots$ but as long as we consider only functions which agree on equivalent points, this may safely be ignored, provided we bear in mind that from all points on e.g. $x_1 = N$, the rates $\tilde{a}$ to the points $(N+1, x_2)$ $[(N-1, x_2)]$ are now $\frac{1}{2N} a_1(-N_1 x_2)$ $[\frac{1}{2N} a_1(N-1, x_2)]$. Denote by $\tilde{X}^N_a(x,t)$ (or simply $\tilde{X}^N(t)$ if no confusion is likely) the Markov process on Σ_N defined by the rates $\tilde{a}_1$ starting from x , and let $X^N(t)$ be the x_1-component of $\tilde{X}^N(t)$. Denote by $Y^N_a(x,t)$ (or simply $Y^N(t)$) the Markov process $\tilde{X}^N(t)/\sim$, where $(x_1, x_2) \sim (\bar{x}_1, \bar{x}_2) \Leftrightarrow \bar{x}_1 - x_1 = 0 \mod 2N$ and $\bar{x}_2 = x_2$. $Y^N(t)$ may be regarded as a Markov process with state space $\tilde{B}_N$, the box B_N with the left side identified with the right (i.e., $(-N, x_2)$ identified with (N, x_2)) . Let m_N be the uniform distribution on $\tilde{B}_N$. Then with respect to m_N , $Y^N(t)$ is clearly a reversible Markov process, in terms of which $\tilde{X}^N(t)$ and hence $X^N(t)$, can be realized. Under this realization

$X^N(t)$ = number of jumps of $Y^N(s)$ in the positive horizontal direction minus number of jumps of $Y^N(s)$ in the negative horizontal direction during the time interval $[0,t]$. (3.8)

(In the above definition the jump from $N-1$ to $-N \equiv N$ [respectively to $N-1$ from $N \equiv -N$] , is considered as a unit positive resp. negative jump). Note that this is well defined if $N \geq 2$ which we will assume. We now need some notation. The rates are most naturally regarded as bond functions $(x,y) \equiv b \to \tilde{a}(b)$ the rate for going from x to y . (Though the original $a(b)$ was symmetric, this may no longer be the case for $\tilde{a}$, because of the redefinition of the rates for leaving $x_1 = \ldots, -N, N, 3N, \ldots$ necessitated by granting infinite "conductivities (i.e. rates) to the vertical bonds on these lines : e.g. $\tilde{a}((N,x_2),(N+1,x_2')) = \tilde{a}((N+1,x_2'),(N,x_2))$ may fail). For any bond $b = (x,y)$, let $\nabla_b f = f(y) - f(x)$. Let $B_x = \{(x,y)\}$ be the set of bonds emanating from x .

Then the generator of the process $\tilde{X}^N_a(x,t)$ is given by

$$L_a f(x) = \sum_{b \in B_x} \tilde{a}(b) \nabla_b f \tag{3.9}$$

while the generator for the process $Y^N_a(x,t)$ may be identified with the restriction of L_a to periodic functions. Here, and from now, "periodic" means periodic with period $2N$ (in the x_1-direction).

Let $\varphi^N_a = L_a x_1$, and note that

$$\varphi_a^N(x) = (\nabla_1^- a_1)(x) \equiv a_1(x) - a_1(x - e_1) \tag{3.10}$$

for $x \in B_N$ $|x_1| \neq N$.

Denote by P_a^N [resp $P_{a,x}^N$] the law of the process $Y^N(t)$ with starting measure m_N [resp. δ_x] .

Let

$$\begin{aligned} D_N(a) &= 2m_N(a_1(x)) - 2\int_o^\infty ds\ m_N(\varphi_a^N \mathbf{E}_{a,x}^N \varphi(Y^N(s))) \\ &= 2m_N(a_1(x)) + 2m_N(\varphi_a^N L^{-1}{}_a\varphi_a^N) \end{aligned} \tag{3.11}$$

REMARKS. It follows easily from Theorem 2.2 and the Remark (1) which follows it, with $Y^N(t)$ playing the role of the Markov process $\xi(t)$, m_N the role of μ and $X^N(t)$ the role of the antisymmetric functional $X_{[0,t]}$ that the process $X^N(t)$ converges to a mixture of Brownian motion with (average) diffusion constant $D_N(a)$, but we don't need this fact.

LEMMA 3.1. Let μ be a translation invariant probability measure on X such that $\mu\{a_i(x) > b\} = 0$ for some $b < \infty$ for all $i = 1,\ldots,d$ .Then the following hold

(i) For any $t > 0$ fixed

$$\lim_{N\to\infty} \mu[m_N(\varphi_a^N \mathbf{E}_{a,x}^N \varphi_a^N(Y^N(t)))] = \mu[\varphi_1(a)\ \mathbf{E}_a\varphi_1(a(t))]$$

where φ_a^N is defined in eq. (3.10) and $\varphi_i(a) = a_i(0)-a_i(-e_i)$

(ii) $D \geq \overline{\lim_{N\to\infty}}\ \mu(D_N(a))$

where the "diffusion coefficient" D for the infinite system is defined in eq. (3.5) and D_N , the one for the finite system in eq. (3.11). □

<u>Proof</u>. Part (ii) follows from (i). In fact, if (i) holds we have

$$\begin{aligned} \int_o^\infty dt\ \mu[\varphi_1(a)\ \mathbf{E}_a\varphi_1(a(t))] &= \int_o^\infty dt\ \lim_{N\to\infty} \mu[m_N(\varphi^N \mathbf{E}_{a,x}^N \varphi^N(Y^N(t)))] \\ &\leq \underline{\lim}_{N\to\infty} \int_o^\infty dt\ \mu[m_N(\varphi^N \mathbf{E}_{a,x}^N \varphi^N)] \\ &= \underline{\lim}_{N\to\infty} \mu[\int_o^\infty dt\ m_N(\varphi^N \mathbf{E}_{a,x}^N \varphi^N)] \end{aligned}$$

where the inequality follows by Fatou's lemma from the fact that the integrand on the right hand side is positive, since m_N is reversible. Furthermore by translation invariance of μ ,

$$\mu[m_N(a_1(x))] = 2\mu[a_1(0)]$$

Now, by the formulas for D and D_N :

$$D = 2\mu[a_1(0)] - 2\int_0^\infty dt\ \mu[\varphi_1(a)\,\mathbf{E}_a\varphi_1(a(t))] \geq$$

$$\geq 2\mu[a_1(0)] - \lim_{N\to\infty} 2\mu[\int_0^\infty m_N(\varphi_a^N\,\mathbf{E}^N_{a,x}\varphi_a^N)] = \lim_{N\to\infty}\mu(D_N(a))$$

Part (i) follows easily from the fact that until the boundary of B_N is reached the periodic random walk and the original random walk agree (provided they start from the same place in the same immediate environment). □

DEFINITION. The effective conductivity.

Let $a \in X$ be fixed. The effective conductivity $\sigma_N(a)$ is the current established across B_N by a unit potential difference. Let V be the corresponding potential. It satisfies

$$L_aV(x) = 0 \qquad -N < x_1 < N \tag{3.12a}$$

$$V(-N,x_2) = 1 \qquad V(N\ ,x_2) = 0 \quad , \quad -N \leq x_2 \leq N-1 \tag{3.12b}$$

Equation (3.12a) merely expresses current balance at x (see eq. 3.13 below). The solution of (3.12) is not unique, but all solutions give rise to the same current

$$\sigma_N(a) \equiv \sum_{x_2=-N}^{N-1} a_1(x_1,x_2)(\nabla_1^+V)(x_1,x_2) \tag{3.13}$$

where

$$(\nabla_1^+V)(x) = V(x + e_1) - V(x) \quad .$$

This expression is independent of x_1 , provided (3.12a) is satisfied, and thus we have

$$\sigma_N(a) = 2N\ m_N(a_1\nabla_1^+V) \tag{3.14}$$

Let $\tilde{V}$ be the natural extension of V to all of Σ_N , not the periodic extension but one producing periodic current :

$$\tilde{V}(x) = V(x) \quad \text{for} \quad x \in B_N \quad . \quad \tilde{V}(x_1 + 2N,x_2) = \tilde{V}(x_1,x_2) - 1 \quad ;$$

Because the left and right boundaries of B_N have been made "superconducting",

$\tilde{V}$ satisfies

$$L_a\tilde{V}(x) = 0 \qquad \text{all } x \in \Sigma_N ,$$

since all this is now required at e.g. $x_1 = -N$ is that the current into this line, from the left, equals the current out, into the right, which is already guaranteed by the fact that the expression (3.13) for $\sigma_N(a)$ is independent of x_1. The relationship between $D_N(a)$ and $\sigma_N(a)$ now follows easily:

Proposition 3.1. For any $a \in X$

$$D_N(a) = 2\sigma_N(a)$$

where D_N is defined in eq. (3.11) and σ_N in eq. (3.13).

Proof. Consider the function $f = x_1 + 2N\tilde{V} - N$ on Σ_N (where the constant $-N$ is added so that, for convenience $f = 0$ on $x_1 = \pm N$). Since $\varphi_a^N = L_a x_1$, it follows that f is a periodic function satisfying

$$L_a f = \varphi_a^N .$$

Thus

$$\begin{aligned} D_N(a) &= 2m_N(a_1) + 2m_N(\varphi_a^N L_a^{-1} \varphi_a^N) \\ &= 2m_N(a_1) + 2m_N(\varphi_a^N f) \\ &= 2m_N(a_1) + 2m_N((\nabla_1^- a_1) f) . \end{aligned}$$

"Integrating" by parts, and using the translation invariance of m_N, (i.e., periodicity) we obtain that

$$\begin{aligned} D_N(a) &= 2m_N(a_1) - 2m_N(a_1 \nabla_1^+ f) \\ &= 2m_N(a_1) - 2m_N(a_1 (1 + 2N\nabla_1^+ V) \\ &= 2 \cdot 2Nm_N(a_1 \nabla_1^+ V) = 2\sigma_N(a) . \qquad \square \end{aligned}$$

Proof of Thm. 3.1 (C). It follows from Lemma 3.1 (ii), Proposition 3.1 and Fatou's lemma that

$$D \geq \overline{\lim_{N\to\infty}}\, \mu(D_N(a) = \overline{\lim_{N\to\infty}}\, \mu(\sigma_N(a)) \geq \mu(\underline{\lim_{N}}\, \sigma_N(a))$$

Moreover it has been proven in [5] and [9] that there exist constants $0 < c_1 < c_2 < \infty$ such that

$$c_1 \leq \underline{\lim}\ \sigma_N(a) \leq \overline{\lim}\ \sigma_N(a) \leq c_2 \tag{3.15}$$

μ a.s. Thus $D \geq c_1 > 0$. □

REMARKS.

1) Independence is required only for eq. (3.15).

2) For boxes other than squares, and for higher dimensions, it remains true that $D_N(a) = 2\sigma_N(a) = m(a_1 \nabla_1^+ V)$, where $V(x_-) = 2N$, $V(x_+) = 0$ and the box extends from $x_1 = x_-$ to $x_1 = x_+$ and $x_+ - x_- = 2N$. Thus $\sigma_N(a)$ is the effective local conductivity : the current per unit cross section per unit average electric field.

CLARIFICATION. Contrary to what was announced in the 1984 Conference on The Mathematics of Phase Transitions (Bowdoin College, June 1984), the problem of the existence of the limit of the expected current, i.e. $\lim_{N\to\infty} < \sigma_N(a) >$ remains open in the percolation case.

ACKNOWLEDGEMENTS. We wish to thank E. Presutti and H. Spohn. Without their support and ideas the present work would not have materialized. Two of us (A.D.M. and P.A.F.) thank the Math. Dept. of Rutgers and J.L. Lebowitz for his friendly hospitality. Finally, we thank the IHES for their hospitality while this was being written, and to Mme J. Martin for typing it.

BIBLIOGRAPHY

1. Anshelevich V.V., Khanin K.M., Sinai G. Ya, Symmetric Random walks in Random Environments, Comm. Math. Phys. 85, 449-470 (1982).

2. De Masi A., Ferrari P.A., Goldstein S., Wick D.W., An invariance principle for reversible Markov processes with applications to random motions in random environments. In preparation.

3. Dürr D., Goldstein S., Remarks on the central limit theorem for weakly depending random variable, To appear in Proceedings BIBOS Conference, Bielefeld 1984.

4. Helland I.S., Central limit theorems for martingales with discrete or continuous time, Scand. J. Stat. 979-994 (1982).

5. Kesten H., Percolation Theory for Mathematicians, Progress in Probability and Statistics, Vol.2, Birkhaüser, Edited by P. Huber and M. Rosenblatt.

6. Kipnis C., Varadhan S.R.S., Central limit theorem for additive functionals of reversible Markov processes and applications to simple exclusion, Preprint 1984.

7. Künnemann R., The diffusion limit for reversible jump processes on $\mathbb{Z}^d$ with ergodic random bond conductivities, Comm. Math. Phys. 90, 27-68 (1983).

8. Figari R., Orlandi E., Papanicolaou G., Diffusive behavior of a random walk in a random medium. Proceedings to Kyoto Conference (1982).

9. Grimmett G., Kesten H., First-passage percolation, net work flows and electrical resistances, Preprint 1984.

10. Guo M. , Limit theorems for interacting particle system, Ph.d. Thesis, Dept. of Math., New-York University, 1984.

11. Lang R., Stochastic Models of Many Particle Systems and their time Evolution, Habilitationsschrift, Universität Heidelberg, 1982.

12. Spohn H., Large scale behaviour of equilibrium time correlation functions for some stochastic Ising models, Springer Lecture Notes in Physics 173 (1982) 304.

13. Spohn H., Equilibrium fluctuations for interacting Brownian particles, Preprint, 1984.

Permanent addresses

For A.D.M. : Università dell'Aquila, Dipartimento di Matematica, 67100 L'Aquila, Italy.

For P.A.F. : Instituto de Matemática e Estatística, Universidade de São Paulo, Cx Postal 20570, São Paulo, Brasil.

For S.G. : Dept of Math., Rutgers University, New Brunswick, N.J. 08903, USA.

For D.W.W. : Math. Dept., Univ. of Washington, GN-50 Seattle, Wa.98115, USA

Contemporary Mathematics
Volume 41, 1985

STOCHASTIC GROWTH MODELS:

TEN PROBLEMS FOR THE 80'S (AND 90'S)

R. Durrett[1]

In this paper we will describe some open problems concerning stochastic growth models, a collection of particle systems that we will describe in the next paragraph, and as the title suggests we suspect many of them will not be resolved before the end of this decade. There are approximately ten of these problems but for convenience of exposition they are not numbered and many are stated as conjectures, questions, or off-hand remarks. Many of these problems can be explained in a few minutes to a person with no prior experience in the area, so with the hope of attracting more people to work on them we have tried to make the discussion below self-contained.

By a stochastic growth model we mean a Markov process whose state at time t ξ_t is a subset of Z^d, which we interpret as the set of occupied sites and which evolves according to the following rules:

(i) particles die at rate δ, i.e. if $x \in \xi_t$

$$P(x \notin \xi_{t+s} | \xi_t) \sim \delta s \quad \text{as} \quad s \to 0$$

(ii) particles are born at rate $b_x(\xi)$, i.e. if $x \notin \xi_t$

$$P(x \in \xi_{t+s} | \xi_t) \sim b_x(\xi_t) s \quad \text{as} \quad s \to 0$$

and the birth rates are supposed to satisfy:

(iii) there is a finite set $S = \{y_1, \ldots, y_n\}$ and a nonnegative function f so that

$$b_x(\xi) = f(\xi(x+y_1), \ldots, \xi(x+y_n))$$

where $\xi(y) = 1$ if $y \in \xi$ and 0 otherwise.

1980 AMS Subject Classification. 60K35.
[1]Partially Supported by NSF Grant MCS83-00836 and an AMS "mid-career" Fellowship, 1984-1986.

(iv) $b_x(\emptyset) = 0$

(v) $\xi \to b_x(\xi)$ is increasing i.e. if $\xi \subset \eta$ then $b_x(\xi) \le b_x(\eta)$.

The first assumption is clearly the most drastic. It implies that each particle lives for an exponential amount of time with mean $1/\delta$ independent of the configuration it finds itself in. This assumption may or may not be reasonable in applications but it allows us to focus our attention on the birth rates which are only required to satisfy three mild conditions. (iii) says that the birth rates are spatially homogeneous and have "finite range", (iv) says there is no "creation from nothing," and (v) says that the birth rates are larger if there are more particles around. One might argue that in some systems the birth rate will be smaller in crowded situations but as the reader will see soon things simplify considerably when (v) is imposed and so we will assume for the rest of the paper that it holds.

Having defined the process, the first problem which must be dealt with is "do these rules specify a unique Markov process?". The answer is yes, of course, but for our purposes below it will be useful to have our processes defined on a special probability space so we will pause now to give an explicit construction. For each $x \in Z^d$ let $\{S_n^x : n \ge 1\}$ and $\{T_n^x : n \ge 1\}$ be the arrival times of two independent Poisson processes having rates 1 and $b^* = b_x(Z^d)$ respectively, and let $\{U_n^x : n \ge 1\}$ be a sequence of independent random variables each with a uniform distribution on $(0,1)$. From the ingredients it should be clear how we are going to construct the process. At times S_n^x $n \ge 1$ we kill a particle at x if one is present. At times T_n^x $n \ge 1$ there will be a birth at x if it is unoccupied and the configuration satisfies $b_x(\xi(T_n^x-)) \ge b^* U_n^x$ where $\xi(T_n^x-)$ denotes the limit of ξ_t as $t \uparrow T_n^x$.

Using properties of the Poisson process it is not hard to show that for any initial configuration A, the construction above gives rise to a well defined Markov process ξ_t^A, $t \ge 0$ with $\xi_0^A = A$, which evolves according to the rules given above. (This is Harris' (1972) construction, details are left as an exercise for the reader.) Notice that as a bonus we have all the processes defined on the same probability space in such a way that if $A \subset B$ then the resulting processes will have $\xi_t^A \subset \xi_t^B$ for all $t \ge 0$.

I. Having constructed our processes the first question we want to ask is: what are their stationary distributions? (iv) implies that the $\emptyset$ is an absorbing state, so $\delta_\emptyset$ = the unit mass on that configuration, is a trivial stationary distribution. To try to construct a nontrivial one we start with $\xi_0^1 = Z^2$ and run the process. An easy argument based on the construction above shows that if $s<t$ we can define ξ_s^1 and ξ_t^1 on the same space in such a way that $\xi_s^1 \supset \xi_t^1$ and from this it follows that as $t\to\infty$ ξ_t^1 converges weakly ($\Rightarrow$) to a limit ξ_∞^1 which is a stationary distribution for the process. If $\xi_\infty^1 \neq \delta_\emptyset$ we have succeeded in constructing a nontrivial equilibrium. If on the other hand $\xi_\infty^1 = \delta_\emptyset$ then we are out of luck. A comparison argument shows that starting from any initial $\xi_0 \subset Z^d$, $\xi_t \Rightarrow \delta_\emptyset$ and hence this is the only stationary distribution.

It is easy to show that if the death rate $\delta >$ the maximum birth rate b^* then

$$P(x \in \xi_t^1) \leq \exp((b^*-\delta)t)$$

and hence $\xi_t^1 \Rightarrow \delta_\emptyset$, so if we let $\delta_c = \sup\{\delta : \xi_\infty^1 \neq \delta_\emptyset\}$ then $\delta_c \leq b^*$. The last result shows δ_c is always finite. It is a much more subtle problem to determine when $\delta_c > 0$. We begin by considering some examples.

Example 1. THE BASIC CONTACT PROCESS IN Z^d.

$$b_x(\xi) = |\xi \cap N_x|$$

where N_x = "the neighbors of x" = $\{y : |x-y| = 1\}$, and $|A|$ = the number of points in A. In words the birth rate is equal to the number of occupied neighbors. Harris (1974) showed that $\delta_c > 0$ in this case.

Example 2. SEXUAL REPRODUCTION IN Z^1.

$$b_x(\xi) = 1 \quad \text{if} \quad x+1 \quad \text{and} \quad x-1 \in \xi$$

If y and $y+1$ are both vacant at some time then they will be for all time, so $\delta_c = 0$.

Example 3. SEXUAL REPRODUCTION IN Z^2.

$$b_x(\xi) = 1 \quad \text{if} \quad x+(1,0) \quad \text{and} \quad x+(0,1) \in \xi$$

It follows from results of Durrett and Gray (1985) that $\delta_c>0$.

Example 4. MENAGE À TROIS IN Z^2.

$$b_x(\xi) = 1 \quad \text{if} \quad |\xi \cap N_x| \geq 3$$

Four unoccupied sites which make a square can never be occupied so $\delta_c=0$.

In the two examples above with $\delta_c=0$ there was a finite set of zeros which could not be destroyed. Given this state of affairs it is natural (if somewhat optimistic) to hope that when this does not happen $\delta_c>0$. We believe this is true and furthermore

Conjecture. $\delta_c>0$ if and only if there is a finite set $S \subset Z^d$ so that $b_0(S)>0$ and 0 is not in the convex hull of S.

It is trivial that the condition is necessary for otherwise a large enough sphere cannot be eaten. A proof of sufficiency can probably be given using a "contour argument" like the ones in Griffeath and Gray (1982) and Durrett and Gray (1985) but the amount of labor required would be enormous (see Toom (1979) for a related result in discrete time) so we leave this as an exercise for a brave or more clever reader.

II. Once one has established that a particular model has $\delta_c>0$, it becomes natural to ask if ξ_∞^1 is the only nontrivial stationary distriubtions when $\delta<\delta_c$, or to be precise: Is there a stationary distribution which is not a convex combination of $\delta_\emptyset$ and ξ_∞^1? In this section we will describe what is known about this question for the basic contact process and related models. To describe the answers which have been obtained in their usual notation we will now and for the rest of the paper change variables $\delta=1/\lambda$ and then speed up time by a factor of λ so the death rate is $\equiv 1$, and the birth rate is $\lambda b_x(\xi)$ $0\leq\lambda<\infty$. With this revision made the birth rates in the basic contact process take on the usual form $b_x(\xi) = \lambda|\xi \cap N_x|$ and the critical value becomes $\lambda_c = \inf\{\lambda:\xi_\infty^1\neq\delta_\emptyset\}$.

In one dimension the answer to our question is: No. This was first proved by Liggett (1978) who showed that the result held for nearest neighbor growth models. In the case of the basic contact process in Z this result also follows from a convergence theorem of Griffeath (1978) and Durrett (1980) (extended to

$\lambda=\lambda_c$ by a remark in Durrett (1984) p. 1013).

<u>Theorem</u> 1. Let $\Omega_\infty = \{\xi_t \neq \emptyset$ for all $t\}$. As $t\to\infty$

$$\xi_t \Rightarrow P(\Omega_\infty^c)\delta_\emptyset + P(\Omega_\infty)\xi_\infty^1.$$

The proof of this theorem relies on two key lemmas.

<u>Lemma</u> 1. Let $r_t = \sup \xi_t^*$, $\ell_t = \inf \xi_t^*$ where $\xi_t^* = \xi_t^{\{0\}}$. If $B \ni \{0\}$ and we construct the processes as we have above, then on $\{\xi_t^* \neq \emptyset\}$

$$\xi_t^* = [\ell_t, r_t] \cap \xi_t^B.$$

Taking $B=Z$ we see that ξ_t^* looks like ξ_t^Z inside $[\ell_t, r_t]$ so all we have to show is that on Ω_∞, $r_t\to\infty$, $\ell_t\to-\infty$ in probability.

<u>Lemma</u> 2. Let $\bar{r}_t = \sup \xi_t^{(-\infty,0]}$. As $t\to\infty$

$$\text{(a)} \quad \frac{\bar{r}_t}{t} \to \alpha(\lambda) \equiv \inf_{m\geq 1} E\,\frac{\bar{r}_m}{m}$$

and

$$\text{(b)} \quad \lambda_c = \inf\{\lambda : \alpha(\lambda) > 0\}.$$

The proof of Lemma 1 relies heavily on the fact that the process is one dimensional and the birth rates are nearest neighbor and breaks down if either of these conditions is dropped. Lemma 2 is only slightly more robust. The proof relies on the fact that we are in one dimension but allows us to conclude (see Durrett (1980)) that (a) holds for growth models with finite range and that (b) if $\alpha(\lambda)<0$ then $\lambda<\lambda_c$. The last fact suggets a natural question which we will state as our next problem. Consider for concreteness

Example 5. <u>THE SECOND NEAREST NEIGHBOR CONTACT PROCESS</u>

$$b_x(\xi) = \lambda|\xi \cap \{x+2, x+1, x-1, x-2\}|.$$

Prove that $\lambda_c = \inf\{\lambda:\alpha(\lambda)>0\}$ and that Theorem 1 holds for this model. These results should be true for any finite range growth model in one dimension.

The state of affairs for the multidimensional contact process is better and worse than that for non-nearest neighbor models on Z^1. Durrett and Griffeath (1982) have shown that Theorem 1 holds for $\lambda>\lambda_c(1)$ = the critical value for the one dimensional model and furthermore that on Ω_∞ ξ_t^* is approximately $\xi_t^Z \cap tC$

when t is large, where C is a fixed convex set (see the paper cited for a precise statement).

The hypothesis in the last result suggests one problem: prove the convergence theorem for all $\lambda \geq \lambda_c$. In order to do this it seems we will need to find a characterization of λ_c like the one given in Lemma 2 which works well in $d>1$. A natural approach in $d=2$ is to let $\lambda_c(1,n)$ = the critical value for the basic contact process on $\{-n,\ldots,n\} \times Z$ and show that as $n\to\infty$ $\lambda_c(1,n) \downarrow \lambda_c(2)$. If you can do this then the techniques in Durrett and Griffeath (1982) and Aizenman et al (1984) will allow you to solve the other problem.

III. The results in the last section were obtained by using coupling methods. A second method for proving convergence theorems is "duality." If the birth rates are of the form

$$(1) \quad b_x(\xi) = \sum_i \lambda_i 1(\xi \cap (x+S_i) \neq \emptyset)$$

then the corresponding growth model ξ_t has a set valued dual process $\tilde{\xi}_t$ which satisfies

$$(2) \quad P(A \cap \xi_t^B \neq \emptyset) = P(B \cap \tilde{\xi}_t^A \neq \emptyset)$$

(for details see Harris (1976), (1978), or Griffeath (1979)). In $\tilde{\xi}_t$ particles die at rate 1 as before but now if $x \in \tilde{\xi}_t$ then at rate λ_i we have events which cause ALL the points of $x+S_i$ to become occupied if they are not already.

It is easy to see that the basic contact process has the form given above with $\lambda_i = \lambda$ $S_i = \{y_i\}$ $1 \leq i \leq 2d$ where $y_1,\ldots,y_{2d}$ are the points in $\{z:|z|=1\}$, and hence this example has the special property that it is self-dual $\xi_t \stackrel{d}{=} \tilde{\xi}_t$.

Not all birth rates can be written in the form given in (1) (Examples 2, 3, and 4 cannot). Those that can are called additive processes and because of the duality equation (2) are much easier to study. In particular we have the following general result due to Harris (1976).

<u>Theorem</u> 2. If the initial distriubtion is translation invariant and puts no mass on the $\emptyset$ configuration, then $\xi_t \Rightarrow \xi_\infty^1$ as $t\to\infty$.

As a corollary we see that any translation invariant stationary distribution is a convex combination of $\delta_\emptyset$ and ξ_∞^1.

The limiting behavior of processes which are not additive is much more complicated. In Example 3 if we start from product measure with density p (i.e. $\{x \in \xi_0^p\}$ are independent and each has probability p) and if $p<p^* = 1$ - the critical probability for oriented percolation then $\xi_t^p \Rightarrow \delta_\emptyset$. The proof is simple: if the initial configuration has a sequence of unoccupied sites $x_1, x_2, \ldots$ with $x_{n+1} \in \{x_n + (1,0),\ x_n + (0,1)\}$ then these sites can never be occupied. The last property is inherited by $x_n - (1,0)$ and $x_n - (0,1)$ when they become empty and gradually spreads to all of Z^2.

The last proof relies only on the existence of a sequence of unoccupied sites with the desired property and it is difficult (but certainly not impossible) to think of a reasonable random set ξ with $P(x \in \xi) = 1$ / (the defense budget) and zero probability of having a sequence of the type described above. (Exercise for experts: Do this under the assumption that ξ has positive correlations in the sense of Harris (1977)). This lack of imagination has led us to make the following:

<u>Conjecture</u>. The density of the largest stationary distribution, $\rho(\lambda) = P(x \in \xi_\infty^1)$, does not go to 0 as $\lambda \downarrow \lambda_c$.

The reader should note that soft arguments show $\lambda \to \rho(\lambda)$ is always right continuous (exercise), so if the conjecture is true then $\delta_\emptyset \neq \xi_\infty^1$ when $\lambda = \lambda_c$. In support of the conjecture we would like to observe that it is true if we replace Z^2 by the binary tree (see Durrett and Gray (1985)).

The proof given above relies heavily on the fact that the sequence of unoccupied sites is immune from attack and breaks down if we consider a more promiscuous model in which any two diagonally adjacent particles can make a baby, i.e.

Example 6. <u>SYMMETRIC SEXUAL REPRODUCTION IN Z^2</u>.

$$b_x(\xi) = \lambda \text{ if there are } y,z \in \xi \cap N_x \text{ with } |y-z|^2 = 2.$$

We conjecture that in this model that there is a P^*, which will now depend on λ, so that if $p<p^*(\lambda)$ then $\xi_t^p \Rightarrow \delta_\emptyset$. The trivial proof above does not work in this case so a new idea is needed to solve the problem.

Having seen $\xi_t^p \Rightarrow \delta_\emptyset$ for $p<p^*(\lambda)$ in Example 3 (and perhaps in Example 6) it becomes natural to let $p_c(\lambda) = \sup\{p : \xi_t^p \Rightarrow \delta_\emptyset\}$ and ask what happens for $p>p_c(\lambda)$. Presumably

$\xi_t^p \Rightarrow \xi_\infty^1$ but we do not even know how to prove this for $p > 1 - \varepsilon(\lambda)$.

IV. In the last three sections our main concern has been with the behavior of the process starting from infinite initial configurations. In this section we will turn our attention to what happens starting from a finite set A and in particular to the question: Is $P(\xi_t^A \neq \emptyset$ for all $t) > 0$? Let $\lambda_f = \inf\{\lambda : P(\xi_t^A \neq \emptyset$ for all $t) > 0$ for some finite set $A\}$. As in Section I the first problem we will consider is finding a necessary and sufficient condition for $\lambda_f < \infty$ and we will lead up to this by considering almost the same examples.

Example 1. <u>THE BASIC CONTACT PROCESS</u>.
Harris (1974) also showed that $\lambda_f < \infty$ in this case. Here the "also" is a little odd, since he used $\lambda_f < \infty$ and the duality equation (2) to conclude that $\lambda_c = \lambda_f < \infty$.

Example 2. <u>SEXUAL REPRODUCTION IN Z^1</u>.
Two consecutive vacant sites can never be occupied so $\lambda_f = \infty$.

Example 3. <u>SEXUAL REPRODUCTION IN Z^2</u>.
If $A \subset [-L,L]^2$ then there can never be births outside $[-L,L]^2$ so $\lambda_f = \infty$ (This is true in the promiscuous version (Example 6) also).

Example 7. <u>2 OUT OF 8 THRESHOLD PROCESS IN Z^2</u>.
Let $M_x = \{y : \sup_i |x_i - y_i| = 1\}$ and let

$$b_x(\xi) = \lambda \quad \text{if} \quad |\xi \cap M_x| \geq 2.$$

We leave it to the reader to show that $\lambda_f < \infty$ in this case (on $\{0,1\} \times Z$ it dominates a discrete time contact process) and to construct other examples with $\lambda_f < \infty$ or $\lambda_f = \infty$. We think that here as in Section I the trivial necessary condition will be sufficient so we

<u>Conjecture</u>. $\lambda_f = \infty$ if and only if there is a finite set $S \subset Z^d$ so that $b_x(S) = 0$ for all $x \notin S$.

The first step in proving a result like this is to recast the hypothesis in a more useable form (like the condition given in the conjecture in Section I). In this case we are not confident that we know the right way to do this so we will leave this (much easier) part of the problem to the reader as well.

Having seen two critical values it is natural to ask if they are equal. Taking $B=Z^d$ in the duality equation (2) gives

$$P(A\cap\xi_t^1\neq\emptyset) = P(\tilde{\xi}_t^A\neq\emptyset)$$

so $\xi_\infty^1=\delta_\emptyset$ if and only if the right hand side is $=0$ for all finite A. Introducing some obvious notation it follows that we have

$$(3) \qquad \lambda_c=\tilde{\lambda}_f.$$

If ξ_t is self dual e.g. the basic contact process then it follows from (3) that $\lambda_c=\lambda_f$. In other cases, we have to do some work to prove this, for instance

Example 8. THE THRESHOLD CONTACT PROCESS IN Z^d.

$$b_x(\xi) = \lambda \quad \text{if} \quad |\xi\cap N_x| \geq 1.$$

This example has the form given in (1) in Section III with $\lambda_1=\lambda$ $S_i=N_x$ and $\lambda_i=0$ $i\geq 2$, so the dual process $\tilde{\xi}_t$ has deaths at rate 1 (as usual) and at rate λ has events which cause all the points of $x+N_x$ to become occupied if they are not already.

In the one-dimensional version of this model (see Durrett and Griffeath (1983)) or, more generally, any nearest neighbor model (see Gray (1985)), it is known that $\lambda_c=\lambda_f$. In the additive case treated in the first reference this is proved by showing that if α and $\tilde{\alpha}$ are the "edge speeds" defined in Lemma 2 in Section II for the process ξ_t and its dual $\tilde{\xi}_t$ then $\alpha(\lambda) = \tilde{\alpha}(\lambda)$ and hence $\lambda_f=\tilde{\lambda}_f=\lambda_c$.

The last argument and a conjecture made in Section II suggest that in one dimensional finite range additive models

$$\begin{array}{ccccc} \lambda_c & = & \inf\{\lambda:\alpha(\lambda)>0\} & = & \lambda_f \\ \| & & \| & & \\ \tilde{\lambda}_f & = & \inf\{\lambda:\tilde{\alpha}(\lambda)>0\} & = & \tilde{\lambda}_c. \end{array}$$

To extend these ideas to $d>1$ we will need to find a suitable way of characterizing λ_c and λ_f. Nonetheless, we

Conjecture. For the threshold contact process $\lambda_c=\lambda_f$ in any dimension.

We think that the last result is true for all additive models but before the reader accepts this on the basis of the mysticism that "all critical values are equal", we would like to point out that Example 3 has $\lambda_c<\infty$ and $\lambda_f=\infty$ and a small perturbation of this example (e.g. add contact process births at rate $\varepsilon\lambda$)

produces a system with $0<\lambda_c<\lambda_f<\infty$. I will leave it to the reader to decide whether it is exciting or depressing that our systems can have two "critical values." In either case the last example raises two questions

(i) Can we have $\lambda_f<\lambda_c$?

(ii) In what circumstances do we have $\lambda_c=\lambda_f$?

We think the answer to (i) is No in Z^d but Yes for the basic contact process on the binary tree. As for (ii) we have already indicated that we think all additive processes will have this property. In addition we think that this will be true for all finite range models in one dimension. We will indicate our (mostly religious) reasons for this belief at the end of the next section. An interesting special case of the last speculation is

Example 9. THE CENTIPEDE (N=100).

$$b_x(\xi) = \lambda \quad \text{if} \quad x+1,\ldots,x+N \quad \text{are all} \quad \in \xi$$

when $N=1$ this is the one-sided contact process of Harris (1974). By comparing with this case one can easily show λ_c, $\lambda_f<\infty$ for all $N<\infty$. Is $\lambda_c=\lambda_f$?

V. In this the last section we will investigate what happens when we add spontaneous births at rate β i.e. change the birth rates to $\beta + b_x(\xi)$. In this case $\delta_\emptyset$ is no longer a stationary distribution but since the flip rates are monotone it can be used to produce one: we run $\xi_t^\emptyset$ and repeat the argument in Section I to conclude that as $t\to\infty$, $\xi_t^\emptyset \Rightarrow \xi_\infty^0$ and the limit is a stationary distribution. It is easy to see that the process has a unique stationary distribution if and only if $\xi_\infty^0 = \xi_\infty^1$, the limit starting from $\xi_0^1 = Z^d$.

If we introduce a superscript β to indicate the value of the parameter, we might hope that $\beta \to \xi_\infty^{1,\beta}$ and $\beta \to \xi_\infty^{0,\beta}$ are continuous at $\beta=0$. It is not hard to show that the first conclusion is correct.

(4) As $\beta\to 0$ $\xi_\infty^{1,\beta} \Rightarrow \xi_\infty^{1,0}$.

Proof. $\beta \to \xi_\infty^{1,\beta}$ is decreasing, so as $\beta\downarrow 0$ $\xi_\infty^{1,\beta}$ converges weakly to a limit $\supseteq \xi_\infty^1$. The limit is a stationary distribution for the process with $\beta=0$, so it cannot be strictly larger.

It is much more difficult to answer the question: "Does $\xi_\infty^{0,\beta} \Rightarrow \delta_\emptyset$ as $\beta \to 0$?", or in words that make a yes answer sound more likely: "Is $\delta_\emptyset$ stable under perturbation?"

In the contact process or more generally additive growth models (see Griffeath (1979)) we have

$$P(x \in \xi_t^1) - P(x \in \xi_t^0) \leq e^{-\beta t}$$

so $\xi_\infty^{0,\beta} = \xi_\infty^{1,\beta}$ for $\beta>0$ and hence $\xi_\infty^{0,\beta} \Rightarrow \xi_\infty^{1,0}$ as $\beta \downarrow 0$. In contrast we have the following result for Example 3 due to Durrett and Gray (1985).

Theorem 3. If λ is such that $\xi_\infty^{1,0} \neq \delta_\emptyset$ and β is chosen so that $6\beta^{1/4}\lambda^{3/4}<1$ then $\xi_\infty^{1,\beta} \neq \xi_\infty^{0,\beta}$.

The result is written in the form above to emphasize (a) the amount of perturbation allowed does not go to 0 as $\lambda \to \lambda_c$, more evidence, we think, for the discontinuity conjectured in Section III, and (b) if it turns out $\xi_\infty^{1,0} \neq \delta_\emptyset$ for $\lambda=\lambda_c$ then the result can be used at the critical value. Unfortunately the formulation completely obscures (c) $\xi_\infty^{0,\beta} \Rightarrow \delta_\emptyset$ as $\beta \to 0$, which also follows from the proof given in the paper cited.

Comparing the results in the last two cases and noticing that in the contact process $\lambda_c=\lambda_f$ while in Example 3. $\lambda_c<\infty, \lambda_f=\infty$ leads to the following wild guess:

<u>Conjecture</u>. $\xi_\infty^{0,\beta} \Rightarrow \delta_\emptyset$ as $\beta \downarrow 0$ if and only if $\lambda \leq \lambda_f$.

This question is in spirit similar to the first one we encountered in Section 1, which may be rephrased as follows: Let $\xi_\infty^{1,\delta}$ be the largest stationary distribution when the death rates is $\equiv \delta$. Find a necessary and sufficient condition for $\xi_\infty^{1,\delta} \Rightarrow Z^2$ as $\delta \downarrow 0$. i.e. When is the trivial equilibrium for the process with $\delta=0$ stable under perturbation? Our conjectured answer was given in Section I . A slightly stronger reformulation of that answer is that if we start from product measure with density $p>1-\varepsilon$ then in the process with $\delta=0$ $\xi_t^p \Rightarrow Z^d$.

Comparing the last statement with the results of Section III:

Ex. 1. $\xi_t^p \Rightarrow \xi_\infty^1$ for all $p>0$

Ex. 3. $\xi_t^p \Rightarrow \delta_\emptyset$ if $p<p^*$

and the state of affairs above

Ex. 1. $\delta_\emptyset$ unstable

Ex. 3. $\delta_\emptyset$ stable

suggests a conjecture which couched in the language of dynamical systems sounds trivially true: if $\delta_\emptyset$ is an attracting fixed point then it is stable under perturbation.

In Section IV we promised to explain why we thought that the centipede and in fact, all one dimensional growth models have $\lambda_c=\lambda_f$. Our reason is probably harder to believe than the conjecture: IF there is an example in one dimension with $\lambda_c<\lambda_f$ and IF the last conjecture we made above was true then by taking λ in (λ_c,λ_f) and adding spontaneous births at a small rate β we get a counterexample to the Positive Rates Conjecture which for growth models says that "all one dimensional finite range models have a unique stationary distribution when $\beta>0$."

The statement in quotation marks is known to hold in the nearest neighbor case (see Gray (1985)) but the general case seems to be a ridiculously hard problem, and hence it is not a member of our list. In closing we should mention a second famous problem which is not on our list for the same reason: Does the basic contact process in Z have $\xi_\infty^1=\delta_\emptyset$ at $\lambda=\lambda_c$? There are many reasons for believing this, see Griffeath (1981) or Durrett (1984) for arguments for oriented percolation, but none of these are strong enough to prove the results and to give the reader an idea how primitive our knowledge is, we leave it as our last problem to decide what happens at $\lambda=\lambda_c$ in any of the models discussed above.

Aknowledgement. Many of the conjectures above and all of the ones concerning the existence of critical values and stability of stationary distributions under various perturbations, were developed in conversations with Larry Gray during the academic year 1983-84 when he visited U.C.L.A.

REFERENCES

M. Aizenman, J.T. Chayes, L. Chayes, J. Fröhlich, and L. Russo (1984). On a sharp transition from area to perimeter law in a system of random surfaces. Comm. Math. Phys., to appear.

R. Durrett (1980). On the growth of one dimensional contact processes. Ann. Prob. 8, 890-907.

R. Durrett (1984). Oriented percolation in two dimensions. Ann. Prob. 12, 999-1040.

R. Durrett and L. Gray (1985). Some peculiar properties of a particle system with sexual reproduction, preprint.

R. Durrett and D. Griffeath (1982). Contact processes in several dimensions. Z. fur Wahr. 59, 535-552.

R. Durrett and D. Griffeath (1983). Supercritical contact processes on Z. Ann. Prob. 11, 1-15.

L. Gray (1982). The positive rates problem for attractive nearest neighbor spin systems on Z. Z. fur Wahr. 61, 389-404.

L. Gray (1985). Duality for general attractive spin systems, with applications in one dimension, preprint.

L. Gray and D. Griffeath (1982). A stability criterion for attractive nearest neighbor spin systems on Z. Ann. Prob. 10, 67-85.

D. Griffeath (1978). Limit theorems for nonergodic set-valued Markov processes. Ann. Prob. 6, 379-387.

D. Griffeath (1979). Additive and Cancellative Interacting Particle Systems. Springer Lecture Notes in Math. 724.

D. Griffeath (1981). The basic contact process. Stoch. Proc. Appl. 11, 151-186.

D. Griffeath (1983). The binary contact path process. Ann. Prob. 11, 692-705.

T. Harris (1972). Nearest neighbor Markov interaction processes on multidimenisonal lattices. Adv. in Math. 9, 66-89.

T. Harris (1974). Contact interactions on a lattice. Ann. Prob. 2, 969-988.

T. Harris (1976). On a class of set-valued Markov processes. Ann. Prob. 4, 175-194.

T. Harris (1977). A correlation inequality for processes in partially ordered spaces. Ann. Prob. 5, 451-454.

T. Harris (1978). Additive set-valued Markov processes and graphical methods. Ann. Prob. 6, 355-378.

R. Holley and T. Liggett (1978). The survival of contact processes. Ann. Prob. 6, 198-206.

H. Kesten (1982). Percolation Theory for Mathematicians. Birkhauser, Boston.

T. Liggett (1978). Attractive nearest neighbor spin systems on the integers. Ann. Prob. 6, 629-636.

T. Liggett (1985). Interacting Particle Systems. Springer-Verlag, New York.

A. Toom (1979). Stable and attractive trajectories in multi-component systems. p. 549-575 in Advances in Probability, Vol. 6, Marcel Dekker, New York.

DEPARTMENT OF MATHEMATICS
U.C.L.A.
LOS ANGELES, CA. 90024

After July 1, 1985:
Department of Mathematics
Cornell University
Ithaca, New York 14853

Contemporary Mathematics
Volume **41**, 1985

LARGE DEVIATIONS AND STATISTICAL MECHANICS

Richard S. Ellis[1]

ABSTRACT. This paper presents an overview of large deviations together with some applications to statistical mechanics. A general large deviation theorem is stated for sequences of random vectors for which a free energy function exists. This theorem is applied to three levels of large deviations for sequences of i.i.d. random vectors and to ferromagnetic models on $\mathbf{Z}^D$.

I. INTRODUCTION. In recent years, the theory of large deviations has undergone an extensive development. To a great measure, this development has been spurred by the important work of Donsker and Varadhan (1975a,1975b,1976,1983). The roots of large deviation theory lie in a number of areas, including statistical mechanics, statistics, probability, and information theory. In statistical mechanics, the first important figure was Boltzmann, who in the 1870's studied the relation between entropy and probability in physical systems. Statistical mechanics also provides a natural context in which large deviation theory may be applied. This is shown, for example, by the lectures of Lanford (1973), who uses large deviation ideas to explain certain foundational questions in statistical mechanics. In my book [Ellis (1985)], I try to bridge the approach of Lanford and the work of Donsker and Varadhan. The present paper highlights some of the main topics in the book. I have not given detailed references here since the book gives them.

II. LARGE DEVIATION PROPERTY FOR RANDOM VECTORS. The theory of large deviations is concerned with the exponential decay of probabilities associated with certain stochastic systems. We first present a general framework for studying such problems. Afterwards, specific examples will be given. Let X be a complete separable metric space and $\{Q_n; n=1,2,\ldots\}$ a sequence of probability measures on the Borel subsets of X. Suppose that as $n \to \infty$ $\{Q_n\}$ converges weakly to the unit point measure concentrated at some point x_0 in X; i.e.,

1980 Mathematics Subject Classification. 60F10, 82A05.
[1]Supported in part by National Science Foundation Grant MCS-8219848.

$Q_n \Rightarrow \delta_{x_0}$ as $n \to \infty$. Thus if A is any Borel set whose closure does not contain x_0, then $Q_n\{A\} \to 0$. In the examples we consider, $Q_n\{A\}$ converges to 0 exponentially fast as $n \to \infty$ with an exponential rate depending on the set A. The situation is abstracted in the following definition.

DEFINITION II.1. Let X be a complete separable metric space, $\mathcal{B}(X)$ the Borel σ-field of X, and $\{Q_n; n=1,2,\ldots\}$ a sequence of probability measures on $\mathcal{B}(X)$. The sequence $\{Q_n\}$ is said to have a large deviation property if there exists a sequence of positive real numbers $\{a_n\}$ which tend to ∞ and a function $I(\cdot)$ which maps X into $[0,\infty]$ such that the following hold.*

(a) $I(\cdot)$ is lower semicontinuous on X.

(b) For each real number b, the set $\{x \in X : I(x) \leq b\}$ is a compact set in X.

(c) $\limsup_{n\to\infty} \frac{1}{a_n} \log Q_n\{K\} \leq -I(K)$ for each closed set K in X.

(d) $\liminf_{n\to\infty} \frac{1}{a_n} \log Q_n\{G\} \geq -I(G)$ for each open set G in X.

$I(x)$ is called an entropy function of $\{Q_n\}$.

One may show that if a large deviation property holds, then the entropy function is unique. Given a Borel set A, $c\ell\, A$ denotes its closure and $\operatorname{int} A$ its interior. The upper and lower large deviation bounds (c) - (d) imply that if $I(c\ell\, A) = I(\operatorname{int} A)$, then

$$\lim_{n\to\infty} \frac{1}{a_n} \log Q_n\{A\} = -I(A). \tag{1}$$

The bounds (c) - (d) with $K = G = X$ also imply that the infimum of $I(x)$ over X equals 0. In many applications, $I(x)$ attains its infimum over X at a unique point x_0; thus

$$I(x_0) = 0 \text{ and } I(x) > 0 \text{ for all } x \neq x_0 \text{ in } X. \tag{2}$$

If A is a Borel set whose closure does not contain x_0, then it follows from hypotheses (a) - (b) of large deviation property that $I(A)$ is positive. If in addition $I(c\ell\, A) = I(\operatorname{int} A)$, then we conclude from (1) that $Q_n\{A\}$ converges to 0 exponentially fast with a_n and that $I(A)$ is the exact exponential rate.

Here is a useful application of the large deviation property to calculate the asymptotics of certain integrals. The theorem, which is due to Varadhan

---

*For A a nonempty subset of X, we define $I(A) = \inf\{I(x) : x \in A\}$. For the empty set, we define $I(\emptyset) = \infty$.

(1966), will be applied later in this paper.

THEOREM II.2. Let X be a complete separable metric space, $\mathcal{B}(X)$ the Borel σ-field of X, and $\{Q_n; n=1,2,\ldots\}$ a sequence of probability measures on $\mathcal{B}(X)$. Assume that $\{Q_n\}$ has a large deviation property with constants $\{a_n\}$ and entropy function I. Let F be a continuous function from X to $\mathbf{R}$.

(a) Assume that $\sup_{x\in X} F(x)$ is finite. Then $\sup_{x\in X}\{F(x) - I(x)\}$ is finite and

(3) $$\lim_{n\to\infty} \frac{1}{a_n} \log \int_X \exp(a_n F(x)) Q_n(dx) = \sup_{x\in X}\{F(x) - I(x)\}.$$

(b) More generally, assume that F is unbounded above and satisfies

(4) $$\lim_{L\to\infty} \limsup_{n\to\infty} \frac{1}{a_n} \log \int_{\{F\geq L\}} \exp(a_n F(x)) Q_n(dx) = -\infty .$$

Then the limit (3) holds and is finite. In particular, if F is bounded above on the union of the supports of the $\{Q_n\}$, then (4) is satisfied and thus the limit (3) holds and is finite.

It is not hard to motivate the limit (3). The large deviation property of $\{Q_n\}$ may be summarized by the heuristic formula

(5) $$Q_n(dx) \approx \exp(-a_n I(x))dx.$$

Hence we may write

(6) $$\lim_{n\to\infty} \frac{1}{a_n} \log \int_X \exp(a_n F(x)) Q_n(dx) \approx \lim_{n\to\infty} \frac{1}{a_n} \int_X \exp[a_n\{F(x) - I(x)\}]dx.$$

By analogy with Laplace's method, one is led to (3).

The main result in this paper is Theorem II.3, which allows one to derive large deviation properties in many situations. Let $\{\Omega_n, \mathcal{F}_n, P_n); n=1,2,\ldots\}$ be a sequence of probability spaces and W_n a random vector on Ω_n which takes values in $\mathbf{R}^d$, some $d \in \{1,2,\ldots\}$. For $n=1,2,\ldots$ and $t \in \mathbf{R}^d$, define the function

(7) $$c_n(t) = \frac{1}{a_n} \log E_n\{\exp\langle t, W_n\rangle\},$$

where $\{a_n\}$ is a sequence of positive real numbers that tend to ∞, E_n denotes expectation with respect to P_n, and $\langle -,- \rangle$ denotes the Euclidean inner product on $\mathbf{R}^d$. We assume that $c_n(t)$ is finite for all $t \in \mathbf{R}^d$ and that the limit

(8) $$c(t) = \lim_{n\to\infty} c_n(t)$$

exists and is finite for all $t \in \mathbf{R}^d$. These assumptions are natural for statistical mechanical applications since $c(t)$ is closely related to the concept of free energy. The function $c(t)$ is called the free energy function of the sequence $\{W_n\}$. Finally, we define the function

$$(9) \qquad I(x) = \sup_{t\in\mathbf{R}^d} \{\langle t,x\rangle - c(t)\} \quad \text{for } x \in \mathbf{R}^d.$$

The next theorem is proved in Ellis (1984).

THEOREM II.3. *Let* Q_n *be the distribution of* W_n/a_n *on* $\mathbf{R}^d$. *The following conclusions hold.*

(a) $I(x)$ *is convex, lower semicontinuous, and non-negative.* $I(x)$ *has compact level sets, and the infimum of* $I(x)$ *over* $\mathbf{R}^d$ *equals* 0.

(b) *The upper large deviation bound is valid:*

$$(10) \qquad \limsup_{n\to\infty} \frac{1}{a_n} \log Q_n\{K\} \le -I(K) \quad \text{for each closed set } K \text{ in } \mathbf{R}.$$

(c) *Assume in addition that* $c(t)$ *is differentiable for all* t. *Then the lower large deviation bound is valid:*

$$(11) \qquad \liminf_{n\to\infty} \frac{1}{a_n} \log Q_n\{G\} \ge -I(G) \quad \text{for each open set } G \text{ in } \mathbf{R}^d.$$

Hence, if $c(t)$ *is differentiable for all* t, *then* (11) *and parts* (a) *and* (b) *imply that* $\{Q_n; n=1,2,\ldots\}$ *has a large deviation property with entropy function* $I(x)$.

Under our assumptions, $c(t)$ is a convex function on $\mathbf{R}^d$. Formula (9) expresses $I(x)$ as the Legendre-Fenchel transform of the convex function $c(t)$. We now motivate this formula. Assuming that $\{Q_n\}$ has a large deviation property with some entropy function $I(x)$ which is convex, we will deduce formula (9) as a consequence. For any $t \in \mathbf{R}^d$

$$(12) \qquad c(t) = \lim_{n\to\infty} \frac{1}{a_n} \log E_n\{\exp\langle t,W_n\rangle\}$$
$$= \lim_{n\to\infty} \frac{1}{a_n} \log \int_{\mathbf{R}^d} \exp(a_n\langle t,x\rangle) Q_n(dx).$$

We apply Theorem II.2 to the continuous function $x \longmapsto \langle t,x\rangle$. Since $c(2t)$ is finite,

$$(13) \qquad \limsup_{n\to\infty} \frac{1}{a_n} \log \int_{\{\langle t,x\rangle \ge L\}} \exp(a_n\langle t,x\rangle) Q_n(dx)$$
$$\le \limsup_{n\to\infty} \frac{1}{a_n} \log\left[e^{-a_n L} \cdot \int_{\mathbf{R}^d} \exp(2a_n\langle t,x\rangle) Q_n(dx)\right]$$
$$= -L + \limsup_{n\to\infty} \frac{1}{a_n} c(2t) = -L .$$

Hence (4) holds, and the formula

(14) $$c(t) = \sup_{x\in\mathbf{R}^d} \{\langle t,x\rangle - I(x)\}$$

follows. Since $I(x)$ is convex and lower semicontinuous, formula (14) may be inverted to yield (9). This completes our motivation of formula (9).

According to Theorem II.3, $I(x)$ is a convex, lower semicontinuous function on $\mathbf{R}^d$, and the infimum of $I(x)$ over $\mathbf{R}^d$ equals 0. If $c(t)$ is differentiable for all t, then $\{Q_n\}$ has a large deviation property with entropy function $I(x)$. It also follows from the theory of Legendre-Fenchel transforms that $I(x)$ attains its infimum over $\mathbf{R}^d$ at the unique point $x_0 = \nabla c(0)$.

What can happen if $c(t)$ is not differentiable for all t? Part (b) of Theorem II.3 states that the upper large deviation bound (10) is valid for all closed sets K. In the following example, $c(t)$ is not differentiable at a single point and the lower large deviation bound (11) fails for a whole class of open sets G. Nevertheless, a large deviation property holds with a nonconvex entropy function. In statistical mechanical applications, a nondifferentiable free energy function corresponds to a phase transition.

EXAMPLE II.4. Let W_n have the distribution $\frac{1}{2}\delta_n + \frac{1}{2}\delta_{-n}$. Then with $a_n = n$

(15) $$c(t) = \lim_{n\to\infty}\frac{1}{n}\log\left[\frac{1}{2}\left(e^{nt} + e^{-nt}\right)\right] = \begin{cases} t & \text{if } t \geq 0, \\ -t & \text{if } t < 0 . \end{cases}$$

Thus $c(t) = |t|$, which is not differentiable at 0, and

(16) $$I(x) = \sup_{t\in\mathbf{R}} \{tx - |t|\} = \begin{cases} 0 & \text{if } |x| \leq 1, \\ \infty & \text{if } |x| > 1 . \end{cases}$$

$I(x)$ attains its infimum not at a single point, but on the whole interval $[-1,1]$. If G is any open subset of $(-1,1)$, then $Q_n\{G\} = 0$ and $\log Q_n\{G\} = -\infty$ while the infimum of $I(x)$ over G equals 0. Thus the lower large deviation bound (11) fails. On the other hand, it is easy to see that the distributions $\{Q_n\}$ of $\{W_n/n\}$ have a large deviation property with entropy function $I_0(x) = 0$ for $x = 1$ and $x = -1$, $I_0(x) = \infty$ for $x \notin \{1,-1\}$. The points 1 and -1 at which $I_0(x)$ attains its infimum are the points at which the distribution Q_n is concentrated for each n. The closed convex hull of $I_0(x)$ is the function $I(x)$ defined in (16).

We end this section by introducing a notion of stochastic convergence that is useful in statistical mechanics. The same notation as in Theorem II.3 is used. We say that W_n/a_n <u>converges exponentially</u> to a constant x_0, and write $W_n/a_n \xrightarrow{\exp} x_0$, if for any $\varepsilon > 0$ there exists a number $N = N(\varepsilon) > 0$ such that

$$(17)\qquad P_n\left\{ \left\|\frac{W_n}{a_n} - x_o\right\| \geq \varepsilon\right\} \leq e^{-a_n N} \quad \text{for all sufficiently large } n.$$

The next theorem is proved in Ellis (1984).

THEOREM II.5. <u>The following statements are equivalent.</u>

(a) $W_n/a_n \xrightarrow{exp} x_o$.

(b) $c(t)$ <u>is differentiable at</u> $t = 0$ <u>and</u> $\nabla c(0) = x_o$.

(c) $I(x)$ <u>attains its infimum over</u> $\mathbf{R}^d$ <u>at the unique point</u> $x = x_o$.

Theorems II.2, II.3, and II.5 will be applied in the next two sections.

III. THREE LEVELS OF LARGE DEVIATIONS FOR I.I.D. RANDOM VECTORS. In this section, $X_1, X_2, \ldots$ is a sequence of i.i.d. random vectors that are defined on a probability space (Ω, F, P) and that take values in $\mathbf{R}^d$. The distribution of X_1 is denoted by ρ. We state three levels of large deviation properties for the sequence $X_1, X_2, \ldots$.

LEVEL-1. Let S_n denote the n'th partial sum $\sum_{j=1}^{n} X_j$, $n = 1,2,\ldots$, and assume that $\int_{\mathbf{R}^d} \|z\| \rho(dz)$ is finite. The strong law of large numbers states that

$$(18)\qquad \frac{S_n}{n} \longrightarrow \mu_\rho \quad \text{P-a.s.,}$$

where μ_ρ is the mean $\int_{\mathbf{R}^d} z\rho(dz)$. If $Q_n^{(1)}$ denotes the distribution of S_n/n on $\mathbf{R}^d$, then it follows that $Q_n^{(1)} \Rightarrow \delta_{\mu_\rho}$ as $n \to \infty$. The next theorem states a large deviation property for the distributions $\{Q_n^{(1)}\}$.

THEOREM III.1. <u>For</u> $t \in \mathbf{R}^d$, <u>define</u>

$$(19)\qquad c_\rho(t) = \log E\{\exp\langle t, X_1\rangle\} = \log \int_{\mathbf{R}^d} \exp\langle t,z\rangle \rho(dz).$$

<u>Assume that</u> $c_\rho(t)$ <u>is finite for all</u> t <u>and define</u>

$$(20)\qquad I_\rho^{(1)}(x) = \sup_{t\in\mathbf{R}^d} \{\langle t,x\rangle - c_\rho(t)\} \quad \text{for } x \in \mathbf{R}^d.$$

<u>Then the following conclusions hold.</u>

(a) $\{Q_n^{(1)}\}$, <u>the distributions of</u> $\{S_n/n\}$ <u>on</u> $\mathbf{R}^d$, <u>have a large deviation property with</u> $a_n = n$ <u>and entropy function</u> $I_\rho^{(1)}(x)$.

(b) $I_\rho^{(1)}(x)$ <u>is a convex function of</u> x. $I_\rho^{(1)}(x)$ <u>measures the discrepancy between</u> x <u>and the mean</u> $\mu_\rho = \int_{\mathbf{R}^d} z\rho(dz)$ <u>in the sense that</u> $I_\rho^{(1)}(x) \geq 0$ <u>with equality if and only if</u> $x = \mu_\rho$.

Part (a) of Theorem III.1 is a consequence of Theorem II.3. Let $a_n = n$ and $W_n = S_n$. Since the random vectors $X_1, X_2, \ldots$ are i.i.d.,

$$(21) \qquad c_n(t) = \frac{1}{n} \log E\{\exp\langle t, S_n\rangle\} = \log E\{\exp\langle t, X_1\rangle\}.$$

Thus $c(t) = \lim_{n\to\infty} c_n(t)$ equals $c_\rho(t)$ for all t. The function $c_\rho(t)$ is differentiable for all t, and so part (a) of Theorem III.1 follows. $I_\rho^{(1)}(x)$ is a convex function of x since it is defined as the Legendre-Fenchel transform of $c_\rho(t)$. According to Theorem II.5, since $c_\rho(t)$ is differentiable at $t = 0$, $I_\rho^{(1)}(x)$ attains its infimum over $\mathbf{R}^d$ at the unique point $x_0 = \nabla c_\rho(0)$. The gradient $\nabla c_\rho(0)$ equals the mean μ_ρ. Part (b) of Theorem III.1 is proved.

Theorem II.5 implies that S_n/n converges exponentially to μ_ρ; i.e., for any $\varepsilon > 0$ there exists a number $N = N(\varepsilon) > 0$ such that

$$(22) \qquad P\left\{ \left\|\frac{S_n}{n} - \mu_\rho\right\| \geq \varepsilon \right\} \leq e^{-nN} \quad \text{for all sufficiently large } n.$$

Since $\sum_{n=1}^{\infty} e^{-nN}$ is finite, the Borel-Cantelli lemma yields the strong law of large numbers: $S_n/n \longrightarrow \mu_\rho$ P-a.s.

We note a special case of Theorem III.1.

EXAMPLE III.2. Let ρ denote the probability measure $\frac{1}{2}\delta_1 + \frac{1}{2}\delta_{-1}$ on $\mathbf{R}$. Then $c_\rho(t) = \log\cosh t$. The distributions $\{Q_n^{(1)}\}$ have a large deviation property with $a_n = n$ and entropy function

$$(23) \qquad I_\rho^{(1)}(x) = \sup_{t\in\mathbf{R}}\{tx - \log\cosh t\} = \begin{cases} \frac{1+x}{2}\log(1+x) + \frac{1-x}{2}\log(1-x) & \text{for } |x| \leq 1, \\ \infty & \text{for } |x| > 1\,. \end{cases}$$

LEVEL-2. $\mathcal{M}(\mathbf{R}^d)$ denotes the space of probability measures on the Borel subsets of $\mathbf{R}^d$ with the topology of weak convergence. $\mathcal{M}(\mathbf{R}^d)$ is a complete separable metric space. For $n = 1,2,\ldots$ and $\omega \in \Omega$, the <u>empirical measure</u> is the measure on $\mathcal{B}(\mathbf{R}^d)$ defined by

$$(24) \qquad L_n(\omega,\cdot) = \frac{1}{n}\sum_{j=1}^{n} \delta_{X_j(\omega)}(\cdot).$$

For A a Borel subset of $\mathbf{R}^d$, $L_n(\omega,A)$ gives the relative frequency with which the sequence $X_1(\omega), X_2(\omega), \ldots, X_n(\omega)$ lies in A. For each ω, $L_n(\omega,\cdot)$ is an element of $\mathcal{M}(\mathbf{R}^d)$. The mean of $L_n(\omega,\cdot)$ is the level-1 sum $S_n(\omega)/n$. The strong law of large numbers implies that

$$(25) \qquad L_n(\omega,\cdot) \Rightarrow \rho \quad P\text{-a.s.}$$

If $Q_n^{(2)}$ denotes the distribution of $L_n(\omega,\cdot)$ on $M(\mathbf{R}^d)$, then it follows that $Q_n^{(2)} \Rightarrow \delta_\rho$ as $n \to \infty$. The next theorem states a large deviation property for the distributions $\{Q_n^{(2)}\}$.

THEOREM III.3. <u>For</u> $\nu \in M(\mathbf{R}^d)$, <u>define the relative entropy of</u> ν <u>with respect to</u> ρ <u>by the formula</u>

$$(26)\qquad I_\rho^{(2)}(\nu) = \begin{cases} \int_{\mathbf{R}^d} \log \frac{d\nu}{d\rho}(x)\,\nu(dx) & \underline{\text{if}}\ \ \nu \ll \rho\ \ \underline{\text{and}}\ \int_{\mathbf{R}^d} \left|\log \frac{d\nu}{d\rho}\right| d\nu < \infty, \\ \infty & \underline{\text{otherwise}}. \end{cases}$$

<u>The following conclusions hold.</u>

(a) $\{Q_n^{(2)}\}$, <u>the distributions of</u> $\{L_n\}$ <u>on</u> $M(\mathbf{R}^d)$, <u>have a large deviation property with</u> $a_n = n$ <u>and entropy function</u> $I_\rho^{(2)}$.

(b) $I_\rho^{(2)}(\nu)$ <u>is a convex function of</u> ν. $I_\rho^{(2)}(\nu)$ <u>measures the discrepancy between</u> ν <u>and</u> ρ <u>in the sense that</u> $I_\rho^{(2)}(\nu) \geq 0$ <u>with equality if and only if</u> $\nu = \rho$.

Theorem III.3 is proved in Donsker and Varadhan (1976).* We sketch a simple proof of part (a) in the case that the random vectors $X_1, X_2, \ldots$ have a finite state space Γ. The reader is referred to Chapter VIII of Ellis (1985) for details. Suppose that $\Gamma = \{x_1, x_2, \ldots, x_r\}$, where $x_1, x_2, \ldots, x_r$ are r distinct real numbers. $M(\Gamma)$ denotes the space of probability measures on the subsets of Γ with the topology of weak convergence. The distribution ρ of X_1 has the form $\sum_{i=1}^r \rho_i \delta_{x_i}$ ($\rho_i = \rho\{x_i\} > 0$), and $Q_n^{(2)}\{M(\Gamma)\} = 1$. $M(\Gamma)$ is homeomorphic to the compact convex subset of $\mathbf{R}^r$ consisting of all vectors $\nu = (\nu_1, \ldots, \nu_r)$ with $\nu_i \geq 0$ and $\sum_{i=1}^r \nu_i = 1$. Let M denote this subset of $\mathbf{R}^r$. For $\nu \in M$,

$$(27)\qquad I_\rho^{(2)}(\nu) = \sum_{i=1}^r \nu_i \log \frac{\nu_i}{\rho_i}\,.$$

where $0 \log 0 = 0$. The vector in M corresponding to the empirical measure $L_n(\omega,\cdot)$ is the vector $L_n(\omega)$ whose i'th component is $L_n(\omega,\{x_i\})$. It suffices to show that the distributions of $\{L_n(\omega)\}$ on $\mathbf{R}^r$ have a large deviation property with $a_n = n$ and entropy function

$$(28)\qquad I(\nu) = \begin{cases} I_\rho^{(2)}(\nu) & \text{for } \nu \in M, \\ \infty & \text{for } \nu \in \mathbf{R}^r \setminus M. \end{cases}$$

*This paper generalizes Theorem III.3 to random vectors taking values in a complete separable metric space.

Using the notation of Theorem II.3, let $a_n = n$ and $W_n(\omega) = nL_n(\omega)$. The free energy function of the sequence $\{nL_n(\omega)\}$ is found to be

$$c(t) = \log \sum_{i=1}^{r} e^{t_i}\rho_i \quad \text{for} \quad t = (t_1,\dots,t_r) \in \mathbf{R}^r. \tag{29}$$

Since $c(t)$ is differentiable for all t, the distributions of $\{L_n(\omega)\}$ on $\mathbf{R}^r$ have a large deviation property with $a_n = n$ and entropy function

$$I(\nu) = \sup_{t\in\mathbf{R}^r} \{\langle t,\nu\rangle - c(t)\}. \tag{30}$$

A short calculation gives (28).

LEVEL-3 (Empirical Pair Measure). We first consider a special case of level-3. A more general level-3 problem is stated below. For simplicity, we assume that the random vectors $X_1, X_2, \dots$ have a finite state space $\Gamma = \{x_1, x_2, \dots, x_r\}$, where $x_1, x_2, \dots, x_r$ are r distinct real numbers. $\mathcal{M}_s(\Gamma^2)$ denotes the space of probability measures on the subsets of Γ^2 with equal one-dimensional marginals. $\mathcal{M}_s(\Gamma^2)$ is topologized by the topology of weak convergence. A measure τ in $\mathcal{M}_s(\Gamma^2)$ has the form $\sum_{i,j=1}^{r} \tau_{ij}\, \delta_{\{x_i,x_j\}}$ with $\tau_{ij} \geq 0$, $\sum_{i,j=1}^{r} \tau_{ij} = 1$, $\sum_{j=1}^{r} \tau_{ij} = \sum_{k=1}^{r} \tau_{ki}$ for each i. For $\omega \in \Omega$ and $n = 2,3,\dots$, define $Y_k^{(n)}(\omega)$ to be the ordered pair $(X_k(\omega), X_{k+1}(\omega))$ if $k \in \{1,2,\dots,n-1\}$ and to be the cyclic ordered pair $(X_n(\omega), X_1(\omega))$ if $k = n$. The <u>empirical pair measure</u> is the measure on the subsets of Γ^2 defined by

$$M_n(\omega,\cdot) = \frac{1}{n}\sum_{k=1}^{n} \delta_{Y_k^{(n)}(\omega)}(\cdot)\,. \tag{31}$$

For A a subset of Γ^2, $M_n(\omega,A)$ gives the relative frequency with which the sequence $Y_1(\omega), Y_2(\omega), \dots, Y_n(\omega)$ lies in A. For each ω, $M_n(\omega,\cdot)$ is an element of $\mathcal{M}_s(\Gamma^2)$. Both one-dimensional marginals of $M_n(\omega,\cdot)$ equal the empirical measure $L_n(\omega,\cdot)$. According to the ergodic theorem

$$M_n(\omega,\cdot) \Rightarrow \pi_2 P_\rho \qquad P\text{-a.s.}, \tag{32}$$

where $\pi_2 P_\rho$ denotes the product measure on the subsets of Γ^2 with identical one-dimensional marginals ρ. If $Q_{n,2}^{(3)}$ denotes the distribution of $M_n(\omega,\cdot)$ on $\mathcal{M}_s(\Gamma^2)$, then it follows that $Q_{n,2}^{(3)} \Rightarrow \delta_{\pi_2 P_\rho}$ as $n \to \infty$. The next theorem states a large deviation property for the distributions $\{Q_{n,2}^{(3)}\}$.

THEOREM III.4. <u>For</u> $\tau = \sum_{i,j=1}^{r} \tau_{ij}\delta_{\{x_i,x_j\}} \in \mathcal{M}_s(\Gamma^2)$, <u>define</u>

$$(\nu_\tau)_i = \sum_{j=1}^{r} \tau_{ij} \quad \underline{\text{for each}} \quad i \in \{1,\dots,r\} \tag{33}$$

and

(34) $$I^{(3)}_{\rho,2}(\tau) = \Sigma \tau_{ij} \log \frac{\tau_{ij}}{(\nu_\tau)_i \rho_j} ,$$

where the sum runs over all i and j for which $(\nu_\tau)_i > 0$. The following conclusions hold.

(a) $\{Q^{(3)}_{n,2}\}$, the distributions of $\{M_n\}$ on $M_s(\Gamma^2)$, have a large deviation property with $a_n = n$ and entropy function $I^{(3)}_{\rho,2}$.

(b) $I^{(3)}_{\rho,2}(\tau)$ is a convex function of τ. $I^{(3)}_{\rho,2}(\tau)$ measures the discrepancy between τ and $\pi_2 P_\rho$ in the sense that $I^{(3)}_{\rho,2}(\tau) \geq 0$ with equality if and only if $\tau = \pi_2 P_\rho$.

In order to prove part (a) of Theorem III.4, it suffices to show a large deviation property for the distributions of the random vectors on $\mathbf{R}^{r^2}$ whose components are given by $M_n(\omega,\{x_i,x_j\})$, $i,j \in \{1,\ldots,r\}$. This large deviation property may be deduced from Theorem II.3. The reader is referred to Chapter IX of Ellis (1985) for the proof of Theorem III.4.

LEVEL-3 (Empirical Process). $(\mathbf{R}^d)^{\mathbf{Z}}$ denotes the set of all sequences $\omega = \{\omega_j; j \in \mathbf{Z}\}$ with each $\omega_j \in \mathbf{R}^d$. $(\mathbf{R}^d)^{\mathbf{Z}}$ is topologized by the product topology. $\mathcal{B}((\mathbf{R}^d)^{\mathbf{Z}})$ denotes the Borel σ-field of $(\mathbf{R}^d)^{\mathbf{Z}}$; T the shift mapping on $(\mathbf{R}^d)^{\mathbf{Z}}$; and $M_s((\mathbf{R}^d)^{\mathbf{Z}})$ the space of strictly stationary probability measures on $\mathcal{B}((\mathbf{R}^d)^{\mathbf{Z}})$ with the topology of weak convergence. $M_s((\mathbf{R}^d)^{\mathbf{Z}})$ is a complete separable metric space. Given a positive integer n and $\omega \in \Omega$, repeat the sequence $X_1(\omega),X_2(\omega),\ldots,X_n(\omega)$ periodically into a doubly infinite sequence, obtaining a point

(35) $$X(n,\omega) = (\ldots,X_n(\omega),X_1(\omega),X_2(\omega),\ldots,X_n(\omega),X_1(\omega),X_2(\omega),\ldots,X_n(\omega),X_1(\omega),\ldots)$$

in $(\mathbf{R}^d)^{\mathbf{Z}}$; $(X(n,\omega))_1 = X_1(\omega)$, $(X(n,\omega))_2 = X_2(\omega)$, etc. For $\omega \in \Omega$, the empirical process is the measure on $\mathcal{B}((\mathbf{R}^d)^{\mathbf{Z}})$ defined by

(36) $$R_n(\omega,\cdot) = \frac{1}{n} \sum_{k=0}^{n-1} \delta_{T^k X(n,\omega)}(\cdot) ,$$

where T^0 is the identity mapping and $T^k = T(T^{k-1})$, $k \geq 1$. For A a Borel subset of $(\mathbf{R}^d)^{\mathbf{Z}}$, $R_n(\omega,A)$ gives the relative frequency with which the sequence $X(n,\omega)$ $TX(n,\omega),\ldots,T^{n-1}X(n,\omega)$ lies in A. Since $X(n,\omega)$ is periodic of period n, $R_n(\omega,\cdot)$ is for each ω an element of $M_s((\mathbf{R}^d)^{\mathbf{Z}})$. All the one-dimensional marginals of $R_n(\omega,\cdot)$ equal the empirical measure $L_n(\omega,\cdot)$ while all the two-dimensional marginals of $R_n(\omega,\cdot)$ equal the empirical pair measure $M_n(\omega,\cdot)$. According to the ergodic theorem

(37) $$R_n(\omega,\cdot) \Rightarrow P_\rho \quad P\text{-a.s.},$$

where P_ρ denotes the infinite product measure on $\mathcal{B}((\mathbf{R}^d)^{\mathbf{Z}})$ with identical one-dimensional marginals ρ. If $Q_n^{(3)}$ denotes the distribution of $R_n(\omega,\cdot)$ on $\mathcal{M}_s((\mathbf{R}^d)^{\mathbf{Z}})$, then it follows that $Q_n^{(3)} \Rightarrow \delta_{P_\rho}$ as $n \to \infty$. The next theorem states a large deviation property for the distributions $\{Q_n^{(3)}\}$.

THEOREM III.5. For $P \in \mathcal{M}_s((\mathbf{R}^d)^{\mathbf{Z}})$, let $\tilde{P}_\omega$ be a regular conditional distribution, with respect to P, of X_1 given the σ-field $\mathcal{F}(\{X_j; j \le 0\})$. We define the mean relative entropy of P with respect to P_ρ by the formula

(38) $$I_\rho^{(3)}(P) = \int_{(\mathbf{R}^d)^{\mathbf{Z}}} I_\rho^{(2)}(\tilde{P}_\omega)\, P(d\omega),$$

where $I_\rho^{(2)}$ denotes relative entropy [Theorem III.3].* The following conclusions hold.

(a) $\{Q_n^{(3)}\}$, the distributions of $\{R_n\}$ on $\mathcal{M}_s((\mathbf{R}^d)^{\mathbf{Z}})$, have a large deviation property with $a_n = n$ and entropy function $I_\rho^{(3)}$.

(b) $I_\rho^{(3)}(P)$ is an affine function of P. $I_\rho^{(3)}(P)$ measures the discrepancy between P and P_ρ in the sense that $I_\rho^{(3)}(P) \ge 0$ with equality if and only if $P = P_\rho$.

Theorem III.5 may be proved by the methods of Donsker and Varadhan (1983). The main result in that paper is a level-3 large deviation theorem for continuous parameter Markov processes taking values in a complete separable metric space. In the special case where the random vectors $X_1, X_2, \ldots$ have a finite state space $\Gamma = \{x_1, x_2, \ldots, x_r\}$, Theorem III.5 may be derived from Theorem II.3. The set Γ is topologized by the discrete topology and the sequence space $\Gamma^{\mathbf{Z}}$ by the product topology. $\mathcal{M}_s(\Gamma^{\mathbf{Z}})$ denotes the space of probability measures on the Borel subsets of $\Gamma^{\mathbf{Z}}$ with the topology of weak convergence. $\mathcal{M}_s(\Gamma^{\mathbf{Z}})$ is compact, and $Q_n^{(3)}\{\mathcal{M}_s(\Gamma^{\mathbf{Z}})\} = 1$. The key step in the proof of Theorem III.5 is to show the large deviation bounds for finite-dimensional closed and open sets in $\mathcal{M}_s(\Gamma^{\mathbf{Z}})$. The proof of the bounds for arbitrary closed and open sets in $\mathcal{M}_s(\Gamma^{\mathbf{Z}})$ then follows from an approximation argument. The proof for finite-dimensional sets depends on the following facts.

(a) For each $\alpha \in \{1,2,\ldots\}$ the distributions of the α-dimensional marginals of $\{R_n\}$ have a large deviation property with entropy function denoted by $I_{\rho,\alpha}^{(3)}$.

---

*There exists a version of $\tilde{P}_\omega$ such that $I_\rho^{(3)}(P)$ is well defined.

(b) For $\alpha \in \{1,2,\dots\}$ and $P \in M_s(\Gamma^{\mathbf{Z}})$, let $\pi_\alpha P$ denote the α-dimensional marginal of P; $\pi_\alpha P$ is a measure on Γ^α. The infimum of $I_\rho^{(3)}(P)$ over all P with $\pi_\alpha P$ fixed to be some measure τ equals $I_{\rho,\alpha}^{(3)}(\tau)$; i.e.,

$$\inf\{I_\rho^{(3)}(P): P \in M_s(\Gamma^{\mathbf{Z}}), \pi_\alpha P = \tau\} = I_{\rho,\alpha}^{(3)}(\tau). \tag{39}$$

This formula is known as a <u>contraction principle</u> relating $I_\rho^{(3)}$ and $I_{\rho,\alpha}^{(3)}$.

Fact (a) follows from Theorem II.3. Fact (b) follows from monotonicity properties of the entropy. The reader is referred to Chapter IX of Ellis (1985) for the proof of Theorem III.5 in the case of a finite state space. For $\alpha = 2$ the entropy function $I_{\rho,2}^{(3)}$ is defined in (34).

The mean relative entropy $I_\rho^{(3)}(P)$ has another interesting characterization. For $\alpha \in \{1,2,\dots\}$ and $P \in M_s(\Gamma^{\mathbf{Z}})$, $I_{\pi_\alpha P_\rho}^{(2)}(\pi_\alpha P)$ denotes the relative entropy of $\pi_\alpha P$ with respect to $\pi_\alpha P_\rho$. We have the formula

$$I_\rho^{(3)}(P) = \lim_{\alpha \to \infty} \frac{1}{\alpha} I_{\pi_\alpha P_\rho}^{(2)}(\pi_\alpha P). \tag{40}$$

$I_\rho^{(3)}(P)$ is closely related to the Kolmogorov-Sinai invariant $h(T)$ of the dynamical system $(\Gamma^{\mathbf{Z}}, \mathcal{B}(\Gamma^{\mathbf{Z}}), P, T)$; e.g., if ρ is the measure $\sum_{i=1}^r \frac{1}{r}\delta_{x_i}$ on Γ, then $h(T) = \log r - I_\rho^{(3)}(P)$.

This completes our discussion of large deviation properties for i.i.d. random vectors. Analogous properties for finite-state Markov chains may also be derived from Theorem II.3.

IV. APPLICATIONS TO STATISTICAL MECHANICS. We now apply the large deviation theorems to ferromagnetic models on the integer lattices $\mathbf{Z}^D$, $D \in \{1,2,\dots\}$. Let Λ be a hypercube in $\mathbf{Z}^D$ centered at the origin. To each site $j \in \Lambda$ there is assigned a variable ω_j, called a <u>spin</u>, which takes the value 1 (spin-up) or -1 (spin-down). The configuration space is the set Ω_Λ of all sequences $\omega = \{\omega_j; j \in \Lambda\}$; thus, $\Omega_\Lambda = \{1,-1\}^\Lambda$. The coordinate functions on Ω_Λ, defined by $X_j(\omega) = \omega_j$, are called the <u>spin random variables</u> at the sites j. Let β be a positive real number, which represents the inverse absolute temperature; $J(\cdot)$ a non-negative real-valued function on $\mathbf{Z}^D$ that satisfies $J(k) = J(-k)$ for each $k \in \mathbf{Z}^D$ and $\sum_{k \in \mathbf{Z}^D} J(k) < \infty$; h a real number, which represents an external magnetic field; and ρ the probability measure $\frac{1}{2}\delta_1 + \frac{1}{2}\delta_{-1}$. The function J is called a <u>ferromagnetic interaction</u>. For each $\omega \in \Omega_\Lambda$, we define the <u>Hamiltonian</u> or <u>interaction energy</u>

(41) $$H_{\Lambda,h}(\omega) = -\frac{1}{2}\sum_{i,j\in\Lambda} J(i-j)\omega_i\omega_j - h\sum_{i\in\Lambda}\omega_i .$$

The ferromagnetic model is defined by the probability measure $P_{\Lambda,\beta,h}$ on the subsets of Ω_Λ which assigns to each $\{\omega\}$, $\omega \in \Omega_\Lambda$, the probability

(42) $$P_{\Lambda,\beta,h}\{\omega\} = \exp[-\beta H_{\Lambda,h}(\omega)]\pi_\Lambda P_\rho\{\omega\}\cdot\frac{1}{Z(\Lambda,\beta,h)} .$$

$\pi_\Lambda P_\rho$ denotes the product measure on the subsets of Ω_Λ with identical one-dimensional marginals ρ. Thus for each $\omega \in \Omega_\Lambda$ $\pi_\Lambda P_\rho\{\omega\}$ equals $2^{-|\Lambda|}$, where $|\Lambda|$ is the number of sites in Λ. $Z(\Lambda,\beta,h)$ is a normalization which is picked so that $\sum_{\omega\in\Omega_\Lambda} P_{\Lambda,\beta,h}\{\omega\} = 1$; i.e.,

(43) $$Z(\Lambda,\beta,h) = \int_{\Omega_\Lambda} \exp[-\beta H_{\Lambda,h}(\omega)]\pi_\Lambda P_\rho(d\omega) = \sum_{\omega\in\Omega_\Lambda} \exp[-\beta H_{\Lambda,h}(\omega)]\frac{1}{2^{|\Lambda|}} .$$

The probability measure $P_{\Lambda,\beta,h}$ is called a <u>finite-volume Gibbs state</u> on Λ and the normalization $Z(\Lambda,\beta,h)$ is called a <u>partition function</u>.

The ferromagnetic model that has been studied the most extensively is the Ising model on $\mathbf{Z}^D$. The corresponding interaction is given by $J(k) = \mathcal{J}$ if $\|k\| = 1$ and $J(k) = 0$ if $\|k\| \neq 1$; $\mathcal{J}$ is a fixed positive number and $\|\cdot\|$ is the Euclidean norm on $\mathbf{Z}^D$.

For all sufficiently large values of β, ferromagnetic models on $\mathbf{Z}^D$ exhibit a phase transition, which is characterized by a positive spontaneous magnetization. The reader is referred to Chapters IV and V of Ellis (1985) for a detailed description of this phase transition and for complete references. Here we only sketch the main facts. Define $S_\Lambda(\omega)$ to be the sum $\sum_{j\in\Lambda} X_j(\omega)$. $S_\Lambda(\omega)$ is called the <u>total spin</u> in Λ. The <u>magnetization</u> is defined as

(44) $$M(\Lambda,\beta,h) = \int_{\Omega_\Lambda} S_\Lambda(\omega)P_{\Lambda,\beta,h}(d\omega)$$

and the <u>Gibbs free energy</u> as

(45) $$\Psi(\Lambda,\beta,h) = -\frac{1}{\beta}\log Z(\Lambda,\beta,h).$$

The magnetization per site, $|\Lambda|^{-1}M(\Lambda,\beta,h)$, and the Gibbs free energy per site, $|\Lambda|^{-1}\Psi(\Lambda,\beta,h)$, are related by the formula

(46) $$\frac{1}{|\Lambda|}M(\Lambda,\beta,h) = -\frac{1}{|\Lambda|}\frac{\partial\Psi(\Lambda,\beta,h)}{\partial h}$$

We will study the magnetization per site and the Gibbs free energy per site in the limit as the hypercubes Λ expand to fill $\mathbf{Z}^D$. This limit is denoted by

$\Lambda\uparrow\mathbf{Z}^D$. The specific magnetization is defined as

(47) $$m(\beta,h) = \lim_{\Lambda\uparrow\mathbf{Z}^D} \frac{1}{|\Lambda|} M(\Lambda,\beta,h)$$

and the specific Gibbs free energy as

(48) $$\psi(\beta,h) = \lim_{\Lambda\uparrow\mathbf{Z}^D} \frac{1}{|\Lambda|} \Psi(\Lambda,\beta,h) .$$

We have the following facts.

(a) $m(\beta,h)$ exists and for each $\beta > 0$, $m(\beta,0) = 0$ and $m(\beta,-h) = -m(\beta,h)$.

(b) The limit $m(\beta,+) = \lim_{h\to 0^+} m(\beta,h)$ exists and is non-negative; $m(\beta,+)$ is called the spontaneous magnetization.

(c) There exists a positive number β_c, which may equal $+\infty$, such that $m(\beta,+) = 0$ for $0 < \beta < \beta_c$ and $m(\beta,+) > 0$ for $\beta > \beta_c$;* β_c is called the critical inverse temperature.

(d) $\psi(\beta,h)$ exists and for each $\beta > 0$ is a continuously differentiable function of $h \neq 0$. For $\beta > 0$ and $h \neq 0$

(49) $$m(\beta,h) = -\frac{\partial\psi(\beta,h)}{\partial h} .$$

(e) For $0 < \beta < \beta_c$, $\psi(\beta,h)$ is differentiable at $h = 0$ and

(50) $$m(\beta,+) = -\frac{\partial\psi(\beta,0)}{\partial h} = 0.$$

For $\beta > \beta_c$, $\psi(\beta,h)$ is not differentiable at $h = 0$; $\partial\psi(\beta,0)/\partial h^+$, the right-hand derivative at $h = 0$, exists and

(51) $$m(\beta,+) = -\frac{\partial\psi(\beta,0)}{\partial h^+} > 0.$$

According to fact (c), there is a positive spontaneous magnetization $m(\beta,+)$ for β exceeding the critical inverse temperature β_c. One may explain this phenomenon as follows. Since the interaction J is non-negative, a positive external field h induces an alignment effect in the finite-volume Gibbs state on Λ, resulting in a positive magnetization $M(\Lambda,\beta,h)$. This gives rise to a positive specific magnetization $m(\beta,h)$. For $0 < \beta < \beta_c$ the interactions among the spins are relatively weak, and as $h \to 0^+$ the alignment effect disappears $(m(\beta,h) \to m(\beta,0) = 0)$. By contrast, for $\beta > \beta_c$ the

---

*The value of $m(\beta_c,+)$ depends upon the model. If $\beta_c = +\infty$, then the statement $m(\beta,+) > 0$ for $\beta > \beta_c$ is vacuous.

interactions among the spins are relatively strong, and the alignment effect persists in the limit $h \to 0^+$ $(m(\beta,h) \to m(\beta,+) > 0 = m(\beta,0))$. Facts (d) and (e) show that for $\beta > 0$, $h \neq 0$ and $0 < \beta < \beta_c$, $h = 0$, the limit $\Lambda \uparrow \mathbf{Z}^D$ and differentiation with respect to h may be interchanged in formula (46), but for $\beta > \beta_c$ and $h = 0$ they may not be interchanged.

The important question is for which interactions J is the critical inverse temperature β_c finite. The answer depends upon the dimension D of the lattice. If $D \geq 2$, then β_c is finite if $J(i)$ and $J(j)$ are positive for two linearly independent vectors i and j in $\mathbf{Z}^D$. In particular, β_c is finite for the Ising model on $\mathbf{Z}^D$, $D \geq 2$. If $D = 1$, then β_c is finite if $J(k) = k^{-2}$ $(k \neq 0)$ or if $J > 0$ and $J(k) \sim |k|^{-\alpha}$, some $1 < \alpha < 2$;* $\beta_c = +\infty$ if $\sum_{k \in \mathbf{Z}} |k| J(k) < \infty$. In particular, $\beta_c = +\infty$ for the Ising model on $\mathbf{Z}$.

We now present four applications of the large deviation theorems stated in Sections II and III.

APPLICATION 1. <u>The Curie-Weiss Model</u>. Let Λ be a symmetric hypercube in $\mathbf{Z}^D$ centered at the origin. For each $\omega \in \Omega_\Lambda = \{1,-1\}^\Lambda$, we define the Hamiltonian

$$(52) \qquad H^{CW}_{\Lambda,h}(\omega) = -\frac{1}{2}\frac{1}{|\Lambda|}\sum_{i,j\in\Lambda}\omega_i\omega_j - h\sum_{i\in\Lambda}\omega_i \, .$$

The <u>Curie-Weiss model</u> is defined by the probability measure $P^{CW}_{\Lambda,\beta,h}$ on the subsets of Ω_Λ which assigns to each $\{\omega\}$, $\omega \in \Omega_\Lambda$, the probability

$$(53) \qquad P^{CW}_{\Lambda,\beta,h}\{\omega\} = \exp[-\beta H^{CW}_{\Lambda,h}(\omega)]\, \pi_\Lambda P_\rho\{\omega\} \cdot \frac{1}{Z^{CW}(\Lambda,\beta,h)} \, .$$

$Z^{CW}(\Lambda,\beta,h)$ is a normalization. Thus the interaction for the Curie-Weiss model is given by

$$(54) \qquad J_\Lambda(i-j) = \begin{cases} |\Lambda|^{-1} & \text{if } i \text{ and } j \in \Lambda, \\ 0 & \text{if } i \text{ or } j \notin \Lambda. \end{cases}$$

Since the interaction depends on the set Λ, the Curie-Weiss model does not belong to the class of ferromagnetic models considered earlier. However, the model is ideal for doing exact calculations, and its behavior gives some insight into the behavior of these ferromagnetic models.

---

*$J(k) \sim |k|^{-\alpha}$ means $\lim_{|k|\to\infty} [\log J(k)/\log|k|] = -\alpha$.

In order to ease the notation, we replace Λ by the set $\{1,2,\ldots,n\}$ and index all quantities by n instead of by Λ. We want to determine the limit of $\langle f(S_n/n)\rangle^{CW}_{n,\beta,h}$ as $n \to \infty$ for any function f in $C(\mathbf{R})$. Here $S_n(\omega)$ is the total spin $\sum_{j=1}^{n} X_j(\omega)$ in $\{1,2,\ldots,n\}$ and $\langle\cdot\rangle^{CW}_{n,\beta,h}$ denotes expectation with respect to $P^{CW}_{n,\beta,h}$. Let $Q_n^{(1)}(dx)$ denote the distribution of the sum $\sum_{j=1}^{n} \omega_j$, $\omega = (\omega_1,\omega_2,\ldots,\omega_n) \in \Omega_n$, with respect to the product measure $\pi_n P_\rho$. Since

$$H^{CW}_{n,h}(\omega) = -n\left[\frac{1}{2}\left(\frac{\sum_{j=1}^{n}\omega_j}{n}\right)^2 + h\,\frac{\sum_{j=1}^{n}\omega_j}{n}\right], \tag{55}$$

we have

$$\begin{aligned}\langle f(\tfrac{S_n}{n})\rangle^{CW}_{n,\beta,h} &= \int_{\Omega_n} f\left(\frac{\sum_{j=1}^{n}\omega_j}{n}\right)\exp[-\beta H^{CW}_{n,h}(\omega)]\pi_n P_\rho(d\omega)\cdot\frac{1}{Z^{CW}(n,\beta,h)} \\ &= \int_{\mathbf{R}} f(x)\exp[n(\tfrac{1}{2}\beta x^2 + \beta h x)]Q_n^{(1)}(dx)\cdot\frac{1}{Z^{CW}(n,\beta,h)}\,,\end{aligned} \tag{56}$$

$$Z^{CW}(n,\beta,h) = \int_{\mathbf{R}} \exp[n(\tfrac{1}{2}\beta x^2 + \beta h x)]Q_n^{(1)}(dx). \tag{57}$$

According to Example III.2, the distributions $\{Q_n^{(1)}\}$ have a large deviation property with $a_n = n$ and entropy function

$$I_\rho^{(1)}(x) = \begin{cases} \frac{1+x}{2}\log(1+x) + \frac{1-x}{2}\log(1-x) & \text{for } |x| \le 1, \\ \infty & \text{for } |x| > 1. \end{cases} \tag{58}$$

We define

$$i_{\beta,h}(x) = -(\tfrac{1}{2}\beta x^2 + \beta h x) + I_\rho^{(1)}(x). \tag{59}$$

For large n, (56) and (57) suggest the heuristic formula

$$\langle f(\tfrac{S_n}{n})\rangle^{CW}_{n,\beta,h} \approx \int_{\mathbf{R}} f(x)\exp[-n i_{\beta,h}(x)]dx \cdot \frac{1}{\int_{\mathbf{R}}\exp[-n i_{\beta,h}(x)]dx}\,. \tag{60}$$

This formula leads one to expect that the limit of $\langle f(S_n/n)\rangle^{CW}_{n,\beta,h}$ as $n \to \infty$ is determined by the points x at which the function $i_{\beta,h}(x)$ attains its infimum. This expectation is in fact correct. Consider, for example, $h = 0$. For $0 < \beta \le 1$, $i_{\beta,0}(x)$ attains its infimum at the unique point $x = 0$. For $\beta > 1$, $i_{\beta,0}(x)$ attains its infimum at the unique points $x = \pm\, x(\beta,+)$, where

$x(\beta,+)$ is the unique positive solution of the equation $di_{\beta,0}(x)/dx = 0$ $(0 < x(\beta,+) < 1)$. We have the limit

$$\text{(61)} \qquad \lim_{n\to\infty} \langle f(\tfrac{S_n}{n})\rangle^{CW}_{n,\beta,0} = \begin{cases} f(0) & \text{for } 0 < \beta \le 1, \\ \tfrac{1}{2}f(x(\beta,+)) + \tfrac{1}{2}f(-x(\beta,+)) & \text{for } \beta > 1. \end{cases}$$

The reader is referred to Theorem IV.4.1 in Ellis (1985) for a proof. We note that if $f(x) = x$ for $-1 \le x \le 1$, then the specific magnetization for the Curie-Weiss model is given by

$$\text{(62)} \qquad m^{CW}(\beta,h) = \lim_{n\to\infty} \langle f(\tfrac{S_n}{n})\rangle^{CW}_{n,\beta,h}.$$

A straightforward analysis shows that $m^{CW}(\beta,+) = \lim_{h\to 0^+} m^{CW}(\beta,h)$ equals 0 for $0 < \beta \le 1$ and is positive for $\beta > 1$. In fact, for $\beta > 1$ $m^{CW}(\beta,+)$ equals the positive number $x(\beta,+)$ at which the function $i_{\beta,0}(x)$ attains its infimum. Thus for the Curie-Weiss model the critical inverse temperature equals 1.

APPLICATION 2. Large Deviations and Exponential Convergence. Let J be a ferromagnetic interaction on $\mathbf{Z}^D$ and $P_{\Lambda,\beta,h}$ the corresponding finite-volume Gibbs state given in (42). For real t, we define

$$\text{(63)} \qquad c_{\beta,h}(t) = \lim_{\Lambda\uparrow\mathbf{Z}^D} \frac{1}{|\Lambda|} \log \int_{\Omega_\Lambda} \exp(tS_\Lambda(\omega))P_{\Lambda,\beta,h}(d\omega).$$

The function $c_{\beta,h}(t)$ is the free energy function of the sequence $\{S_\Lambda\}$ with respect to the measures $\{P_{\Lambda,\beta,h}\}$. Since

$$\text{(64)} \qquad tS_\Lambda(\omega) - \beta H_{\Lambda,h}(\omega) = -\beta H_{\Lambda,h+t/\beta}(\omega),$$

it follows that

$$\text{(65)} \qquad \begin{aligned} c_{\beta,h}(t) &= \lim_{\Lambda\uparrow\mathbf{Z}^D} \frac{1}{|\Lambda|} \log \left[Z(\Lambda,\beta,h+t/\beta)\cdot\frac{1}{Z(\Lambda,\beta,h)}\right] \\ &= -\beta[\psi(\beta,h+t/\beta) - \psi(\beta,h)], \end{aligned}$$

where $\psi(\beta,h)$ is the specific Gibbs free energy. For $0 < \beta < \beta_c$ and any real h, $c_{\beta,h}(t)$ is a differentiable function of $t \in \mathbf{R}$. According to Theorem II.3, for $0 < \beta < \beta_c$ and any real h, the $P_{\Lambda,\beta,h}$-distributions of $\{S_\Lambda/|\Lambda|\}$ have a large deviation property with $a_\Lambda = |\Lambda|$ and entropy function

$$\text{(66)} \qquad I_{\beta,h}(x) = \sup_{t\in\mathbf{R}} \{tx - c_{\beta,h}(t)\}.$$

By contrast, for $\beta > \beta_c$ and any real h, $c_{\beta,h}(t)$ is not differentiable at $t = -\beta h$, and so we are unable to apply Theorem II.3 to conclude that the $P_{\Lambda,\beta,h}$-distributions of $\{S_\Lambda/|\Lambda|\}$ have a large deviation property. The validity of the large deviation property for $\beta > \beta_c$ is an open question.

A related problem involves exponential convergence properties of $\{S_\Lambda/|\Lambda|\}$. For $\beta > 0$, $h \neq 0$ and $0 < \beta < \beta_c$, $h = 0$, $c_{\beta,h}(t)$ is differentiable at $t = 0$. Theorem II.5 implies that for these values of β and h

$$\frac{S_\Lambda}{|\Lambda|} \xrightarrow{\exp} c'_{\beta,h}(0) = -\frac{\partial\psi(\beta,h)}{\partial h} = m(\beta,h) \quad \text{w.r.t.} \quad \{P_{\Lambda,\beta,h}\} \tag{67}$$

as $\Lambda\uparrow\mathbf{Z}^D$. In other words, in the limit $\Lambda\uparrow\mathbf{Z}^D$, all but an exponentially small set of configurations ω have essentially the same value $m(\beta,h)$ of spin per site $S_\Lambda(\omega)/|\Lambda|$. The situation is different for $\beta > \beta_c$ and $h = 0$. Since $c_{\beta,0}(t)$ is not differentiable at $t = 0$, the exponential convergence of $S_\Lambda/|\Lambda|$ to a constant fails [Theorem II.5].

APPLICATION 3. Evaluation of the Specific Gibbs Energy for the Ising Model on $\mathbf{Z}$. In order to ease the notation, we consider only $h = 0$. A similar evaluation holds for $h \neq 0$. For the Ising model on $\mathbf{Z}$, the interaction is given by

$$J(k) = \begin{cases} J & \text{if} \quad |k| = 1, \\ 0 & \text{if} \quad |k| \neq 1, \end{cases} \tag{68}$$

where J is a fixed positive number. Let Γ be the set $\{1,-1\}$, Ω the sequence space $\Gamma^{\mathbf{Z}}$, and P_ρ the infinite product measure on the Borel subsets of Ω with identical one-dimensional marginals ρ. If $\Lambda = \{-N,-N+1,\dots,N-1,N\}$ for some positive integer N, then we may write the partition function $Z(\Lambda,\beta,0)$ as

$$Z(n,\beta) = \int_\Omega \exp[\beta J \sum_{j=1}^{n-1} \omega_j\omega_{j+1}] P_\rho(d\omega), \tag{69}$$

where $n = 2N + 1 = |\Lambda|$. The specific Gibbs free energy $\psi(\beta,0)$ is given by

$$-\beta\psi(\beta,0) = \lim_{n\to\infty} \frac{1}{n} \log Z(n,\beta), \tag{70}$$

and it is easy to see that

$$\lim_{n\to\infty} \frac{1}{n} \log Z(n,\beta) = \lim_{n\to\infty} \frac{1}{n} \log Z_0(n,\beta), \tag{71}$$

where $Z_0(n,\beta) = \int_\Omega \exp[\beta J(\sum_{j=1}^{n-1} \omega_j\omega_{j+1} + \omega_n\omega_1)]P_\rho(d\omega)$. Let $X_j(\omega) = \omega_j$ be the coordinate functions on Ω and $M_n(\omega,\cdot)$ the corresponding empirical pair measure defined in formula (31). $M_n(\omega,\cdot)$ takes values in the space $\mathcal{M}_s(\Gamma^2)$, which is the space of probability measures on the subsets of Γ^2 with equal one-dimensional marginals. For each $\omega \in \Omega$,

$$\sum_{j=1}^{n-1} \omega_j\omega_{j+1} + \omega_n\omega_1 \tag{72}$$
$$= n[M_n(\omega,\{1,1\}) - M_n(\omega,\{1,-1\}) - M_n(\omega,\{-1,1\}) + M_n(\omega,\{-1,-1\})],$$

and thus

$$Z_0(n,\beta) = \int_{\mathcal{M}_s(\Gamma^2)} \exp[n\beta J(\tau_{1,1} - \tau_{1,-1} - \tau_{-1,1} + \tau_{-1,-1})]P_\rho\{\omega: M_n(\omega,\cdot) \in d\tau\}. \tag{73}$$

According to Theorem III.4, the P_ρ-distributions of $M_n(\omega,\cdot)$ on $\mathcal{M}_s(\Gamma^2)$ have a large deviation property with $a_n = n$ and entropy function $I^{(3)}_{\rho,2}$ defined in (34). We conclude from Theorem II.2 that

$$-\beta\psi(\beta,0) = \lim_{n\to\infty} \frac{1}{n} \log Z(n,\beta) = \lim_{n\to\infty} \frac{1}{n} \log Z_0(n,\beta) \tag{74}$$
$$= \sup_{\tau\in\mathcal{M}_s(\Gamma^2)} \{\beta J(\tau_{1,1} - \tau_{1,-1} - \tau_{-1,1} + \tau_{-1,-1}) - I^{(3)}_{\rho,2}(\tau)\}.$$

This formula for the specific Gibbs free energy is an example of the <u>Gibbs variational formula</u>. We will return to (74) after considering the Gibbs variational formula for arbitrary ferromagnetic models on $\mathbf{Z}$.

APPLICATION 4. <u>Evaluation of the Specific Gibbs Free Energy for Ferromagnetic Models on</u> $\mathbf{Z}$. Let J be a ferromagnetic interaction on $\mathbf{Z}$ that satisfies $J(k) = 0$ for all sufficiently large k. The smallest number L with the property that $J(k) = 0$ whenever $|k| > L$ is called the <u>range</u> of the interaction. For example, for the Ising model on $\mathbf{Z}$, the range equals 1. Let Γ be the set $\{1,-1\}$, Ω the sequence space $\Gamma^{\mathbf{Z}}$, and P_ρ the infinite product measure on the Borel subsets of Ω with identical one-dimensional marginals ρ. If $\Lambda = \{-N,-N+1,\ldots,N-1,N\}$ for some positive integer N, then we may write the partition function $Z(\Lambda,\beta,h)$ as

$$Z(n,\beta,h) = \int_\Omega \exp[\beta\{\frac{1}{2} \sum_{i,j=1}^{n} J(i-j)\omega_i\omega_j + h \sum_{i=1}^{n} \omega_i\}]P_\rho(d\omega), \tag{75}$$

where $n = 2N + 1$. The specific Gibbs free energy $\psi(\beta,h)$ is given by

(76) $$-\beta\psi(\beta,h) = \lim_{n\to\infty} \frac{1}{n} \log Z(n,\beta,h).$$

Let $X_j(\omega) = \omega_j$ be the coordinate functions on Ω and $R_n(\omega,\cdot)$ the corresponding empirical process defined in formula (36). $R_n(\omega,\cdot)$ takes values in the space $M_s(\Omega)$, which is the space of strictly stationary probability measures on the Borel subsets of Ω. Given a measure P in $M_s(\Omega)$, we define

(77) $$u(h;P) = -\frac{1}{2} \sum_{k\in\mathbf{Z}} J(k) \int_\Omega \bar{\omega}_1 \bar{\omega}_{k+1} P(d\bar{\omega}) - h \int_\Omega \bar{\omega}_1 P(d\bar{\omega}).$$

The function $P \longrightarrow u(h;P)$ is bounded, continuous, and affine. A short calculation shows that since $J(k) = J(-k)$ and J has range L,

$$\sup_{\omega\in\Omega} \Big| \frac{1}{2} \sum_{i,j=1}^{n} J(i-j)\omega_i\omega_j + h \sum_{i=1}^{n} \omega_i + nu(h;R_n(\omega,\cdot)) \Big| = O(L)$$

for all $n > L$. Thus,

(78) $$\begin{aligned} -\beta\psi(\beta,h) &= \lim_{n\to\infty} \frac{1}{n} \log Z(n,\beta,h) \\ &= \lim_{n\to\infty} \frac{1}{n} \log \int_\Omega \exp[-n\beta u(h;R_n(\omega,\cdot))] P_\rho(d\omega) \\ &= \lim_{n\to\infty} \frac{1}{n} \log \int_{M_s(\Omega)} \exp[-n\beta u(h;P)] P_\rho\{\omega: R_n(\omega,\cdot) \in dP\}. \end{aligned}$$

According to Theorem III.5, the P_ρ-distributions of $R_n(\omega,\cdot)$ on $M_s(\Omega)$ have a large deviation property with $a_n = n$ and entropy function $I_\rho^{(3)}$ defined in (38). We conclude from Theorem II.2 that

(79) $$-\beta\psi(\beta,h) = \sup_{P\in M_s(\Omega)} \{-\beta u(h;P) - I_\rho^{(3)}(P)\}.$$

If J is an arbitrary ferromagnetic interaction on $\mathbf{Z}$ ($J \geq 0$, $J(k) = J(-k)$, and $\sum_{k\in\mathbf{Z}} J(k) < \infty$), then one may prove the same formula by approximating J by interactions of range L and then by taking $L \to \infty$.

Formula (79) is called the <u>Gibbs variational formula</u>. Let $G_{\beta,h}$ denote the subset of $M_s(\Omega)$ at which the supremum in (79) is attained. Since the function

(80) $$P \longrightarrow -\beta u(h;P) - I_\rho^{(3)}(P)$$

is upper semicontinuous and affine and since the space $M_s(\Omega)$ is compact,

$\mathcal{G}_{\beta,h}$ is nonempty, closed, and convex. Elements of $\mathcal{G}_{\beta,h}$ are called <u>infinite-volume Gibbs states</u>.* The set $\mathcal{G}_{\beta,h}$ has an interesting relation to the set of finite-volume Gibbs states $\{P_{\Lambda,\beta,h}; \Lambda \subseteq \mathbf{Z}\}$ defined in (42). One modifies the definition of $P_{\Lambda,\beta,h}$ by incorporating external conditions, which are fixed configurations of spins on $\mathbf{Z}\backslash\Lambda$ that interact with the spins on Λ via the given interaction. Let $\mathcal{G}^0_{\beta,h}$ denote the subset of $\mathcal{M}_s(\Omega)$ consisting of all weak limits of convergent subsequences of these modified measures as $\Lambda\uparrow\mathbf{Z}$. We have the following facts.

(a) $\mathcal{G}_{\beta,h}$ coincides with the closed convex hull of $\mathcal{G}^0_{\beta,h}$.

(b) For $\beta > 0$, $h \neq 0$ and $0 < \beta < \beta_c$, $h = 0$, $\mathcal{G}_{\beta,h}$ consists of a unique measure.

(c) For $\beta > \beta_c$ and $h = 0$, $\mathcal{G}_{\beta,0}$ contains at least all measures of the form

$$P = \lambda P_{\beta,0,+} + (1-\lambda)P_{\beta,0,-}, \quad 0 \le \lambda \le 1. \tag{81}$$

The measure $P_{\beta,0,+}$ (resp., $P_{\beta,0,-}$) equals the weak limit, as $\Lambda\uparrow\mathbf{Z}$, of the finite-volume Gibbs states with the external configurations 1 (resp., -1) at each site of $\mathbf{Z}\backslash\Lambda$.

Fact (a) is known as the <u>Gibbs variational principle</u>. The existence of more than one measure in $\mathcal{G}_{\beta,0}$ for $\beta > \beta_c$ is called a <u>level-3 phase transition</u>.

We note that for the Ising model on $\mathbf{Z}$ with $h = 0$,

$$u(0;P) = -J\int_\Omega \bar{\omega}_1\bar{\omega}_2 \, P(d\bar{\omega}) \quad \text{for } P \in \mathcal{M}_s(\Omega). \tag{82}$$

This integral involves only the two-dimensional marginal $\pi_2 P$ of P, and $\pi_2 P$ is an element of $\mathcal{M}_s(\Gamma^2)$. Hence we may rewrite the Gibbs variational formula (79) as

$$-\beta\psi(\beta,0) = \sup_{\tau\in\mathcal{M}_s(\Gamma^2)} \{\beta J\int_{\Gamma^2} \bar{\omega}_1\bar{\omega}_2\tau(d\bar{\omega}) - \inf\{I^{(3)}_\rho(P) : P \in \mathcal{M}_s(\Omega), \pi_2 P = \tau\}\} . \tag{83}$$

According to the contraction principle (39),

---

*In Ellis (1985), the set of infinite-volume Gibbs states is defined differently, but the two definitions are equivalent.

(84) $\inf\{I_\rho^{(3)}(P) : P \in M_s(\Omega), \pi_2 P = \tau\} = I_{\rho,2}^{(3)}(\tau)$ for $\tau \in M_s(\Gamma^2)$.

Thus (83) reduces to

(85) $$-\beta\psi(\beta,0) = \sup_{\tau\in M_s(\Gamma^2)} \{\beta J \int_{\Gamma^2} \bar{\omega}_1\bar{\omega}_2 \tau(d\bar{\omega}) - I_{\rho,2}^{(3)}(\tau)\} .$$

This formula has already been derived in (74) via a large deviation analysis involving the empirical pair measure.

For the Ising model on $\mathbb{Z}$, the critical inverse temperature β_c equals $+\infty$. For this model with any $\beta > 0$ and $h = 0$, the supremum in the Gibbs variational formula (79) is attained at a unique measure $P \in M_s(\Omega)$. P is a Markov chain, and it equals the weak limit of the finite-volume Gibbs states $P_{\Lambda,\beta,0}$ as $\Lambda\uparrow\mathbb{Z}$. Let τ denote the two-dimensional marginal of this Markov chain. Then τ is the unique measure at which the supremum in (85) is attained. Similar statements are valid for the Ising model on $\mathbb{Z}$ with $\beta > 0$ and $h \neq 0$.

Our derivation of the Gibbs variational formula via level-3 large deviations is valid only for $D = 1$. For arbitrary $D \geq 1$, one may derive the Gibbs variational formula by other methods. The facts about the set $G_{\beta,h}$ stated above are also valid for any $D \geq 1$.

V. SUMMARY. In Theorem II.3 we stated a general large deviation theorem for sequences of random vectors for which a free energy function exists. This theorem was applied in Section III to three levels of large deviations for sequences of i.i.d. random vectors. Level-1 involves the sum S_n/n; level-2 the empirical measure L_n; level-3 the empirical pair measure M_n and more generally the empirical process R_n. The theorems in Sections II and III were applied in Section IV to ferromagnetic models on $\mathbb{Z}^D$. The first application treated the Curie-Weiss model by means of level-1 large deviations for i.i.d. random variables. The second application treated arbitrary ferromagnetic models by means of the general large deviation theorem, Theorem II.3. The third and fourth applications used level-3 large deviations for i.i.d. random variables in order to derive the Gibbs variational formula for models on $\mathbb{Z}$. It is clear from these applications that statistical mechanics provides a natural context in which large deviation theory may be applied. The use of large deviation methods should lead to further insight into ferromagnetic and other statistical mechanical models in the future.

BIBLIOGRAPHY

M. D. Donsker and S. R. S. Varadhan (1975a). Asymptotic evaluation of certain Markov process expectations for large time, I. Comm. Pure Appl. Math. 28, 1-47.

M. D. Donsker and S. R. S. Varadhan (1975b). Asymptotic evaluation of certain Markov process expectations for large time, II. Comm. Pure Appl. Math. 28, 279-301.

M. D. Donsker and S. R. S. Varadhan (1976). Asymptotic evaluation of certain Markov process expectations for large time, III. Comm. Pure Appl. Math. 29, 389-461.

M. D. Donsker and S. R. S. Varadhan (1983). Asymptotic evaluation of certain Markov process expectations for large time, IV. Comm. Pure Appl. Math. 36, 183-212.

R. S. Ellis (1984). Large deviations for a general class of random vectors. Ann. Probab. 12, 1-12.

R. S. Ellis (1985). Entropy, Large Deviations, and Statistical Mechanics. Grundlehren der Mathematischen Wissenschaften, Volume 271. Springer-Verlag, New York.

O. E. Lanford (1973). Entropy and equilibrium states in classical statistical mechanics. In: Statistical Mechanics and Mathematical Problems, 1-113. Edited by A. Lenard. Lecture Notes in Physics 20. Springer, New York.

S. R. S. Varadhan (1966). Asymptotic probabilities and differential equations. Comm. Pure Appl. Math. 19, 261-286.

DEPARTMENT OF MATHEMATICS AND STATISTICS
UNIVERSITY OF MASSACHUSETTS
AMHERST, MASSACHUSETTS 01003

Contemporary Mathematics
Volume 41, 1985

Nonergodic one-dimensional media and reliable computation

Péter Gács*

Abstract. We construct a one-dimensional array of cellular automata on which arbitrarily large computations can be implemented reliably, even though each automaton at each step makes an error with some constant probability. In statistical mechanics, this construction refutes the "positive probability conjecture", which states that any one-dimensional infinite particle system with positive transition probabilities is ergodic, and our approach takes its origin from Kurdyumov's ideas for this refutation.

Earlier designs for reliable computation were circuits whose intricate interconnection pattern (arising from the error-correcting organization) he had to assume to be immune to errors. In a uniform cellular medium, the error-correcting organization exists only in "software", therefore errors threaten to disable it. The real technical novelty of the paper is therefore the construction of a *self-repairing organization.*

1. Introduction

Can we avoid the accumulation of errors in arbitrarily large computations using unreliable components? Earlier answers to this question (see [vN 52], [D 77], [Ta 68], [Kz 73]) assumed computing devices whose connection pattern was not local but error probability did not depend on the length of the wires used. If local connections are required then computing devices can be modeled as *particle systems with nearest-neighbor interaction*, or shortly, *media.* Such systems were the subject of extensive research in the last 15 years in statistical physics (see [Li 77], [Gri 79], [KS 80]). This paper is devoted to the case of *discrete time*, though at the end, we make some remarks about the case of continuous time. Discrete-time deterministic media are known in Computer Science under the names "cellular automata", "cellular space", "iterative array", "systolic array", etc.

One of the main questions asked in connection with any medium is whether it is *ergodic.* Without going into the formal definition, it is clear that an ergodic infinite medium

1979 Mathematics Subject Classification 60K35, 68C99.

* Supported in part by the National Science Foundation grants No. MCS 8110430, MCS 8104008, MCS 8302874 and DCR 8405270, and in part by the DARPA grant No. N00014-82-K0193 monitored by the ONR. Part of the work was done while the author was at the University of Rochester.

0271-4132/85 $1.00 + $.25 per page

can not be used for computation since sooner or later, it forgets almost all information about its initial configuration in a finite number of steps. Besides being nonergodic, a computing device should be *stable*, in the sense that if its probability distribution is perturbed a little, it will still work reliably, and will certainly stay nonergodic.

Toom was the first to construct stable nonergodic media (see [To 74], [To 80]). The set of sites is the two-dimensional lattice $\mathbf{Z}^2$. Each cell has two states, 0 and 1. The medium works according to a deterministic transition function. In one of the examples, to determine its state in the next moment, each cell computes the majority of the current states of the three cells consisting of its Northern and Eastern neighbor and itself. In a next paper, joint with J.Reif, we will apply Toom's construction (with a new proof) to obtain a *three-dimensional* reliable computing device.

It has been conjectured that if all local transition probabilities are positive then a *one-dimensional* medium is ergodic. This is the so-called Positive Probability Conjecture. If this conjecture held then there would be no stable nonergodic one-dimensional medium.

The Positive Probability Conjecture implies that there is no "simple" one-dimensional reliable memory. Here, simplicity means spatial uniformity and local error-correction i.e. that the memory is a one-dimensional finite medium (with, say, wraparound at the ends). The designer is free to specify the transition rule and an error probability bound independent of the size of the device, but the actual work of each cell at each step will deviate from the rule with some probability within the bound. Suppose that we want to store one bit of information, i.e. there are two possible starting configurations, u_0 and u_1. For a memory of size K, let m_K be the maximum number of steps after which it is still possible to find out which of u_0 and u_1 was the initial content, with a probability of mistake less than $2/3$. According to the Positive Probability Conjecture, the value m_K is bounded by some number depending only on the error-correcting rule and the error probability but not on K. Thus if the conjecture is true then *no amount of redundancy can make a one-dimensional memory reliable.*

It is instructive to try simple error-correcting rules in a one-dimensional memory and see them fail. For definiteness, let us assume that we started with u_0 and the error probability is ρ. It seems natural to choose the K-fold repetition of 0 for u_0. The first idea for an error-correcting rule is majority voting among the three neighbors of a cell (including itself). However, this rule will not eliminate any island of 1's longer than one cell. As more and more of these islands are brought in by errors, the content of the memory will lose all similarity to u_0 in about ρ^{-2} steps. (Due to the technical complexity of the probability model, the failure of this voting rule is not really proved yet, though the methods of [Gray 83] are believed to eventually lead to its proof.)

Thus "local voting" is ruled out. The rules considered in [G 78] carry information from one end of a long island to the other end and thus seem to be better than local voting. But since no measures are taken to protect this information against new errors, the rules eliminate a finite set of islands only if no new errors occur.

The above reasoning suggests that the simple task of protecting a bit of information requires the capability of carrying information to large distances reliably. It is hard to imagine how to do this without setting up some structure, "division of labor" among cells: assigning roles to cells in a way varying in space and time. Tsirel'son did this (see [Ts 76]) but he sacrificed homogeneity: components of three different kinds are present, and the component kind changes in both space and time according to a grand plan not subject to errors.

Thus, if we are not willing to give up uniformity in space-time, the task of protecting one bit of information leads us to the task of setting up a non-local (e.g. hierarchical) organization and a rule that will continuously restore this organization from the damages caused by errors.

Using the above insights, G.Kurdyumov made in [K 78] some valuable suggestions for the construction of a one-dimensional stable medium. The presentation was too tentative to be taken seriously by most researchers in the field. Using Kurdyumov's ideas as a starting point, in the next pages we will show the construction of a one-dimensional stable medium that is also a reliable and (asymptotically) economical computing device. This result is a refutation of the Positive Probability Conjecture.

I hope that after the above discussion, the reader expects a complicated construction and proof. Unfortunately, the complexity of the proof goes beyond these expectations. My main encouragement to the reader is that I think the effort to understand the proof is worth-while. The problem it solves is simple (preserving information in a noisy environment), and some of its principles seem to be ones of a biological or social organization. If it turns out that no simpler solution exists then this proof contributes to a deeper understanding of the significance of these principles.

2. Statement of the result

Our definitions are an adaptation of the ones in [To 80]. A (one-dimensional) *medium (cellular automaton, cellular space)* D is given by a finite set $S = S_D$ of states and a transition function $D : S^3 \to S$. The set of *cells*, or *sites* is the set $\mathbf{Z}$ of integers. The set of time-steps is the set $\mathbf{Z}^+$ of nonnegative integers. Put $A = \mathbf{Z}^+ \times \mathbf{Z}$. Any function $x : A \to S_D$, is an *evolution*. We write $x[t,n]$ for the state of cell n at time t. An evolution x is a *trajectory* of the medium D if for all t, n we have

$$x[t+1,n] = D(x[t,w-1], x[t,w], x[w+1]). \tag{1}$$

We say that an *error* occurred in x at $(t+1,w)$ (with respect to D) if (1) does not hold. If y is a trajectory then the event $x[v] \neq y[v]$ is a *deviation*. Intuitively, a medium is reliable if it can keep the number of deviations small despite the occasional occurrence of errors.

Let D be a medium and y a trajectory. Let $\big(\xi[t,w] \ : \ (t,w) \in A\big)$ be a system of random variables. We say that ξ is a ρ-perturbation of y if

(a) $\xi[0,n] = y[0,n]$ for all n with probability 1, and

(b) for each subset B of A the probability that for *each* $v \in B$ an error occurs in ξ at v is not more than $\rho^{|B|}$.

Let $m(\rho)$ be the maximum probability of $\xi[t,n] \neq y[t,n]$ for all ρ-perturbations ξ of y and all t, n. The trajectory y is *stable* if $\lim \rho \to 0 = 0$. It is easy to find a medium with one stable trajectory. What we are after is *at least two stable trajectories.*

Theorem *There is a deterministic medium M with several stable trajectories.*

The rest of the paper is devoted to an outline of the proof.

3. The construction

To speak of intervals of integers more concisely, let us write

$$[m \dots n] = \{m, m+1, \dots, m\},$$

and by analogy with real analysis, $[m \dots n-1] = [m \dots n)$. For a space-time rectangle $[a \dots b] \times [m \dots n]$, we will call the segment $\{a\} \times [m \dots n]$ the *input* of the rectangle. If $x[t,n]$ is an evolution and $[m \dots n]$ is an interval then we will write

$$x[t,[m \dots n]] = (x[t,m], x[t,m+1], \dots, x[t,n]).$$

Sparsity. Two constants, P and T, will play a central role in the definition of M. When M is on one of the stable trajectories, the cells will be organized into *blocks*: intervals of length P. These blocks will be grouped into 2-blocks: intervals of length P^2, etc. Similarly, the trajectory divides the time axis into *working periods* of length T, and these periods are hierarchically grouped into k-periods of length T^k for all k. Let us denote by $\mathsf{P}^k[n]$ the k-block $[nP^k, \dots (n+1)P^k)$. The time period $\mathsf{T}[i]$ is defined similarly. A k*-rectangle* is any rectangle in space-time whose time-projection and space-projection have lengths T^k and P^k respectively.

A set E of errors is called 0–sparse if it is empty. It is k*-sparse* if for every rectangle I of size $3T^k \times 4P^k$ there is a $k-1$–rectangle J such that $E \cap I \setminus J$ is $k-1$–sparse. (Here $\setminus$ is set-theoretical subtraction.) There is a similar definition of k–sparsity for subsets of the set $\mathbf{Z}$ of sites. If for some t and interval I the set of sites $n \in I$ with $y[t,n] \neq x[t,n]$ is k–sparse then we will say that x is a k*-distortion* of y over $\{t\} \times I$. It is routine to prove:

Lemma 1 *Let ξ be any ρ-perturbation of a trajectory y. For any k-rectangle R, the probability that the set of errors is not k-sparse on R is less than $\rho^{2^{k-1}+1/2}$.*

This lemma suggests to introduce the notion of k*-trajectory*. An evolution x is a k–trajectory on the rectangle R if set of errors is k–sparse in x over R.

We introduce two families of rectangles.

$$W_0^k[h,i] = [(h-1)T^k \dots (h+2)T^k] \times [(i-2)P^k \dots (i+2)P^k),$$
$$W_1^k[h,i] = [hT^k \dots (h+2)T^k] \times [(i-1)P^k \dots (i+1)P^k).$$

Our plan is to construct M and several trajectories y in such a way that the following lemma holds.

Lemma 2 *For all k, h, i, if x is a $k-1$-trajectory over $W_0^k[h,i]$, and either*

(a) x is a k-distortion of y over the input of $W_0^k[h,i]$, or

(b) for some $t < T^k$, we have $x = y$ over $\{hT^k - t\} \times [(i-2)P^k \dots (i+2)P^k)$, then x is a $k-1$-distortion of y over the input of any $W_0^{k-1}[t,n] \subset W_1^k[h,i]$.

Proof of the Theorem using Lemmas 1 and 2: We show that y is a stable trajectory. Let (t,n) be a point in space-time where we want to estimate the probability that $\xi[t,n] \neq y[t,n]$. Let us define the increasing sequence $R_0, R_1, \dots$ of rectangles as follows. Put $R_0 = W_0^0[t+1,n]$. Notice that (t,n) is in the input of R_0. For $k > 0$, let R_k have the form $W_0^k[h,i]$ with $R_{k-1} \subset W_1^k[h,i]$. Such a sequence obviously exists.

Let U be the event that for all $k > 0$, the evolution ξ is a $k-1$-trajectory over R_k. By Lemma 1, for any $k > 0$, the probability that this is not so is less than $\rho^{2^{k-2}+1/2}$. The sum of these terms for all $k > 0$ is $O(\sqrt{\rho})$. Therefore U holds with probability $1 - O(\sqrt{\rho})$.

Let x be a realization of ξ for which U holds. It remains to show that $x[t,n] = y[t,n]$. Let r be the first k such that R_k intersects the start line $\{0\} \times \mathbf{Z}$. If $r = 0$, there is nothing to prove.

Otherwise, notice that $x = y$ over $\{0\} \times \mathbf{Z}$. It follows from *(b)* of Lemma 2 that x is an $r-1$-distortion of y on the input of R_{r-1}. Using *(a)* of Lemma 2 we conclude from here that x is a $r-2$-distortion of y on the input of R_{r-2}. Using the same conclusion repeatedly, one finds that x is a 0-distortion of y on the input of R_0. In particular, we have $x[t,n] = y[t,n]$. ∎

A proof with the above structure can also be given for the result in [Ts 76] and important cases of [To 80]. The proof for Tsirel'son's result is more transparent than the original one. Toom's original proof in [To 80] uses a topological construction that does not lend itself easily to good probability estimates for finite media. The new proof gives these estimates (they will appear in a subsequent paper).

A 1–sparse error set. To satisfy Lemma 2 we need a rule M that, for all k, when applied over the rectangle $W_0^k[h,i]$, copes with a $k-1$–sparse set of errors, as well as with a k–sparse set of deviations in the input. Let us look first at small k.

A 1–sparse set of errors is one in which whenever two errors are closer than $4T$ steps in time they are at least $3P$ steps apart in space. If the trajectory y consists of the infinite repetition of the same symbol s in space and time then e.g. majority voting among the three neighbors could protect this trajectory from a 1–sparse set of errors. But for later use, we have to organize a general *computation* reliably in the presence of a 1–sparse set of errors.

By a computation, we mean a trajectory of an arbitrary 1–dimensional medium. It is known that some of these media are *universal* in that they can *simulate* any other medium D. Medium U simulates a medium D if with blocksize P_1, work period T_1 and decoding function $\phi : S_U^{P_1} \to S_D$ if for any trajectory x of D there is a trajectory z of U such that

$$\phi(z[T_1 h, \mathsf{P}_1[i]]) = x[h,i].$$

Let us fix a universal medium U and show how it can be simulated by another medium M_1 resistant to a 1–sparse set of errors. By this we mean the following. There is a decoding function γ such that if z is a trajectory of U that we want to simulate then there is a trajectory y of M_1 such that for any 1–trajectory x of M_1 that is equal to y at time 0, it is possible to decode z from x by γ.

We divide the cells of medium U (sofar arbitrarily) into blocks of size P_1. The time for U is divided into work periods of size T_1. For simplicity, let us suppose $T_1 = P_1$. (In general, P_1 and T_1 should be such that for the trajectories of interest, the outcome of a working period $\mathsf{T}_1[h]$ of a block $\mathsf{P}_1[i]$ should only depend on z at time hT_1 in the three blocks $\mathsf{P}_1[i-1]$, $\mathsf{P}_1[i]$, $\mathsf{P}_1[i+1]$.) We choose P to be larger than $\max(P_1, T_1)$.

The cells of M_1 will also be organized into blocks. These blocks have size $P > 3P_1$. There will be a working period $T > 9P$. The *computational* function of block $\mathsf{P}[i]$ of M_1 is to simulate, during its working period $\mathsf{T}[h]$, the working period $\mathsf{T}_1[h]$ of block $\mathsf{P}_1[i]$ in z. The state of each cell in the medium M_1 is determined by the value of several *variables*. Two of these deserve special mention. Variable π shows the position of the cell within its own block, and variable τ shows the number of steps elapsed since the beginning of the current working period. Neighbor cells are *consistent* if they have the same τ value, and their π value differs by 1 mod P. Variables τ and π are called *control variables*. Since the transition function depends on τ, we view it as a *program* telling us what to do at the various points of the working period. If x is an evolution then in it, the value of e.g. the variable π in cell n at time t will be denoted by $\pi(x)[t,n]$. The dependence on x will not always be shown.

Each block $\mathsf{P}[i]$ in M_1 contains three disjoint subblocks $\mathsf{K}_1[i], \mathsf{K}_2[i], \mathsf{K}_3[i]$ of length P_1. The variable X is used to represent the trajectory z of U in triple repetition. In the trajectory y of M_1 simulating z, each of the three strings $X(y)[hT, K_j[i]]$, $(i = 1,2,3)$, is equal to the string $z[hT_1, \mathsf{P}_1[i]]$. The decoding ϕ uses majority vote among the three strings.

The work period of M_1 consists of three similar parts.

In the first part, input is first shipped in from the two neighbor blocks. They will be stored into temporary variables $Input_j$ for $j = 1,2,3$. Especially, (treating our variables like ones in a computer program and ignoring their time dependence in the notation), the string $X[\mathsf{K}_j[i+s]]$ will be copied into $Input_j[\mathsf{K}_s[i]]$ for $s = 1,2,3$. For $s = -1$ this means that the three thirds of the left neighbor block of $\mathsf{P}[i]$ send their X content into the left third of $\mathsf{P}[i]$, into the three "tracks" provided by the three variables $Input_j$. Now a majority vote decides among the three variables $Input_j$ in each cell, and the result is stored in the variable Y of the cell.

Now the medium M_1 applies the transition rule of U to the variables Y for T_1 steps. This is the central part of the simulation. After the computation, the result is in the string $Y[\mathsf{K}_2[i]]$. It is sent out in three identical copies into the strings $Output_1[\mathsf{K}_j[i]]$, $j = 1,2,3$. Here $Output_s$, $s = 1,2,3$ are three new variables. This is the end of the first third of the working period.

The whole reading-decoding-computation-encoding cycle is repeated two more times, with results stored in $Output_2$ and $Output_3$. The final result is obtained by majority vote from the outputs of the three identical computations, and is written into the variable X at the end of the working period.

The various information transfer operations are performed with the help of "mailbox variables". The above organization (a proper selection of variables, coding and repetition) provably corrects a 1-sparse set of errors *provided* that, by some magic, the control variables restore themselves quickly after the errors. Indeed, these variables show for each cell what to do at the present time with the information entrusted to it. The information-destroying effect of the errors is confined to one third of the working period and one third of the block: therefore it will be corrected by majority vote.

The restoration of the control variables is thus crucial. In the simple case of a 1-sparse set of errors this is also easily done by local majority vote.

A 2-sparse error set. A set of errors is 2-sparse if there can be some "bad" 1-rectangles outside which the set of errors is 1-sparse, but these bad 1-rectangles are several 2-rectangles apart. Let us try now to find a medium M_2 that also resists a 2-sparse set of errors. The construction is based on M_1. After a bad rectangle "happened", the structure of the cells in the one or two blocks affected will be lost. Especially, the control variables of the cells lose their values, hence these cells no longer form blocks. These control variables must be restored.

For the task of restoration, the *dead* state is introduced. In this state, the values of the control variables are irrelevant. A dead cell is inconsistent with any live cell. Four procedures deal with the restoration of the control variables. Two of them, *Purge* and *Shrink*, kill cells inconsistent with their neighbors. *Purge* does this for local error correction, while *Shrink* does it with the aim of killing partial blocks. The two other procedures, *Heal* and *Grow*, resurrect the dead neighbors of live cells, setting the control variables of the resurrected cell in accordance with the live neighbor. *Heal* does this for local error correction, while *Shrink* does it with the aim of resurrecting a block of dead cells next to a live block. They depend on certain constants C_j chosen appropriately. As a hint on the choice of the constants, let us remark that if we choose the constants C_1, C_2, C_3, C_4, P, T, in this order, the choice of any one of them is restricted only by a lower bound expressed in terms of the previously defined ones.

The procedure *Grow* resurrects some dead cells next to a group of live cells. But *Grow* will be organized to be "tentative" because we do not want that one error enable a bad block to kill a good one. Thus, *Grow* has $2C_4$ steps of "advance", C_4 steps of "retreat" (when the resurrected cells will be killed if they have an inconsistent neighbor), and $2C_4$ idle steps. Repetitions of this procedure will be inserted in the first part of the working period, before any computation. We put in enough repetitions to resurrect slightly more than a whole block of dead cells. Let T_0 denote the time when the repetitions of *Grow* end. The resurrected cells will be distinguished from the rest as the "extension" of the block, but the distinction will be erased after all applications of *Grow* have been completed.

Procedure *Shrink* kills some cells next to an inconsistency inside a block. It consists of two parts. In the first C_1 steps, it kills a cell if the cell is not at the end of its block and is inconsistent with one of its neighbors. In the remaining $C_4 - C_1$ steps, it stays idle. We begin applying *Shrink* after T_0 and repeat it to the end of the working period, running in parallel with the computational part.

Procedure *Heal* resurrects some dead cells if they form a small gap. To avoid uncontrollable growth, the first C_3 steps are the resurrecting ones, marking the resurrected cells temporary. In the remaining C_3 steps, the temporary cells are killed if they have an inconsistent neighbor and are not at the end of the block. *Heal* will be inserted after every C_4 steps of the previously described program.

Procedure *Purge* kills C_2 or fewer cells if they are surrounded by inconsistencies. In the first c_2 steps, it propagates an error message from a place of inconsistency to the right. In the next C_2 steps, any cell who has the message and is inconsistent with its right neighbor will be killed. *Purge* will be inserted after every C_1 steps of the previously defined program (including *Heal*).

Let us denote by M_1' the above extension of the definition of M_1. Even if M_1' might recover the block structure from a 2-sparse set of errors, the *information content* of the blocks affected by the bad rectangle would be lost. Let us see how that can be restored. We are still free to vary the *content* of the computation of the medium U. We can use the universality of U and have each block $\mathcal{P}_1[i]$ of cells in U simulate a cell i of some trajectory of the medium M_1'. In this way, a bad rectangle can be handled by the 1-blocks like one error by the original cells (of course, only after the block structure in the damaged blocks has been restored).

Let g and γ be the encoding and decoding function of the simulation of M_1' by U. We want to make sure that whenever the control variables are correct the simulation will be performed, independently of the values of the other variables. Such a simulation can be called a *forced simulation*. To achieve it, we do the following. Before applying the rules of U to the Y variables, the medium casts the strings $Y[\mathcal{K}[j]]$, $(j = 1, 2, 3)$, into a code: it replaces $Y[\mathcal{K}[j]]$ with $g(\gamma(Y[\mathcal{K}[j]]))$.

There are two more parts to be added to the program. They implement the idea that "dead" means dead on all levels.

(a) if at the end of the computational part, it is discovered that the computed new X string is the code of a dead cell then all cells of the block will be killed.

(b) The procedure *Grow* writes into the X variables of the resurrected block the code of a dead cell.

This new medium will be called M_2.

The medium M_2 will indeed correct a 2-sparse set of errors, though this is rather tedious to prove. A simple way tho carry the idea of M_2 further to 3-sparse sets of errors and beyond is to use a trick in the definition of M_2. It is possible to have U simulate not only M_1' but rather M_2. This seems first circular since M_2 is the rule we are just defining, but the technique of the "fixpoint theorem" ("recursion theorem") of recursion theory makes this possible.

Thus the medium $M = M_2$ simulates itself (by the mediation of the universal medium U). This gives rise to an infinite hierarchy of simulations of M by itself. The effect of rule (a), of course, *ripples down* through all levels, while the effect of rule (b) *ripples up* through all levels.

After we fixed the content of the U-computation to be a simulation of M, the reader may wonder where is our freedom to define several different trajectories. Though any extra variables can be added, a simple possibility is to consider different trajectories depending on the choice of space-time origin (the content of the control variables on all levels). This is a countable number stable trajectories, but it is not difficult to have a continuum of stable trajectories either.

The idea of the proof. The crucial property of the above organization is that in any gap of size P^k caused by errors, the structure of the hierarchical simulation will be spontaneously restored, provided the set of errors is k-sparse for about T^k steps. Indeed, any partial blocks are eliminated by the procedure *Shrink*. Due to the forced simulation, live (consistent) blocks of cells simulate the behavior of cells of M (small errors are corrected by *Heal* and the voting). Thus they also apply *Shrink* and *Purge* in simulation, etc. In this way, after less than $2T^k$ steps, the gap does not contain "almost" any live cells that do not belong to a level k organization consistent with the one around the gap. Now the procedures *Grow* and *Heal* can refill the gap.

The assertion to be proved inductively is that in a rectangle where x is a k-trajectory, the whole rectangle will be divided into disjoint "k-traces" and "no man's land". The traces are space-time domains in which the cells are organized up to the kth level. Different traces correspond to different possible ways of organization, especially, different possible positions of the origin of the coordinates (this determines the values of the control variables on all levels of simulation). Thus a k-trace is determined by a pair (a, b) of numbers where $a \in [0 \dots T^{k+1})$, $b \in [0 \dots P^{k+1})$. The trace $(0,0)$ can be called the *canonical trace*. The "no man's land" is the area that was said in the previous paragraph to contain "almost" no live cells. Formally, it will simply be the area not belonging to any trace.

For the canonical trace, as well as for any other trace, the function $x^k[h,i]$ is defined. It is the trajectory represented by k-blocks of the trace. It and the notion of the k-trace are defined simultaneously, inductively. The k-rectangle $\daleth[h] \times \text{P}[i]$ belongs to the k-trace of $(0,0)$ if

(a) all but possibly an adjacent pair of the $k-1$-rectangles in its input belong to the $k-1$-trace of $(0,0)$,

(b) the string $x^{k-1}[hT, \text{P}[i]]$ represented by its input differs only at most at a pair of places from a code of some value $x^k[h,i]$,

(c) The control variables of $x^k[h,i]$ have the values h mod $\daleth$ and i mod P respectively.

An evolution will be called *k-organized* if its traces are disjoint, the evolution $x^k[h,i]$ as well as the evolutions corresponding to other traces are trajectories, further if any trace will grow into the no man's land whenever its program demands so and no other trace is nearby. The *key lemma* says that a k-trajectory is k-organized.

In the inductive proof, we can assume that except in one error rectangle, x is a $k-1$-trajectory. Hence it is $k-1$-organized everywhere outside the error rectangle. Moreover, the $k-1$-blocks in the $k-1$-traces behave in almost every respect as individual cells. The case of a general k is therefore not much more difficult than the case $k = 2$.

4. Conclusion

Simplification. It is an important task to simplify the construction and proof in [G 83] conceptually as much as possible. Some progress has been achieved here and the new version of the paper will reflect it.

The problem of quantitative simplification is no less pressing. The medium described in [G 83] has millions of states. In order to lend the medium physical credibility, reducing the number of states (while retaining nearest-neighbor interaction) is crucial. A great amount of reduction is possible within the present framework: e.g. most variables could be shared by a few neighbor cells. This was avoided in [G 83] only for reasons of clarity. But further attempts at reduction bring up principial problems since "forced self-simulation", a crucial feature of the medium, is expensive.

Positive memory capacity. In [G 83], one needs $K \log^c(KL)$ cells to store K bits of information for L steps. The time delay is $\log^c(KL)$ per step. A memory used for a fixed number L of steps is a special communication channel of the type considered in Information Theory, and one can ask whether it has a positive capacity C in the sense that only $KC/\log s$ cells are needed to store K bits for L steps (at least if K is not smaller than $\log L$). Here, s is the number of cell-states. This problem has a positive solution but several details need to be elaborated. Besides the positive capacity, the time delay will also be decreased to $\log^{1+\epsilon}(TK)$ with ϵ arbitrarily small.

Here are some of the changes to be made on [G 83] for positive capacity. A block of level k has now $Pk^2 2^{k-1}$ subblocks of level $k-1$. (The new proof of Toom's result also uses these block sizes.) In the inductive definition of k-sparsity we allow 2^{k-1} exceptional subrectangles instead of just one. The simple idea of self-simulation must be abandoned for a more direct definition of the work of a $k+1$-block in terms of the k-blocks comprising it. The spatial repetition must be replaced with an algebraic code of rate $1 - 1/k^2$. Thus some P_k/k^2 checkbits are added to every k-block of size P_k. The control, mailbox and other service variables are shared by several cells to cut their contribution to the redundancy. The temporal repetition is about 2^k times on level k.

Tradeoffs involving redundancy, dimension and simplicity. John Reif of Harvard University noticed in October 1984 that Toom's media can be used for the creation of an extremely simple *three-dimensional* medium simulating any perfect one-dimensional cellular medium D in *real time*. Indeed, one uses Toom's voting rule in parallel planes and the rule D across the planes.

The three-dimensional reliable medium simulates any perfect one-dimensional medium in real time, essentially without encoding, but using repetition $\log^{2+\epsilon}(KT)$ times, i.e. with information storage capacity $1/\log^{2+\epsilon}(KT)$. The one-dimensional medium described in the previous subsection has logarithmic time-delay, but positive constant information storage capacity. It remains open whether there is any *tradeoff* between these possibilities. The two-dimensional case is intriguing, since a three-dimensional cellular medium is *not physically realizable*. Indeed, each component turns free energy into heat at a constant rate, hence needs a long private wire for feeding.

Continuous time. Most interactive particle systems considered by probabilists and physicists work in continuous time as a Markov process. Since such systems are physically more realistic the question arises whether reliable computation is achievable in them. As Bennett remarked in 1983, continuous time brings in the additional problem of *synchronization*. Still, this problem seems to be solvable. The hierarchical organization used for error-correction can also be used for synchronization. Each block of cells is supposed to be synchronized to a close tolerance and is periodically resynchronized. Higher order blocks are synchronized to progressively looser tolerances. The details have to be worked out.

Self-organization. The one-dimensional cellular space described in the previous section works reliably only if the starting configuration contains already the (input-independent) hierarchical organization. It would sound more natural that if the input is one bit then the starting configuration should consist just of the repetition of this one bit. If error-correction requires structure then the medium should be able to build up this structure, out of "nothing". In probabilistic terms, this would correspond to the existence of several invariant measures that are also *translation-invariant*.

This goal is the *most intriguing* among the ones proposed. We run into several new problems. First: Without errors, no structure can arise, since all cells have the same state and the same rule. Hence whatever structures may arise will be random (e.g. if there are blocks, their position will be random). Second: killing a small organized island surrounded by inconsistencies was one of the main principles of the construction in [G 83]. Self-organization requires that we permit these islands to grow. A possible new rule replacing the old one could be that islands die if they are prevented from growth for longer time. Assigning different speeds to different kinds of growth (making "self-organizing" growth the slowest) is another trick that might work.

REFERENCES

[D 77] R.L.Dobrushin, S.I.Ortyukov: Upper Bounds on the Redundancy of Self-correcting Arrangements of Unreliable Elements, *Problems of Inf.Transm.* **13/3** 201-218

[G 78] P.Gács, L.A.Levin, G.Kurdyumov: One-dimensional homogeneous media dissolving finite islands, *Problems of Inf. Transm.* **14/3** 92-96

[G 83] P.Gács: Reliable Computation with Cellular Automata, *Proceedings of the ACM Symposium on the Theory of Computing*, Boston 1983
Journal version submitted to the *Journal of Computer and System Sciences.*

[Gray 82] L.F.Gray: The Positive Rates Problem for Attractive Nearest Neighbor Spin Systems on **Z**, *Z. Wahrscheinlichkeitstheorie verw. Gebiete* **61** (1982) 389-404

[Gri 79] D.Griffeath: Additive and Cancellative Interacting Particle Systems, *Lecture Notes in Math.* **724**, Springer, New York 1979

[KS 80] R.Kindermann, L.Snell: Markov Random Fields and their Applications. *Contemporary Mathematics* **1** Amer.Math.Soc. 1980

[Ku 78] G.L.Kurdyumov: An Example of a Nonergodic Homogeneous One-dimensional Random Medium with Positive Transition Probabilities *Soviet Math.Dokl.***19** (1978/1) 211-214

[Kz 73] A.V.Kuznietsov: Information Storage in a Memory Assembled from Unreliable Components *Problems of Information Transm.* **9/3** (1973) 254-264

[Li 77] T.M.Liggett: The Stochastic Evolution of Infinite Systems of Interactive Particles, *Lecture Notes in Math.* **598** Springer, New York 1977, 187-248

[Ta 68] M.C.Taylor: Reliable Information Storage in Memories Designed from Unreliable Components, *Bell Syst. Tech. J.* **47/10** (1968) 2299-2337
Reliable Computation in Computing Systems Designed from Unreliable Components, *ibid.* (1968) 2339-2366

[To 74] A.L.Toom: Nonergodic Multidimensional Systems of Automata, *Problems of Information Transm.* **10** 239-246

[To 80] A.L.Toom: Stable and Attractive Trajectories in Multicomponent Systems, *Advances in Probability* **6** (Multicomponent Systems, ed. R.L.Dobrushin) Dekker, NY 1980, 549-575

[Ts 76] B.S. Tsirel'son: Reliable Information Storage in a System of Unreliable Components with Local Interactions, *Lecture Notes on Math.* **653** Springer 1978, 15-30 (Interactive Markov-processes in Biology, ed.Dobrushin,Kryukov,Toom)

[vN 53] J. von Neumann: Probabilistic Logics and the Synthesis of Reliable Organisms from Unreliable Components. *Automata Studies* (Shannon, McCarthy, eds.) Princeton Univ.Press NJ 1956

Department of Computer Science
Boston University
111 Cummington Street
Boston, Massachusetts, 02215

Contemporary Mathematics
Volume **41**, 1985

DISORDERED ISING FERROMAGNETS AND PERCOLATION

H. O. Georgii

ABSTRACT. We consider Ising spin systems in two dimensions with ferromagnetic random interactions. We investigate the behaviour of the critical temperature near the threshold for percolation through active bonds by looking at a correlated site-bond percolation problem.

1. PRELIMINARIES ON ISING FERROMAGNETS

The two-dimensional Ising system is well-known to be the simplest mathematical model showing a ferromagnetic phase transition. It is specified as follows. Let $S = \mathbb{Z}^2$ be the square lattice and $B = \{ \{i,j\} \subset S : |i-j| = 1 \}$ the set of all "bonds" between adjacent sites of S. At each site $i \in S$ there is a random variable σ_i (called "spin") taking the values ± 1 only. The spins are coupled by a ferromagnetic nearest-neighbour interaction which is specified by a family $J = (J_b)_{b \in B}$ of coupling constants $J_b \geq 0$ and a scaling parameter $\beta \geq 0$, the "inverse absolute temperature". More precisely, for each bond $b = \{i,j\} \in B$ the interaction energy of σ_i and σ_j is $- \beta J_b \sigma_i \sigma_j$. As $J_b \geq 0$, σ_i and σ_j have minimal energy iff $\sigma_i = \sigma_j$, i.e. the interaction is ferromagnetic. (The classical choice is $J \equiv 1$.) We further assume that non-adjacent spins do not interact. By means of the Gibbs ansatz these hypotheses are transformed into the following requirement on the joint distribution of the σ_i's:

For each finite $\Lambda \subset S$ the conditional distribution of $\sigma_\Lambda = (\sigma_i)_{i \in \Lambda}$ given $\sigma_{S \setminus \Lambda} = (\sigma_i)_{i \in S \setminus \Lambda}$ has density

$$(1.1) \qquad Z(\sigma_{S\setminus\Lambda})^{-1} \exp [\beta \sum_{b=\{i,j\} \in B,\ b \cap \Lambda \neq \emptyset} J_b \sigma_i \sigma_j]$$

with respect to the equidistribution on $\{-1,1\}^\Lambda$; here $Z(\sigma_{S\setminus\Lambda})$ is a normalizing constant.

1980 Mathematics Subject Classification. 60K35, 82A25, 82A43, 82A57.

We let $G(\beta J)$ denote the set of all distributions of spin systems $\sigma = (\sigma_i)_{i\in S}$ satisfying the preceding requirement. Each $\mu \in G(\beta J)$ is called a Gibbs measure for the potential βJ and should be thought of as an equilibrium distribution of a spin system with interaction J at temperature β^{-1}. If $|G(\beta J)| > 1$ then the system has a "free choice" to pick its equilibrium state. Therefore the phenomenon of non-uniqueness of the Gibbs measure is interpreted as a phase transition.

As J is ferromagnetic, the correlation inequalities of FKG-Holley [1] and Griffiths [11] may be applied and give the following results [7]:

(1.2) As $\Lambda \uparrow S$, the weak limit $\mu_+^{\beta J}$ of the distributions defined by (1.1) and the "boundary condition" $\sigma_{S\setminus\Lambda} \equiv 1$ exists and belongs to $G(\beta J)$.

(1.3) $|G(\beta J)| > 1$ iff $\sup_{i\in S} \int \sigma_i \, d\mu_+^{\beta J} > 0$.

Thus phase transition is equivalent to spontaneous magnetization.

(1.4) For each $i \in S$, $\int \sigma_i \, d\mu_+^{\beta J}$ is nonnegative and increasing in β and J. Hence for each J there is a "critical inverse temperature" $\beta_c(J) \in [0,\infty]$ such that

$$|G(\beta J)| \begin{cases} = 1 & \text{if } \beta < \beta_c(J) \\ > 1 & \text{if } \beta > \beta_c(J) . \end{cases}$$

$\beta_c(J)$ is a decreasing function of J.

As for the existence of phase transitions, we should mention the famous result of Onsager [8]:

(1.5) If $J \equiv 1$ then $\sinh 2\beta_c(J) = 1$.

2. DISORDERED ISING MODELS. MAIN RESULT.

As stated above, the standard choice of the interaction is $J \equiv 1$. However, in real ferromagnetic materials the spin interaction is usually disordered because of lattice defects or impurities, and the latter may be considered as being random. For instance, this holds for quenched binary alloys of a ferromagnetic and a non-magnetic metal such as FeAu. Therefore it is realistic to look at Ising models with random coupling variables J_b, $b \in B$. We make the usual assumption that the random variables J_b, $b \in B$, are independent and have the same distribution, say α. The underlying probability measure will be denoted by P_α, and the corresponding expectation by E_α. We also assume that the random interaction is ferromagnetic, i.e. $\alpha([0,\infty[) = 1$. Then (1.2) implies that $\sup_{i\in S} \int \sigma_i \, d\mu_+^{\beta J}$ is invariant under translations of the J_b's and therefore constant P_α - almost surely. Combining this observation with (1.3) and (1.4) we obtain the following basic result.

(2.1) For each probability measure α on $[0,\infty[$ there is a critical inverse temperature $\beta_c(\alpha) \in [0,\infty]$ such that

$$P_\alpha(|G(\beta J)| = 1) = 1 \, , \; E_\alpha \int \sigma_o \, d\mu_+^{\beta J} = 0 \qquad \text{if} \quad \beta < \beta_c(\alpha) \, ,$$

$$P_\alpha(|G(\beta J)| > 1) = 1 \, , \; E_\alpha \int \sigma_o \, d\mu_+^{\beta J} > 0 \qquad \text{if} \quad \beta > \beta_c(\alpha) \, .$$

$\beta_c(\alpha)$ is an increasing function of the distribution function of α.

We are going to investigate $\beta_c(\alpha)$. More precisely, we ask how $\beta_c(\alpha)$ depends on the atom $\alpha(\{0\})$ of α at 0. To this end we write $\alpha = (1-p)\delta_o + p\alpha_+$. Here δ_o is the Dirac measure at 0, $p = \alpha(]0,\infty[) = P_\alpha(J_b > 0)$ is the probability that a bond $b \in B$ is "active", and $\alpha_+ = \alpha(\,\cdot\,|]0,\infty[)$ is the conditional distribution of J_b given b is active. (Thus α_+ is a probability measure on $]0,\infty[$.) We fix α_+ and ask for the properties of the decreasing function

$$(2.2) \qquad p \to \beta_c(\alpha_+, p) = \beta_c(p\alpha_+ + (1-p)\delta_o)$$

on $[0,1]$. This question is intimately connected to the Bernoulli bond percolation problem on (S,B). For let $\xi_b = 1_{\{J_b>0\}}$ be the indicator function of the event that b is active. Then the ξ_b, $b \in B$, are Bernoulli random variables with $P_\alpha(\xi_b = 1) = p$. For each realization of $\xi = (\xi_b)_{b\in B}$ we define the random subgraph $G(\xi) = (S, \{b \in B : \xi_b = 1\})$ of (S,B). $G(\xi)$ is the subject of Bernoulli bond percolation theory on (S,B). A famous theorem of Harris and Kesten [6] states that

$$P_\alpha(\text{each cluster of } G(\xi) \text{ is finite}) = 1 \qquad \text{if} \quad p \leq 1/2 \, ,$$

$$P_\alpha(G(\xi) \text{ contains a unique infinite cluster}) = 1 \quad \text{if} \quad p > 1/2 \, .$$

In particular, if $p \leq 1/2$ then P_α - a.s. each spin σ_i is correlated to at most finitely many other spins and therefore $\int \sigma_i \, d\mu_+^{\beta J} = 0$ for all $\beta \geq 0$.

Consequently, we may state:

(2.3) <u>If</u> $p \leq 1/2$ <u>then</u> $\beta_c(\alpha_+, p) = \infty$.

If $p > 1/2$ then the existence of an infinite cluster of correlated spins should imply that $\beta_c(\alpha_+, p)$ is finite, but the continuity of the percolation probability suggests that $\beta_c(\alpha_+, p) \to \infty$ as $p \downarrow \frac{1}{2}$. We therefore expect a shape of $\beta_c(\alpha_+, \cdot)$ as in the next figure.

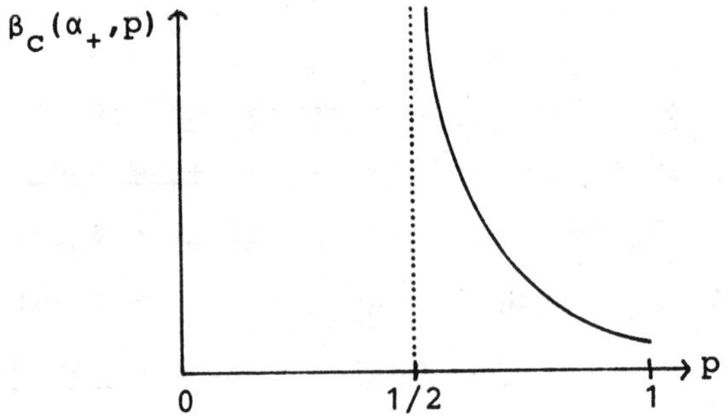

This picture is confirmed and made precise by Theorem 1 below. Let

$$\hat{\alpha}_+(t) = \int e^{-tx} \alpha_+(dx) = E_\alpha(e^{-tJ_b} \mid J_b > 0)$$

denote the Laplace transform of α_+ . $\hat{\alpha}_+$ is a continuous strictly decreasing function from $[0,\infty]$ onto $[0,1]$ and therefore has a continuous decreasing inverse $\hat{\alpha}_+^{-1} : [0,1] \to [0,\infty]$. We write $x_+ = \max(x,0)$.

THEOREM 1. <u>Let</u> $c_1 = \sqrt{8}$, $c_2 = 1/4$, <u>and let</u> $c_3 > 0$ <u>be sufficiently small and</u> $c_4 > 0$ <u>sufficiently large. Then</u>

$$\frac{1}{2}\hat{\alpha}_+^{-1}(c_1(p-\tfrac{1}{2})_+^{c_2}) \leq \beta_c(\alpha_+, p) \leq \hat{\alpha}_+^{-1}(c_3(p-\tfrac{1}{2})_+^{c_4})$$

<u>for all</u> $0 \leq p \leq 1$ <u>and all probability measures</u> α_+ <u>on</u> $]0,\infty[$.

Unfortunately, our proof only gives poor estimates of c_3 and c_4 . Nevertheless, Theorem 1 contains a reasonable amount of information on the kind of the singularity of $\beta_c(\alpha_+, \cdot)$ at 1/2 . Let us discuss some examples.

EXAMPLES. (a) $\alpha_+ = \delta_1$, the Dirac measure at 1 . Then $\hat{\alpha}_+(t) = e^{-t}$. Hence Theorem 1 shows that there are constants A_1 , $A_2 > 0$ such that

$$-A_1 \log(p-\tfrac{1}{2})_+ \leq \beta_c(\delta_1, p) \leq -A_2 \log(p-\tfrac{1}{2})_+ .$$

(b) Let $q > 1$ and α_+ the stable distribution with index $1/q$ and $\operatorname{supp} \alpha_+ = [0,\infty[$. Then $\hat{\alpha}_+ = e^{-t^{1/q}}$. Hence

$$A_1 \mid \log (p - \tfrac{1}{2})_+ \mid^q \leq \beta_c(\alpha_+,p) \leq A_2 \mid \log (p - \tfrac{1}{2})_+ \mid^q$$

for suitable constants A_1 , $A_2 > 0$.

(c) Suppose $\alpha_+(]0,x]) \sim cx^{1/q}$ as $x \downarrow 0$ for some $c > 0$, $q > 0$. Then an Abelian theorem gives the power estimates

$$A_1 (p - \tfrac{1}{2})_+^{-qc_2} \leq \beta_c(\alpha_+,p) \leq A_2 (p - \tfrac{1}{2})_+^{-qc_4} ,$$

where A_1 , $A_2 > 0$ are suitable constants.

(d) Let α_+ be any probability measure on $]0,\infty[$. Then there is a probability measure $\tilde{\alpha}_+$ on $]0,\infty[$ such that $\hat{\tilde{\alpha}}_+(t) = \hat{\alpha}_+(\log (1+t))$. From Theorem 1 we conclude that

$$2\beta_c(\tilde{\alpha}_+,p) + 1 \geq \exp \beta_c(\alpha_+ , \tfrac{1}{2} + a_1 (p - \tfrac{1}{2})_+^{a_2})$$

with suitable constants a_1 , $a_2 > 0$. Iterating this procedure we find probability measures α_+ on $]0,\infty[$ with an arbitrary fast divergence of $\beta_c(\alpha_+,\cdot)$ at $p = \frac{1}{2}$.

3. A CORRELATED SITE - BOND PERCOLATION PROBLEM .

Here we sketch the main ideas for a proof of Theorem 1 . The complete proof is given in [4] . We define a random subgraph $G(\xi,\sigma)$ of (S,B) as follows: The set of vertices of $G(\xi,\sigma)$ is the set $\{i \in S : \sigma_i = 1\}$ of positive spins, and the set of edges is the set

$$\{b = \{i,j\} \in B : \xi_b = \sigma_i = \sigma_j = 1\}$$

of active bonds between positive spins. Here $\xi_b = 1_{\{J_b > 0\}}$, and the joint distribution of (J,σ) is specified by the law $P_{\alpha,\beta} = P_\alpha \times \mu_+^{\beta J}$. More precisely, J is governed by P_α , and the conditional distribution of $\sigma = (\sigma_i)_{i \in S}$ given J is $\mu_+^{\beta J}$. A theorem of Coniglio et al. [3] implies that

$$P_{\alpha,\beta} \text{ (0 belongs to an infinite cluster of } G(\xi,\sigma)) > 0$$

whenever $\beta > \beta_c(\alpha)$. On the other hand, standard arguments show that $\beta \geq \beta_c(\alpha)$ whenever

$$P_{\alpha,\beta} \text{ (}G(\xi,\sigma)\text{ contains an ocean)} = 1 .$$

Here an "ocean" is an infinite cluster with the property that all clusters of its complement (the "islands") are finite. Thus Theorem 1 follows from

THEOREM 2. Let $c_1, \ldots, c_4$ be as in Theorem 1 . Then (i) $P_{\alpha,\beta}$ (0 belongs to an infinite cluster of $G(\xi,\sigma)) > 0$ implies $2\beta \geq \hat{\alpha}_+^{-1}(c_1(p-\frac{1}{2})_+^{c_2})$, and (ii) $\beta > \hat{\alpha}_+^{-1}(c_3(p-\frac{1}{2})_+^{c_2})$ implies

$P_{\alpha,\beta}(G(\xi,\sigma)$ contains an ocean) = 1 .

We sketch the main stages in the proof of Theorem 2 .

PROOF OF (i) . For fixed J , the event

$$C_o(\xi) = \{0 \text{ belongs to an infinite cluster of } G(\xi,\sigma)\}$$

is a monotone limit of increasing cylinder events. The FKG - Holley inequalities therefore imply that

$$\mu_+^{\beta J}(C_o(\xi)) \leq \nu^{\beta J}(C_o(\xi)) .$$

Here $\nu^{\beta J}$ is the unique Gibbs measure with vanishing interaction and an "external field"

$$h_i = \beta\, \Sigma_{b\in i} J_b$$

acting on σ_i for all $i \in S$. $\nu^{\beta J}$ is an inhomogeneous product measure. By a comparison of product measures we obtain for the conditional expectation with respect to ξ

$$E_\alpha(\nu^{\beta J}(C_o(\xi)) \mid \xi) \leq \nu^{\beta,\alpha_+}(C_o(\xi)) ,$$

where $\nu^{\beta,\alpha_+} = E_{\alpha_+}\nu^{\beta J}$. Unfortunately, ν^{β,α_+} is not a product measure except when α_+ is degenerate. This is because $J_{\{i,j\}}$ contributes to both h_i and h_j . Therefore each J_b is replaced by two i.i.d. random variables J_b^1 and J_b^2 such that $J_b = J_b^1 \wedge J_b^2$ in distribution. Again exploiting the stochastic order between product measures we thus obtain

$$\nu^{\beta,\alpha_+}(C_o(\xi)) \leq \mu^{\beta,\alpha_+}(C_o(\xi)) ,$$

where μ^{β,α_+} is the Bernoulli measure with probability $1-(\hat{\alpha}_+(2\beta)/2)^4/2$ of "success". Thus

$$E_\alpha\, \mu^{\beta,\alpha_+}(C_o(\xi)) \geq E_\alpha\, \mu_+^{\beta J}(C_o(\xi)) > 0 .$$

This means: If σ_i, $i \in S$, and ξ_b, $b \in B$, are independent $\{0,1\}$-valued random variables with $P(\sigma_i = 1) = 1 - 2^{-5}\hat{\alpha}_+(2\beta)$ and $P(\xi_b = 1) = p$ then with positive probability the origin belongs to an infinite cluster of $G(\xi,\sigma)$. In other words, $(1 - 2^{-5}\hat{\alpha}_+(2\beta), p)$ belongs to the percolative region of the Bernoulli mixed site-bond percolation problem on (S,B). Now a theorem of Hammersley [6] applies, stating that $p(1 - 2^{-5}\hat{\alpha}_+(2\beta))$ exceeds the threshold $\frac{1}{2}$ of the pure bond problem. This gives part (i) of Theorem 2.

PROOF OF (ii). Let T be the upper half-plane of S, $W(\xi)$ the cluster of $(T, \{b \in B : b \subset T, \xi_b = 1\})$ containing 0, and $W(\xi,\sigma) \subset W(\xi)$ the cluster of

$$(T, \{b = \{i,j\} \in B : b \subset T, \xi_b \sigma_i \sigma_j = 1\})$$

containing 0. Using a 0-1 law for $P_{\alpha,\beta}$ and some standard methods of percolation theory we may show that

$$P_{\alpha,\beta}(G(\xi,\sigma) \text{ contains an ocean}) = 1$$

provided $P_{\alpha,\beta}(|W(\xi,\sigma)| = \infty) > 0$. A power estimate of Kesten [6] states that

$$P_\alpha(|W(\xi)| = \infty) \geq a_1 (p - \tfrac{1}{2})_+^{a_2}$$

for suitable a_1, $a_2 > 0$. We therefore look at the event

$$C = \{|W(\xi)| = \infty, \ |W(\xi,\sigma)| < \infty\}.$$

Assertion (ii) will be proved once we have found constants c_3, $c_4 > 0$ such that

$$P_{\alpha,\beta}(C) < a_1 (p - \tfrac{1}{2})^{a_2} \tag{3.1}$$

when $p > \frac{1}{2}$ and $\hat{\alpha}_+(\beta) < c_3 (p - \frac{1}{2})^{c_4}$.

Suppose C occurs. Look at the set $\Delta(\xi,\sigma)$ of bonds connecting a site $i \in W(\xi,\sigma)$ and a site j in the infinite cluster of $W(\xi) \setminus W(\xi,\sigma)$. $\Delta(\xi,\sigma)$ is "the outer boundary of $W(\xi,\sigma)$ in $W(\xi)$" and may be visualized as follows. For each $\{i,j\} \in \Delta(\xi\ \sigma)$ we draw a line segment of unit length with center $(i+j)/2$ perpendicular to the line through i and j. The resulting line segments form a "ξ-<u>contour</u>", i.e. a non-empty finite set of finite polygons which surround 0 and have their initial and final points on the "islands" in $W(\xi)$ or on the boundary line of T. We ignore all polygons which return to their starting island. In the figure below we have indicated five islands in $W(\xi)$ and a contour consisting of three polygons.

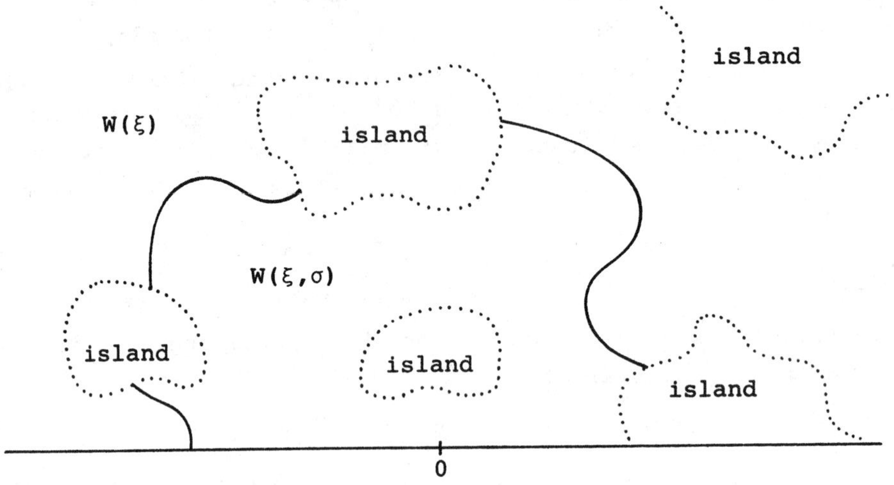

Each bond $b = \{i,j\} \in \Delta(\xi,\sigma)$ is critical in the sense that $\xi_b \sigma_i \sigma_j = -1$. Looking at (1.1) we see that each critical bond requires an energy amount of $2\beta J_b^+$, where J_b^+ has distribution α_+. This fact provides us with the estimate

$$P_{\alpha,\beta}(\text{all bonds in } \Delta \text{ are critical} \mid \xi) \leq \hat{\alpha}_+(\beta)^{|\Delta|}$$

for each finite $\Delta \subset \{b \in B : \xi_b = 1\}$. Consequently,

$$P_{\alpha,\beta}(C) \leq E_\alpha \sum_{\text{all possible } \xi\text{-contours } \Delta} \hat{\alpha}_+(\beta)^{|\Delta|}$$

$$= \sum_{\ell \geq 1} \hat{\alpha}_+(\beta)^\ell \; E_\alpha N(\ell,\xi) ,$$

where $N(\ell,\xi)$ is the number of all possible ξ-contours of total length ℓ. Clearly, $N(\ell,\xi)$ may be large if $W(\xi)$ contains large islands near the origin. Fortunately, a second power estimate of Kesten [6] guarantees that the islands are small in the sense

$$P_\alpha(\text{i and j belong to the same island}) \leq a_3 \exp[-a_4(p-\tfrac{1}{2})^{a_5}|i-j|],$$

where $a_3, a_4, a_5 > 0$. This estimate can be extended to the case of finitely many disjoint islands, which together with a counting of contours gives

$$P_{\alpha,\beta}(C) < a_6 \, \hat{\alpha}_+(\beta) \, (p-\tfrac{1}{2})^{-a_7}$$

with constants $a_6, a_7 > 0$. Choosing $c_3 = a_1/a_6$ and $c_4 = a_2 + a_7$ we obtain inequality (3.1).

It should be noted that the preceding argument is a refined version of the famous argument of Peierls for the existence of a phase transition in the classical Ising model.

4. THE CASE OF BOND DILUTION

The bond diluted Ising ferromagnet is the random Ising ferromagnet with coupling distribution $\alpha = p\delta_1 + (1-p)\delta_0$. This case was already considered in Example (a) to Theorem 1. However, a much sharper result is available.

THEOREM 3. <u>For all</u> $0 \leq p \leq 1$,

$$\frac{1}{2}\log\frac{1}{2} \leq \beta_c(\delta_1, p) - \frac{1}{2}\log(p-\tfrac{1}{2})_+^{-1} \leq \frac{1}{2}\log\frac{3}{2}.$$

Theorem 3 relies on an argument of Bergstresser [2] and the fact that $\beta_c(\delta_1,p) < \infty$ for $p > \frac{1}{2}$; the latter follows from Theorem 1. As Bergstresser considers a slightly different model, we will sketch a proof.

PROOF. Let $\Lambda \subset S$ be finite, $w = \tanh\beta$, and for a fixed realization of J we let $\langle\cdot\rangle = \langle\cdot\rangle^{\beta J}_{\Lambda,+}$ denote the expectation with respect to the distribution at (1.1) with $\sigma_{S\setminus\Lambda} \equiv 1$. We take any $b = \{i,j\} \in B$ with $b \cap \Lambda \neq \emptyset$ and $J_b = 1$, define an interaction J^b by $J^b_{b'} = J_{b'}$ for $b' \neq b$ and $J^b_b = 0$, and put $\langle\cdot\rangle^b = \langle\cdot\rangle^{\beta J^b}_{\Lambda,+}$. Writing $\sigma_b = \sigma_i\sigma_j$ and using the identity $\exp(-J_b\sigma_b) = \cosh\beta(1-w\sigma_b)$ we obtain

$$\langle\sigma_0\rangle^b = \langle\sigma_0(1-w\sigma_b)\rangle / \langle 1-w\sigma_b\rangle$$

and therefore

$$\langle\sigma_0\rangle - \langle\sigma_0\rangle^b = (\langle\sigma_0\sigma_b\rangle - \langle\sigma_0\rangle\langle\sigma_b\rangle)\,w/(1-w\langle\sigma_b\rangle).$$

Griffiths' inequalities imply $w \leq \langle\sigma_b\rangle \leq 1$ and $\langle\sigma_0\sigma_b\rangle \geq \langle\sigma_0\rangle\langle\sigma_b\rangle$. Furthermore,

$$\frac{d}{d\beta}\langle\sigma_0\rangle^{\beta J}_{\Lambda,+} = \sum_{b\cap\Lambda\neq\emptyset} J_b(\langle\sigma_0\sigma_b\rangle - \langle\sigma_0\rangle\langle\sigma_b\rangle).$$

Similarly, if $E_p = E_\alpha$ for $\alpha = p\delta_1 + (1-p)\delta_0$ then a short computation shows

$$\frac{\partial}{\partial p} E_p < \sigma_o >_{\Lambda,+}^{\beta J} = p^{-1} \sum_{b \cap \Lambda \neq \emptyset} E_p[J_b(< \sigma_o > - < \sigma_o >^b)] .$$

Combining these results we end up with the estimate

$$\frac{w}{1-w^2} \frac{\partial}{\partial \beta} E_p < \sigma_o >_{\Lambda,+}^{\beta J} \leqq p \frac{\partial}{\partial p} E_p < \sigma_o >_{\Lambda,+}^{\beta J} \leqq \frac{w}{1-w} \frac{\partial}{\partial \beta} E_p < \sigma_o >_{\Lambda,+}^{\beta J} .$$

To exploit the first inequality we choose some $p_o > 1/2$ and let $u(p) = \text{ar tanh}\ p_o / p$ for $p > p_o$. Then

$$\frac{d}{dp} E_p < \sigma_o >_{\Lambda,+}^{u(p) J} \geqq 0 .$$

Integrating over p and letting $\Lambda \uparrow S$ we see that the function $p \to E_p \int \sigma_o \, d\mu_+^{u(p) J}$ is increasing. Griffiths' inequalities and Theorem 1 imply

$$\lim_{p \downarrow p_o} E_p \int \sigma_o \, d\mu_+^{u(p) J} \geqq \lim_{\beta \to \infty} E_{p_o} \int \sigma_o \, d\mu_+^{\beta J} > 0 .$$

Hence $u(p) \geqq \beta_c(\delta_1, p)$. Letting $p_o \to 1/2$ we obtain the upper bound of $\beta_c(\delta_1, p)$.

For the lower bound we choose $p_o < 1/2$ and let $\ell(p) = \text{ar tanh}\ p_o / (2p - p_o)$ for $p \geqq 1/2$. Then a similar argument gives

$$E_p \int \sigma_o \, d\mu_+^{\ell(p) J} \leqq E_{1/2} \int \sigma_o \, d\mu_+^{\ell(1/2) J} = 0 ,$$

and in the limit $p_o \uparrow 1/2$ we obtain the lower bound of $\beta_c(\delta_1, p)$.

5. REMARKS AND QUESTIONS

(5.1) THE THINNESS OF THE OCEAN. Let $p > 1/2$, $\xi = (\xi_b)_{b \in B}$ a system of Bernoulli variables for p, and $\mathcal{O}(\xi)$ the unique infinite cluster in $G(\xi)$. $\mathcal{O}(\xi)$ is an ocean. It is easily seen that, for all α_+ and almost all ξ, $\beta_c(\alpha_+, p)$ is the critical inverse temperature of the random Ising ferromagnet with coupling distribution α_+ on the graph $(\mathcal{O}(\xi), \{b \in B : b \subset \mathcal{O}(\xi), \xi_b = 1\})$. In particular, $\beta_c(\delta_1, p)$ is the critical temperature of the standard Ising model on $\mathcal{O}(\xi)$. Therefore $\beta_c(\alpha_+, p)$ may be thought of as a quantity which measures the "thinness" of $\mathcal{O}(\xi)$.

(5.2) GENERAL SPINS. Theorems 1 and 2 (but not Theorem 3) extend to the case of spins taking values in a bounded symmetric interval [4]. The extension to unbounded or vector - valued spins is an open problem.

(5.3) HIGHER DIMENSIONS. The lower bound in Theorem 1 carries over to the case $S = \mathbb{Z}^d$ for $d > 2$ [4]. Of course, the constants c_1, c_2 vary with d, and $(p - 1/2)_+$ has to be replaced by $(p - p_d)_+$, where p_d is the threshold for Bernoulli bond percolation on $\mathbb{Z}^d$. Unfortunately, our proof of the upper bound relies on a number of two-dimensional arguments and requires a knowledge of Bernoulli percolation which so far is only available in two dimensions. Therefore the extension of the upper bound to dimensions $d > 2$ is a major open problem. We even do not know whether

$$p_d(\alpha_+) = \inf \{p : \beta_c(\alpha_+, p) < \infty\}$$

coincides with p_d for all α_+ and $d > 2$. In this respect it is interesting to note that Theorem 3 extends to all d with $p_d(\delta_1) = p_d$. L. and J. Chayes (private communication) claim to have a proof of $p_3(\delta_1) = p_3$.

(5.4) LINES OF CONSTANT MAGNETIZATION. The proofs of Theorems 1 and 3 also give upper and lower bounds for the level lines of the magnetization $E_\alpha \int \sigma_0 \, d\mu_+^{\beta J}$.

(5.5) SITE DILUTION. Let $(\eta_i)_{i \in S}$ be a family of $\{0,1\}$-valued Bernoulli variables with parameter p, and consider the dependent random variables $J_b = \eta_i \eta_j$, $b = \{i,j\} \in B$. The random Ising ferromagnet with interaction $(J_b)_{b \in B}$ is called the site diluted Ising model. The proofs of Theorems 1 and 2 (but not 3) carry over to this case and give the same result as in the case of bond dilution except that $1/2$ has to be replaced by the threshold for Bernoulli site percolation on (S,B).

(5.6) NONFERROMAGNETIC INTERACTIONS. Our use of correlation inequalities relies on the assumption $\alpha([0,\infty[) = 1$. This hypothesis excludes Ising systems with random interactions of arbitrary sign which are of great interest in the theory of spin-glasses. However, many difficulties arise when the J_b's are i.i.d. random variables of a.s. non-constant signs. First of all, it is unknown if the sets $\{|G(\beta J)| > 1\}$ are measurable. Secondly, there is no monotonicity in β. Therefore $\beta_c(\alpha)$ has to be defined as the infimum over all β which are such that, for all $\beta' \geq \beta$, $\{|G(\beta' J)| > 1\}$ has positive inner probability. It is easily seen that $\beta_c(\alpha)$ remains invariant under reflection of α at 0. Here are some questions. Pick two probability measures α_+ on $]0,\infty[$ and α_- on $]-\infty,0]$ and let

$$p_c(\alpha_+,\alpha_-) = \inf\{p : \beta_c(p'\alpha_+ + (1-p')\alpha_-) < \infty \text{ for all } p' \geq p\} .$$

What is the value of $p_c(\alpha_+,\alpha_-)$? Is $p_c(\alpha_+,\alpha_-)$ related to the threshold of some percolation problem as, for instance, percolation of unfrustrated plaquettes? What is the behaviour of $\beta_c(p\alpha_+ + (1-p)\alpha_-)$ near $p_c(\alpha_+,\alpha_-)$? So far there is only an upper bound of $p_c(\delta_1,\delta_{-1})$ which differs greatly from the available Monte Carlo estimates [9] .

(5.7) PHASE TRANSITION AND PERCOLATION. It is tempting to conjecture that percolation and the occurrence of a phase transition are equivalent in the sense that

$$P_\alpha(\,|\,\mathcal{G}(\beta J)\,| > 1) = P_{\alpha,\beta}(G(\xi,\sigma) \text{ contains an infinite cluster}).$$

In two dimensions, this should be true. For in dimension 2 it is believed that each infinite cluster of $G(\xi,\sigma)$ is an ocean $P_{\alpha,\beta}$-almost surely. The latter has been proved in the classical Ising model $(\alpha = \delta_1)$ [3,10] .

(5.8) NUMBER OF PHASES. A well-known theorem of Russo, Aizenman and Higuchi [5] states that $\mathcal{G}(\beta J)$ has two extreme elements when $J \equiv 1$ and $\beta > \beta_c(J)$ (and $d = 2$) . What is the effect of disorder on the number of phases? Is it still true that $\mathcal{G}(\beta J)$ has two extreme points P_α-almost surely for all α on $[0,\infty[$ and all $\beta > \beta_c(\alpha)$?

ACKNOWLEDGEMENT. I am indebted to J.W. Halley for pointing out reference [2] to me.

REFERENCES

1. Batty, C.J.K. and Bollmann, H.W., "Generalized Holley-Preston inequalities on measure spaces and their products", Z. Wahrscheinlichkeitsth. 53 (1980), 157-173.

2. Bergstresser, T.K., "Rigorous upper and lower bounds on the critical temperature in Ising models with random, quenched, broken-bond disorder", J. Phys. C 10 (1977), 3831-3849.

3. Coniglio, A., Nappi, C.R., Peruggi, F. and Russo, L., "Percolation and phase transitions in the Ising model", Commun. math. Phys. 51 (1976), 315-323.

4. Georgii, H.O., "On the ferromagnetic and the percolative region of random spin systems", Adv. Appl. Prob. 16 (1984), to appear.

5. Higuchi, Y., "On the absence of non - translation invariant Gibbs states for the two - dimensional Ising model", in: J. Fritz, J.L. Lebowitz, D. Szász (eds.), Random Fields, Esztergom (Hungary) 1979, North Holland, Amsterdam, 1981, Vol. I, 517-534.

6. Kesten H., Percolation Theory for Mathematicians, Birkhäuser, Boston, 1982.

7. Lebowitz, J.L. and Martin-Löf, A., "On the uniqueness of the equilibrium state for Ising spin systems", Commun. math. Phys. 25 (1972), 276-282.

8. McCoy, B.M. and Wu, T.T., "The two - dimensional Ising model", Harvard Univ. Press, Cambridge, Mass., 1973.

9. Roepstorff, G., "The Peierls-Griffiths argument for disordered Ising systems", J. Math. Phys. 22 (1981), 3002-3005.

10. Russo, L., "The infinite cluster method in the two - dimensional Ising model", Commun. math. Phys. 67 (1979), 251-266.

11. Sylvester, G.S., "Inequalities for continuous - spin Ising ferromagnets", J. Statist. Phys. 15 (1976), 327-341.

MATHEMATISCHES INSTITUT
LUDWIG-MAXIMILIANS-UNIVERSITÄT
THERESIENSTR. 39
D - 8000 MÜNCHEN 2
WEST GERMANY

Contemporary Mathematics
Volume 41, 1985

THE CRITICAL BEHAVIOR OF A CLASS OF SIMPLE INTERACTING SYSTEMS -- A FEW ANSWERS AND A LOT OF QUESTIONS

Lawrence Gray[1]

ABSTRACT. A class of Markovian interacting particle systems is defined which is relatively simple and yet rich enough to provide examples of many types of equilibrium behavior, as well as fuel for some intriguing conjectures. The systems may be viewed as simplified statistical mechanical models with parameters which play the roles of various physical quantities. Several questions are raised concerning the way in which the parameters influence critical behavior. Some answers are given, mostly in the nearest neighbor one-dimensional case. These answers are compared and contrasted with predictions based on common statistical mechanical rules of thumb.

1. INTRODUCTION. It has not been too many years since a conference that took as its theme "The mathematics of phase transitions" would feel itself obligated to stick with discussions of the equilibrium behavior of statistical mechanical models for which certain physical assumptions (such as time-reversibility and the relationship between the interaction potential and some Hamiltonian) are almost automatic. However today we hear side by side with the term "phase transition" other mathematical buzzwords such as "infinite cluster size", "reliable information storage", "population survival" and more generally, "non-ergodicity". Time reversibility and potentials have nothing to do with these (indeed the first phrase comes from percolation theory, with time playing no role at all), and yet the critical phenomena that they describe are closely connected with those from more "physical" theories.

It has thus become apparent that there is an entire complex of models which have in common that they are all systems of infinitely many interacting components in which some sort of critical phenomenon occurs. The study of one class of models often sheds light on others which on first consideration may not seem closely related. It is the purpose of this article to define and discuss one class of systems that are very much a part of this complex,

1980 Mathematics Subject Classification. 60 K 35.
[1]Partially supported by NSF Grant DMS 83 - 01080.

in the hopes that some math physicist or percolation expert can be enticed to take a look at some of the problems that have occupied some of us for years, to our mutual benefit.

2. POLLING MODELS. To make visualization easier, we will speak in terms of a population of people who each hold one of two opinions on some controversial issue. Each person is continually polling a certain finite set of other people, to see how their opinions compare with his/her own. Thus each person x is influenced by the members of his/her reference set, which we denote by R_x. If too many members of R_x hold the opposite opinion from x, then x has a tendency to change his/her mind. Such a change will often be called a flip. In our simple models, all people will have identical behavior patterns (i.e. there will be a translation invariance assumption).

We now give a more precise description. We assume that the people can be arranged at the sites of the d-dimensional integer lattice $\mathbb{Z}^d$ in such a way that for each x in $\mathbb{Z}^d$, $R_x = R + x$, where R is a finite subset of $\mathbb{Z}^d \setminus \{0\}$. To simplify matters, we will assume here that R is either a punctured ball centered at 0, or the non-negative part of such a ball. Thus R is either $B_N = \{x : 0 < |x| \leq N\}$, or $B_N^+ = \{x : 0 < |x| \leq N$ and $x_k \geq 0$ for $k = 1, 2, \ldots, d\}$, where $N \geq 1$ and $|x|$ is the usual Euclidian norm of x. If $R = B_N$, we call the model spatially symmetric; if $R = B_N^+$, we call it one-sided.

We now introduce some parameters: two noise parameters β and δ, which are non-negative; a stubbornness factor $\theta \in [0,1]$; and a bias $\alpha \in [0,1]$. These parameters determine the infinitesimal rates at which flips occur, as given in the following table. We label the two opinions + and -, and use $|R|$ for the cardinality of the set R.

	If at most $\theta\|R\|$ members of R_x disagree with x	If more than $\theta\|R\|$ members of R_x disagree with x
- to +	β	$1 + \beta$
+ to -	δ	$1 - \alpha + \delta$

Rates at which flips occur

It can be seen from this table that β and δ are the rates at which spontaneous flips occur, and that θ measures how thickheaded the population is (the larger θ is, the more people it takes to influence a given individual). The parameter α introduces an asymmetry between the +'s and -'s. Note that these models are attractive in the sense that "likes attract likes": the more people there are who agree with person x, the less tendency there is for x to change his/her opinion.

There are many ways of describing what is meant by "infinitesimal rates." One brief attempt will be made here. One does not need to master its details in order to follow the sequel. Let $r_x(t)$ be the rate given in the table for person x at time t. For example, if at time t person x has opinion + and at least $\theta|R|$ members of R_x have opinion -, then $r_x(t) = 1 - \alpha + \delta$. Let

$$T_x(t) = \int_0^t r_x(s)\,ds.$$

Also let $t_1, t_2, \ldots$ be a Poisson location of points on the positive half of the real line with density 1 (i.e. the values t_1, $t_2 - t_1$, $t_3 - t_2$, ... are iid exponentially distributed random variables with mean 1). Then at each time t that $T_x(t)$ takes one of the values t_k, person x changes his/her mind. This procedure is carried on at each site, with the Poisson locations being chosen independently for each x. The interaction comes from the way in which the rate $r_x(t)$ is affected by the opinions held by members of R_x at time t. It can be shown that there is a Markov process which conforms to this description and that it is unique in an appropriate sense.

The "polling models" defined above form a small subclass within the class of Markovian spin-flip systems. They have been chosen for three reasons. First, although they are relatively simple, they include examples of all of the types of critical behavior known for those members of the larger class which have attractive local translation invariant interactions on $\mathbb{Z}^d$. Proofs of theorems for general systems seem to be no easier for polling models, and the open questions appear equally difficult to answer. Second, certain features have been built into the class of polling models that have ready analogies in physical models, and the spatially symmetric case may be used as a sort of mock statistical mechanics without any special assumptions about time-reversibility or interaction potentials. In particular, the noise parameters represent excitations due to thermal energy, the stubbornness factor relates to the strength of the interaction potential and the bias corresponds to an external positive field. These four factors, together with the dimension d and the interaction range N provide all the essential material needed to produce "phase transitions." The third and final reason is that polling models also include many interesting population growth models, such as the contact process (Example 2 below). These models have applications in a variety of fields, and they provide some of the simplest dynamical models which exhibit critical behavior.

3. EXAMPLES. We give here a few simple examples of polling models, together with a short description of equilibrium behavior in each case. When a system has a unique equilibrium, it is called ergodic. Non-ergodicity corresponds to the existence of a phase transition. A critical phenomenon occurs if there is a switch from ergodicity to non-ergodicity at some (non-zero) parameter value. As might be expected, one is interested in estimating critical values, determining rates of convergence to equilibrium and decay of correlations, and in behavior near critical values.

EXAMPLE 1. The stubborn case: $\theta = 1$. In this case, there is no interaction. The people behave independently and the system is ergodic for all choices of β, δ, α, N and d. It converges exponentially fast to equilibrium from any initial state. The equilibrium measure is product measure.

EXAMPLE 2. The contact process: $\theta = 0$, $\alpha = 1$, $\beta = 0$. This is a model for population growth. The sites with +'s are called occupied, and those with -'s are called vacant. (The contact process was originally described in terms of the spread of disease, with + = infected and - = healthy. Healthy sites get the disease from infected ones in their reference set: in other words, the disease spreads through contact.) Occupied sites become vacant at rate δ (since $\alpha = 1$). Vacant sites do not become occupied (since $\beta = 0$), unless at least one member of the reference set is occupied (since $\theta = 0$). Since it takes only one "parent" to produce an offspring at a vacant site, reproduction is asexual in this model (compare Example 4). For all non-empty choices of R, the contact process is ergodic for large δ and non-ergodic for sufficiently small $\delta > 0$, so a critical phenomenon occurs. This process is the continuous time version of a discrete time information transmission model of Stavskaya and Pyatetskii-Shapiro (1968). Harris (1974) first showed the existence of critical behavior in continuous time.

EXAMPLE 3. The majority vote model: $\theta = 1/2$, $\alpha = 0$, $\beta = \delta$. A person changes his/her opinion at rate δ if there is agreement, at rate $1 + \delta$ if there is disagreement with the majority opinion of the reference set (a strict majority is required). In the spatially symmetric case, this model makes a good analogy to the classical Ising model. In fact if $d = N = 1$, it is a stochastic Ising model with a time-reversible equilibrium which is a Gibbs state for some nearest neighbor pair potential. It is believed that the equilibrium behavior of the majority vote model mirrors that of the Ising model in all dimensions, but only the $d = N = 1$ case has been rigorously analyzed. This model is equally frustrating and intriguing for other values of $\theta < 1 - 1/|R|$.

EXAMPLE 4. A population model with sexual reproduction: $d = 2$, $R = B_1^+ = (1,0) \cup (0,1)$, $\theta = 1/2$, $\alpha = 1$, $\beta = 0$. This model is like the contact process (Example 2) except that a vacant site can become occupied only if both sites in the reference set are occupied (instead of at least one). It is ergodic for large δ and non-ergodic for small $\delta > 0$, but it also has some interesting features which distinguish it from the contact process. Furthermore, some version of this model (perhaps in higher dimensions) looks like a good candidate for a model which violates one of the statistical mechanical rules of thumb that will be discussed in the next section (see Question 4). Its critical behavior is analyzed in Durrett and Gray (1985). A description of some of its features can also be found in the article by Durrett in this volume.

4. A FEW ANSWERS FOR A LOT OF QUESTIONS.

There are many rules of thumb in statistical mechanics, such as "no phase transitions at high temperatures," which are assumed to be true in some generality, although many of them can be proved only in special cases. We will list some of these here, rephrasing them as questions about polling models. We think of polling models as test cases for the validity of these rules for general critical phenomena. In the few cases where the answers to the questions are known, we will state them and briefly describe some of the tools used in their derivation. Our first question is the only easy one of the bunch:

QUESTION 1. Does ergodicity always occur when β and δ are large?

ANSWER 1. Yes, always.

Thus the rule "no phase transitions at high temperatures" is valid for polling models. In fact there is a general ergodicity criterion due to Dobrushin (1971) and improved by Sullivan (1974) which implies that any polling model with $\beta + \delta > |R|$ is ergodic, with exponential convergence to equilibrium. There are various proofs, but one of the most appealing uses a technique known as coupling. This is a method for comparing two Markov processes by defining them jointly on some probability space ("coupling them") in such a way that the comparison becomes easy. To prove ergodicity, two arbitrary initial states are chosen, and then two versions of the process, one for each initial state, are coupled together. Under the above conditions on the parameters it can be shown that as time increases, the two processes come arbitrarily close together in a way that implies that the initial states have a vanishing effect on their evolution as time goes to infinity. Ergodicity follows from a standard argument. This technique is widely used throughout the field of Markov processes and should be in everyone's arsenal. The proof used in Holley's extension of the FKG inequality is a good example (see Holley [1974]).

QUESTION 2. Can non-ergodicity exist in dimension $d = 1$ if $\beta > 0$ and $\delta > 0$? (This is called the "positive rates problem".)

ANSWER 2. No for nearest neighbor systems (i.e. those with range $N = 1$). The answer is otherwise unknown.

The question rephrases the rule "no phase transitions in one-dimensional systems with local, translation invariant interactions." The condition that β and δ be positive corresponds to the fact that in physical systems the temperature is always positive. It is somewhat surprising that Question 2 is unanswered for polling models, except in the nearest neighbor case. Even this case was solved only recently (Gray [1983]). It is even more surprising that if we leave the realm of polling models for a moment, there appears to be a system which violates the rule (see Gacs' article in this volume). This model does not have attractive interactions, but otherwise fits into the general framework of Markovian interacting systems of which polling models are a part. Nevertheless, one still believes that the answer to Question 2 is negative for attractive systems. The difficulty of showing this (when one does not have available the special tricks of statistical mechanics) points up weaknesses in our understanding of the role played by spatial dimension in critical phenomena.

The tools used to solve the nearest neighbor case include coupling (see above), and another useful set of methods called graphical techniques. These have a long and venerable history, beginning with the old Peierls' argument. Until recently, they have been used mainly to prove the existence of critical phenomena (see the discussion of Answer 3 below), but in this problem, they are the key to proving that critical behavior does not occur. In one dimension, one can use an appropriate space-time graph to picture the time evolution as a sort of struggle between +'s and -'s. In order for a system to be non-ergodic, the outcome of this struggle must depend on the initial state. However, in the nearest neighbor case, it can be shown that there is not enough room in one dimension for both sides to be able to win: either one side always wins, no matter what the initial state, or there is a type of stand-off. Either way, ergodicity is the result. Graphical methods seem to be useful in exploiting the lack of room in one dimension and would thus seem important to any further progress on this problem. For further discussion of the role of dimension, see Question 9 below.

QUESTION 3. What can be said about rates of convergence to equilibrium (or rates of decay of correlations in equilibrium) at non-critical parameter values?

ANSWER 3. They are exponential for nearest neighbor, one-dimensional systems in which either β or δ is 0. Otherwise the answer is unknown.

The statistical mechanical rule of thumb is that "convergence and decay rates are exponentially fast at non-critical parameter values in systems with a finite number of states at each site." As with Question 2, the actual cases in which this rule can be verified for polling models are surprisingly few. Answer 3 does not even cover all the nearest neighbor, one-dimensional cases. In particular, one would expect, based on the result in Answer 2, that exponential convergence would hold in nearest neighbor one-dimensional models with positive β and δ (most proofs of ergodicity give exponential convergence almost for free), but this is not known.

Let us state Answer 3 in more detail. In all cases where we assert that convergence to equilibrium is exponentially fast, it is also known that the equilibrium measure itself has exponentially decaying correlations.

ANSWER 3. (Expanded). Assume that $d = 1$ and $N = 1$. We ignore the case $\theta > 1 - 1/|R|$, since with $d = N = 1$, this is equivalent to the trivial case $\theta = 1$ found in Example 1.

a. If $0 \leq \theta \leq 1 - 1/|R|$ and $\delta = 0$, then the system converges exponentially fast in distribution to the point mass at "all +'s" for all $\beta > 0$.

b. If $\alpha = 0$, then the system converges exponentially fast to "all +'s" if $\delta = 0$ and $\beta > 0$, or to "all -'s" if $\beta = 0$ and $\delta > 0$.

c. If $\alpha > 0$, $\beta = 0$, and $0 \leq \theta \leq 1 - 1/|R|$, then there is a finite critical value $\delta_c > 0$ such that the system is ergodic for $\delta > \delta_c$ and non-ergodic for $\delta < \delta_c$. In the ergodic case, the system converges exponentially fast to "all -'s" from any initial state. In the non-ergodic case, there is a non-trivial equilibrium measure (not a point mass) which the system approaches exponentially fast if it starts at the initial state "all +'s".

Parts a. and b. are quite easy to show. One concentrates on strings of consecutive +'s if $\delta = 0$ or on strings of consecutive -'s if $\beta = 0$ and shows that such strings grow at a linear rate under the conditions stated, so that they quite rapidly take over the whole system. Unfortunately, this argument does not generalize to the non-nearest neighbor case or to higher dimensions.

Part c. is much more interesting. The existence of a critical phenomenon was shown by Harris (1974) for $\alpha = 1$ and by Gray and Griffeath (1982) for $0 < \alpha < 1$. A combinatorial argument known as the "contour method" was used. It is a graphical technique which is related to the Peierls' argument, and to early proofs of the existence of infinite clusters in percolation models, but which stems more directly from methods developed by various Russian mathematicians for studying discrete time systems (see Toom [1979]). It seems to be adaptable to a wide variety of situations and is one of the first things to try when looking for critical behavior.

The exponential convergence results in c. are due to Durrett and Griffeath (1983) for $\alpha = 1$ and to Gray (1985) for $0 < \alpha < 1$. Again the method is graphical, but this time there are two steps. The first one is known as a "block argument". It is used to transform a problem about a system which is close to critical into a similar problem about another system which is far from critical. In this case, the polling model is transformed into an oriented percolation model. The percolation model is far enough from critical so that contour methods can be used to prove exponential mixing results (this is the second step in the graphical argument). These are then turned around to imply the desired convergence behavior in the polling model. The block argument is a kind of poor man's renormalization group method. It seems to be ideally suited to proving results that work for parameter values right up to critical. In percolation theory, the idea goes back to Russo and plays a role in Kesten's famous $p_c = 1/2$ result (see Kesten [1982]). For another example, look for the work (yet to appear) of Newman in collaboration with Aizenman and Schulman on "one-dimensional $1/|x - y|^s$ percolation". For a discussion of several problems related to Question and Answer 3, see the article by Durrett elsewhere in this volume.

QUESTION 4. What can be said about rates of convergence to equilibrium at critical parameter values?

ANSWER 4. Nothing at this time.

There is not a single example of a polling model which exhibits critical behavior in which it can be determined what happens at the critical parameter value. The rule of thumb is "systems are ergodic at critical parameter values, with slow convergence to equilibrium." However there is evidence that some version of the model in Example 4 is not ergodic at the critical value, so the question is really up in the air (and very difficult).

For the remaining questions, the most appropriate setting is the spatially symmetric case with $\beta = \delta$. With these symmetry assumptions, we obtain those polling models that one would expect to behave most like physical systems, with the parameters β, θ and α taking the roles of temperature, potential, and external field strength (caution: in statistical mechanics, β is often used for reciprocal temperature). Based on experience with physical systems, one would hope to predict the ways in which these three parameters, together with the range N and the dimension d, affect critical behavior.

Five questions are listed, one for each of the parameters. The predicted answer, based on physical intuition, is given in a short discussion appended to each question. The author is not aware of any work with polling models that has been done on any of these problems. In spite of this, the first two at least seem tractable and worth thinking about. Solutions to them would probably involve some useful ideas. The last three questions are probably very hard, but also well worth some thought, since any progress on them, even in special cases, would lead to important insights into the causes of critical behavior. The last one especially is the author's personal favorite. It involves the role that dimension plays (see also Question 2). Some speculations are offerred on that particular question, for lack of anything more substantial.

Assume for Questions 5 - 9 that R is spatially symmetric and that $\beta = \delta > 0$.

QUESTION 5. Can non-ergodicity exist if $\alpha > 0$?

The answer should be "No", since in statistical mechanical systems like the Ising model, for example, phase transitions disappear in the presence of an external field. If $\alpha > 0$, then +'s should have enough of an advantage over -'s so that even if the initial state is "all -'s", islands of +'s which appear spontaneously due to noise will grow and spread throughout the system. Eventually, the islands will join together in large masses that will be indistinguishable in behavior and shape from the groups of -'s with which they intermingle. Such a scenario should lead to ergodicity. The reader should compare the discussion following Question 9.

If Question 5 has the answer we expect, then non-ergodicity can only exist when $\alpha = 0$. How should the remaining parameters be chosen if one is trying to produce a non-ergodic system? Our physical intuition says that the range should be large, the noise small, and the interaction strong. Thus we ask the following questions, expecting positive answers:

QUESTIONS 6 - 8. Assume that $\alpha = 0$ and that the remaining parameters are such that the system is non-ergodic. Will it still be non-ergodic if

6. the range N is increased?

7. the noise β is decreased?

8. the interaction parameter θ is made closer to $1/2$? (we naively assume that the strongest interaction occurs at $\theta = 1/2$. Perhaps there is some other value of θ which is optimum for causing non-ergodicity. If so, what is it? Does it depend on the other parameters?)

Of the three, Question 7 is probably the most interesting. As long as it remains unanswered, one does not know that there is a unique critical value β_c (the "critical temperature"), even if one shows that the system is non-ergodic for some value of β. To answer Question 7, one probably needs to prove that the correlations between sites are decreasing functions of β. This is known to be true for many statistical mechanical systems.

We finally come to the last question, the most intriguing and difficult of them all:

QUESTION 9. Is the following statement true? Let R be spatially symmetric, $\beta = \delta$, $\alpha = 0$, $\theta = 1/2$ (or θ close to $1/2$), and $N \geq 1$. Then the corresponding polling model is

a. ergodic for all $\beta > 0$ if $d = 1$.

b. non-ergodic for all sufficiently small $\beta > 0$, with no non-translation invariant equilibria, if $d = 2$.

c. non-ergodic for all sufficiently small $\beta > 0$, with both translation invariant and non-translation invariant equilibria, if $d \geq 3$. (The existence of non-translation invariant equilibria is called "phase seperation".)

The first part is just a special case of Question 2, but perhaps the symmetry that we have here makes for a more appropriate conjecture. The other parts are known to be true for certain statistical mechanics models (see Aizenman [1979] for $d = 2$ and Dobrushin [1973] for $d = 3$). The question has been put here in an ambitious form. No one has yet shown that non-ergodicity can exist at all for these models, let alone prove theorems about phase seperation.

Besides the analogies with statistical mechanics, is there any good reason to believe the statements in Question 9? Is there some decisive difference between the cases $d = 1$, 2 and 3 that one can link to critical behavior? We offer here some ideas about dimensions one and two, leaving higher dimensions to others more ambitious.

Consider the simplest case, namely $N = 1$. Start the system off with all -'s. In both cases, $d = 1$ or 2, we expect the following to happen. Since $\beta > 0$, spontaneous flips from - to + will occur at isolated sites. Occasionally, enough of these will occur near each other to form a small island of +'s that is somewhat self-sustaining, in the sense that each member of this island finds that at least half of the people in its reference group also belong to the island. This requires at least two sites in one dimension and at least four in two dimensions. For small $\beta > 0$, the size of such an island changes in a way that approximates a simple birth and death process. The size increases roughly at a rate proportional to the number of -'s that border the island and decreases roughly at a rate proportional to the number of +'s along the "shoreline". Of course this approximation is not exact, since the island can also grow by joining other islands or decrease when "lakes" of -'s form inside of it. We are interested in the total area (or length when $d = 1$) that is enclosed by the shoreline, so we ignore lakes until they actually break through the shoreline, forming "bays". Due to the symmetry between +'s and -'s, the formation of bays from lakes and the growth of the island due to joining with other islands tend to compensate each other as far as the total area enclosed by the shoreline is concerned, especially as the island gets large. Since we are only speculating, we will assume that we can ignore these "second-order" effects.

Thus we visualize an island as a connected set of +'s, ignoring any holes formed by lakes, where we define "connected" in the obvious way in terms of nearest neighbors. The island is surrounded by a sea of -'s, with no other islands in sight (truly a desert island). In this simplified picture, an island in one dimension is merely a string of consecutive +'s, with a shoreline consisting of two non-neighboring sites. The size increases at the same rate that it decreases, so that even though it will eventually disappear with probability one, the expected time until it does so is infinite. Because of the translation invariance of the model, it is not hard to imagine that islands will cover larger and larger areas and eventually take over half the system.

The situation in two dimensions is different. The shoreline is a closed path which continually changes shape. We will assume that all the sites along the shoreline are stable, in the sense that at least half of the members of the reference set of any shoreline site are contained in the island. Under this assumption, the shoreline will be a simple closed curve, and direction changes along the curve occur at right angles, or "corners". These corners are the only places where significant flips occur. Flips at other sites result in unstable sites which we are ignoring. There are two kinds of corners, which we call right corners and left corners. If we walk around the shoreline so that the island lies at our left, the right corners are those at which we make a right turn. Significant flips from - to + occur only at right corners, while those from + to - occur only at left corners. We are finally ready for the crucial fact: there are always four more left than right corners, so the sea always has an advantage over the island. We expect the number of corners to be roughly proportional to the length of the shoreline, which is roughly proportional to the square root of the total area enclosed (this last estimate is not important). Therefore, the area should behave like a birth and death process on the non-negative integers, with birth rate $c\sqrt{x}$ and death rate $c\sqrt{x} + 4$. Such a process hits 0 in finite expected time. Is this enough for the sea to keep the islands under control? If so, then the proportion of +'s will stay bounded below 1/2, producing an equilibrium in which there is a strict majority of -'s. By symmetry, there is another equilibrium in which +'s are in the majority, so the system is non-ergodic. The reader is invited to attempt to fill in the details.

BIBLIOGRAPHY

1. Aizenman, M. "Instability of phase coexistence and translation invariance in two dimensions", Phys. Rev. Letters, 43 (1979), 407 - 409.

2. Dobrushin, R.L. "Markov processes with a large number of locally interacting components", Problems of Information Transmission, 7 (1971), 149 - 164 and 235 - 241.

3. Dobrushin, R.L. "Investigation of Gibbsian states for lattice systems", Theor. Probab. Appl., 18 (1973), 253 - 271.

4. Durrett, R. and Gray, L. "Some peculiar properties of a particle system with sexual reproduction", to appear (1985).

5. Durrett, R. and Griffeath, D. "Supercritical contact processes on ℤ", Ann. Probab., 11 (1983), 1 - 15.

6. Gray, L. "The positive rates problem for attractive nearest neighbor spin systems on ℤ", Z. f. Wahr. verw. Geb., 61 (1983), 389 - 404.

7. Gray, L. "Duality for general attractive spin systems, with applications in one dimension", Ann. Probab., to appear (1985).

8. Gray, L. and Griffeath, D. "A stability criterion for attractive nearest neighbor spin systems on ℤ", Ann. Probab., 10 (1982), 67 - 85.

9. Harris, T. "Contact interactions on a lattice", Ann. Probab., 2 (1974), 969 - 988.

10. Holley, R. "Remarks on the FKG inequalities", Commun. Math. Phys., 36 (1974), 227 - 231.

11. Kesten, H. Percolation Theory for Mathematicians, Birkhauser, Boston (1982).

12. Stavskaya, O.N. and Pyatetskii-Shapiro, I.I. "On homogeneous networks of spontaneously active elements", Prob. kibernetiki, Nauka Moscow, 20 (1968), 91 - 106.

13. Sullivan, W.G. "A unified existence and ergodic theorem for Markov evolution of random fields", Z. f. Wahr. verw. Geb., 31 (1974), 47 - 56.

14. Toom, A. "Stable and attractive trajectories in multicomponent systems", Adv. in Probab., 6 (1979), 549 - 573.

SCHOOL OF MATHEMATICS
UNIVERSITY OF MINNESOTA
MINNEAPOLIS, MN 55455

Contemporary Mathematics
Volume 41, 1985

PHASE TRANSITION FOR A CLASS OF MARKOV PROCESSES ON $(\mathbb{N})^S$

Andreas Greven

ABSTRACT. We consider a class of Markov processes on $(\mathbb{N})^S$ $(S=\mathbb{Z}^n)$ with transitions of the type: motion of a particle, birth and death of a particle, extinction of all particles at a site and splitting of all particles at a site.

We compare the different processes in our class from the point of view of phase transition. We will define a notion of "phase transition" also in models with diverging or decaying (to 0) particle density, by looking at a renormalised respectively rescaled process.

Then we show how presence, absence and strength of the interdependence of the fate of different particles influences the longterm behaviour of the process and shapes the appearance of the phase transition.

Finally we raise the question how robust these features are against random perturbations of the evolution mechanism and also here our notion of phase transition for systems with diverging or decaying particle density proves to be useful.

1. INTRODUCTION TO THE PROBLEM. DESCRIPTION OF THE MODEL.

In this paper we analyse a class of infinite particle systems on $\mathbb{Z}^n$ from the point of view of phase transition. Here we mean by phase transition that the longterm behaviour of our system changes drastically, if the values of the parameters of the evolution or the dimension of the underlying lattice changes. Drastic change in the longterm behaviour means in this context that the system changes from: "forgetting" it's initial state as $t\to\infty$ to :"remembering" certain properties of the initial state (typically the particle density) for all times.

Our attempt to understand the phenomenon described above proceeds by comparing different models in our class under the perspective of phase transition. It will turn out that the phenomenon of phase transition is closely related to the degree to which the evolutions of different particles influence each other. This relation

AMS Subject Classification: 60 K 35

between interaction and phase transition will be analysed in all detail. The feature exhibited in our class of processes are fairly representative. Later we will also discuss the question: how robust are these features connected with the phase transition if we replace the spatially homogeneous evolution mechanism by evolutions taking place in a random environment?

Phase transition for systems on $\{0,1\}^S$ have been studied extensively in the literature, a survey can be found in [2]. Another example for the phenomenon of phase transition for a system on $(\mathbb{R}^+)^S$ can be found in [6].

The models we consider in this paper are Markov processes with state space $(\mathbb{N})^S$ $(S=\mathbb{Z}^n)$. $(\eta \in (\mathbb{N})^S$ is interpreted as follows: $\eta(x)=$ number of particles at a site x). The structure of our processes is motivated by population dynamics. We have the following kinds of transitions: motion of individual particles, birth and death of individual particles, extinction of all particles at a site and splitting of all particles at a site. So the "classical" systems of infinitely many Random Walks or Branching Random Walks belong also to our class of processes.

Let us now introduce our models precisely: We consider Markov processes $(\eta_t^\mu)_{t\in\mathbb{R}^+}$ on $(\mathbb{N})^S$ with initial distribution μ and time evolution according to the table below. About μ we will always assume that:

(i) μ is translation invariant and shiftergodic.

(ii) $\int \eta(x)\,d\mu < \infty$.

In the following table we list the possible transitions and the exponential rate at which they occur (the ingredients are explained below)

rate	transition	intuitive interpretation
$mp(x,y)\eta(x)$	$\eta \to \eta + (\delta_y - \delta_x)$	motion $x \to y$
$b_1\eta(x)$	$\eta \to \eta - \delta_x + \sum_y N_y \delta_y$	birth with displacement
$d_1\eta(x)$	$\eta \to \eta - \delta_x$	death
$b_2\tilde{\tilde{q}}(x,y)$	$\eta \to \eta + \eta(x)\delta_y$	splitting at a site, displacement
d_2	$\eta \to \eta - \eta(x)\delta_x$	extinction at a site

Here $p(x,y)$, $\tilde{q}(x,y)$, $\tilde{\tilde{q}}(x,y)$ are transition matrices, which are assumed homogeneous. We also assume that a certain convex combination (namely (8)) of these kernels is irreducible. b_1,b_2,d_1,d_2 are nonnegative real numbers.

To describe the births we have random variables N , giving the size of the offspring of a certain particle at a certain time and furthermore we have a random variable N_y which gives the number of those of the N children who move (instantaneous with their birth) from their place of birth, say x, to the site y. ($N=\sum_y N_y$). More precisely: Suppose the parent particle is located at site x. Then N_y denotes the number of outcomes y in an i.i.d. experiment of random length N, where the occurence of y has the probability $\tilde{q}(x,y)$. We will always assume $k:=E(N)<\infty$; $\mathrm{Prob}(N\in\{0,1\})=0$.

The existence and uniqueness (in the sense of being the limit of appropriate finite systems [8]) of processes fitting the given description can be established using slight modifications of techniques developed by Liggett and Spitzer in [8].

Later we discuss some effects occuring if we introduce a spatial structure of the evolution mechanism. This is done by replacing the constants $b_1,b_2,d_1,d_2,L(N)$ by a collection of random variables $\{b_1(x)\}_{x\in S},\dots,\{L(N(x))\}_{x\in S}$ $(L(\cdot)=La\omega(\cdot))$. These random variables are realised at time $t=0$ and with this realisation the process evolves for all times.

The models above include, or are continuous time versions of models, discussed earlier by Spitzer [10], Kallenberg [7], Griffeath [5], Greven [3],[4], Dawson and Fleischmann [1].

2. HOW TO STUDY (η_t^μ) FOR $t\to\infty$ VIA A NEW PROCESS (ξ_t^μ)

a) What are the interesting questions to ask about our process $(\eta_t^\mu)_{t\in\mathbb{R}^+}$? This depends of course on the behaviour of the particle density $E(\eta_t^\mu(x))$, namely wether we have diverging, decaying (to 0) or stable particle density. Therefore let us observe first: (compare [4], [3])

<u>PROPOSITION 1</u>: $E(\eta_t^\mu(x)) = E^\mu(\eta(o))\exp(b_1k+b_2-d_1-d_2-b_1), x\in S, t\in\mathbb{R}^+$ (1)

From now on we will use the abbreviation: $\delta = b_1k+b_2-d_1-d_2-b_1$.

Obviously our situation here is different from what we are used to with spin flip systems. In order to see interesting features of our process (η_t^μ) one has to consider a renormalised process in the case $\delta>0$ and a rescaled version of the process in the case $\delta<0$. In section 3a,b and 4 we will see that it is not appropriate to consider only the case where $\delta = 0$. In order to study the long-term behaviour of (η_t^μ) one should look at the following new process (ξ_t^μ) with statespace $(\mathbb{R}^+)^S$:

$$\xi_t^\mu(x) := \begin{cases} \eta_t^\mu(x)\ E(\eta_t^\mu(x))^{-1} & \text{if } \delta > 0 \\ \eta_t^\mu(x) & \text{if } \delta = 0 \\ \sum_{z \in I_t} \eta_t^\mu([\tfrac{1}{2} e^{|\delta| t/n} x] + z) & \text{if } \delta < 0 \end{cases} \tag{2}$$

Here I_t is a cube of the form $[a,b)^n$ such that $|I_t| \sim e^{|\delta| t}$.

This new process (ξ_t^μ) tells us how our process (η_t^μ) grows respectively dies out, it describes the fluctuations of the system around the dynamics of the averages (i.e. the particle density).

b) We are going to study the longterm behaviour of (ξ_t^μ) now. We know from Proposition 1 that:

$$E(\xi_t^\mu(x)) \underset{t\to\infty}{\to} E^\mu(\eta(o)) \qquad \sup_t E^\mu(\xi_t(x)) < \infty \tag{3}$$

Therefore the set $\{L(\xi_t^\mu)\}_{t \in \mathbb{R}^+}$ has (for $t \to \infty$) weak limit points in the space of probability distributions on $(\mathbb{R}^+)^S$. ($(\mathbb{N})^S$ for $\delta \leq 0$). The structure of the set of these limit distributions will lead us to the phenomenon of phase transition, but first we have to investigate another question, namely wether there is exactly one limit point in $\{L(\xi_t^\mu)\}_{t \in \mathbb{R}^+}$ for $t \to \infty$ and wether this limit point is the same at least for all initial distributions μ with the same particle density (remember μ is assumed shiftergodic!).

<u>THEOREM 1.</u> Define $\rho = \int \eta(x) d\mu$ and suppose that a weak limit point of $\{L(\xi_t^\mu)\}_{t \in \mathbb{R}^+}$ inherits from μ the property:

$$\lim_{t\to\infty} (E^\nu |\sum_y u_t(x,y)\eta(y) - \rho|) = 0 \tag{4}$$

(where $u_t(x,y) = \sum_{n=0}^\infty e^{-t} \frac{t^n}{n!} u^n(x,y)$ with a irreducible homogeneous kernel $u(x,y)$). Then we have:

a) $$L(\xi_t^\mu) \underset{t\to\infty}{\overset{w}{\to}} \nu_\rho \quad \forall \mu\text{: translation invariant, shift ergodic, } \int \eta(x) d\mu = \rho \tag{5}$$

b) In the case $\delta = 0$, ν_ρ is an invariant measure of the original process. ν_ρ is extremal in the set of all translation-invariant invariant measures with finite particle density.⌋

The role of (4) here is to show that successful couplings exist for certain initial distributions. The question is of course in which cases (4) is fulfilled. This is a very common, but usually difficult, problem in the theory of infinite particle systems. (Compare for example the situation for generalised potlatch processes or the exclusion process. In both these cases the problem has not been solved completely so far). Probably (4) holds always in our class of processes, but we can prove only the following reduction to a statement which covers many (but not all) situations we are interested in:

PROPOSITION 2:

$$\left(\sup_{t\in\mathbb{R}^+} (E^{\mu}(\xi_t^2(x)))<\infty\right) \Rightarrow \left((4) \quad \text{holds}\right) \tag{6}$$

We make a short remark on the techniques of proof used to obtain theorem 1. The proof involves a coupling argument as a main step: For two translation-invariant, shiftergodic measures with equal particle density we construct a process $(\hat{\eta}_t^{\bar{\mu},\bar{\bar{\mu}}})_{t\in\mathbb{R}^+}$ on $(\mathbb{N})^S \times (\mathbb{N})^S$ such that the marginals are versions of $(\eta_t^{\bar{\mu}})$ and $(\eta_t^{\bar{\bar{\mu}}})$, the initial distribution is $\bar{\mu} \otimes \bar{\bar{\mu}}$ and that as many transitions as possible occur simultaneously in both components. Then one shows that $L(\hat{\eta}_t^{\bar{\mu},\bar{\bar{\mu}}})$ for $t \to \infty$ concentrates (weakly) fast enough on the diagonal. Details can be found in [4].

3. PHASE TRANSITION

a) What do we mean by phase transition? We say a phase transition occurs if for certain values of the parameters (or "effective" dimension) of the model we have that a weak limit point ν_ρ of $\{L(\xi_t^{\mu})\}$ $(t\to\infty)$ equals $\delta_{\{\xi\equiv 0\}}$ and for others ν_ρ has the property $\int \xi(x)d\nu_\rho = \rho$. Another important change in the longterm behaviour of our system is of the form that we obtain measures ν_ρ with $\int \xi^2(x)d\nu_\rho = +\infty$ instead of $\int \xi^2(x)d\nu_\rho < \infty$ by changing the values of parameters of the evolution.

A first attempt to investigate this problem would be to study the second moments $E^{\mu}(\xi_t^2(x))$ and their behaviour for $t \to \infty$. If these moments are bounded by a constant for all times then we would know that there exists a unique limit point ν_ρ depending only on the particle density ρ of μ. This ν_ρ would also have the property $\int \xi(x)d\nu_\rho = \rho$. ((5),(6)).

In order to be able to do calculations with $E^{\mu}(\xi_t^2(x))$ we assume that our initial distribution μ has the property $\int \eta^2(x)d\mu < \infty$.

In our theorem about $E^{\mu}(\xi_t^2(x))$ we will need a continous time random walk (X_t) on $\mathbb{Z}^n$. (X_t) has the jumprate 1 and the transition kernel $\hat{r}(x,y)$ where:

$$\hat{r}(x,y) = 1/2(r(x,y)+r(y,x)) \tag{7}$$

$$r(x,y) = \left(\frac{m}{m+b_1k+b_2}\, p(y,x) + \frac{b_1k}{m+b_1k+b_2}\, \tilde{q}(y,x) + \frac{b_2}{m+b_1k+b_2}\, \tilde{\tilde{q}}(y,x)\right) \tag{8}$$

Furthermore we need the following quantities connected with (X_t):

$$G_{\hat{r}} = \sum_{n=0}^{\infty} \hat{r}^n \tag{9}$$

$$\lambda : (b_2+d_2)(m+b_1k+b_2)^{-1}\int_0^\infty e^{-\lambda t}\mathbb{E}\,1_{\{X_t^{(o)}=o\}}dt = 1 \tag{10}$$

Note that $(b_2+d_2)(m+b_1k+b_2)^{-1}\, G_{\hat{r}}(o,o) > 1$ implies that $\lambda > 0$. In terms of the random walk (X_t) we obtain a complete description of the behaviour of $E^\mu(\xi_t^2(x))$ for $t \to \infty$:

<u>THEOREM 2:</u>

<u>case I</u>: $b_1+b_2+d_2 = 0$

$$\sup_{t\in\mathbb{R}^+} E^\mu(\xi_t^2(x))) < \infty \tag{11}$$

<u>case II</u>: $b_2+d_2 = 0$ but $b_1 > 0$

$$\sup_{t\in\mathbb{R}^+} (E^\mu(\xi_t^2(x))) < \infty \quad \text{if: } \delta \neq 0 \text{ or } (X_t) \text{ is transient} \tag{12}$$

$$\lim_{t\to\infty} (E^\mu(\xi_t^2(x))) = +\infty \quad \text{if: } \delta = 0 \text{ and } (X_t) \text{ is recurrent} \tag{13}$$

<u>case III</u>: $b_2+d_2 > 0$

$$\sup_{t\in\mathbb{R}^+} (E^\mu(\xi_t^2(x))) < +\infty \text{ if} \begin{cases} \delta \geq 0 & \text{and } \frac{1}{2}\frac{b_2+d_2}{m+b_1k+b_2} < (G_{\hat{r}}(o,o))^{-1} \\ \delta < 0 & \text{and } \lambda < |\delta| \end{cases} \tag{14}$$

$$\lim_{t\to\infty} (E^\mu(\xi_t^2(x))) = +\infty \text{ if} \begin{cases} \delta \geq 0 & \text{and } \frac{1}{2}\frac{b_2+d_2}{m+b_1k+b_2} \geq (G_{\hat{r}}(o,o))^{-1} \\ \delta < 0 & \text{and } \lambda \geq |\delta| \end{cases} \tag{15}$$

We see that: if there is no interdependence between particles and we have also "conservation of the number of particles", then no phase transition occurs.

This is different in the presence of interdependence between different particles. The interdependence between particles is in our class of processes such, that they cause the buildup of clusters. (Compare part c) of this section), that is we have sets of sites with high numbers of particles and others with only few particles (always compared to the averages). We see also that a certain amount of cluster buildup is created already by the fact that particles are created and destroyed in a randomly fashion, this latter mechanism determines the behaviour in low dimensions while in higher dimensions the effects of the interdependence between particles are shaping the longterm behaviour of (ξ_t^μ).

<u>There is a very significant difference in the behaviour of systems with interdependence between particles</u> $(b_2+d_2 > 0)$ <u>and those</u>

systems without this interdependence. In the case of no interdependence a phase transition can occur only in the case $\delta = 0$ and the relevant parameter is essentially the dimension of the lattice. In the case of interdependence between particles a phase transition in the moment behaviour occurs for all values of $\mathrm{sign}(\delta)$, if $G_{\hat{r}}(o,o) \in (1,2)$. Here the interesting dimension is typically ≥ 3 where depending on the values of $G_{\hat{r}}(o,o)$ the behaviour of (ξ_t^{μ}) undergoes a drastic change if $(b_2+d_2)(m+b_1k+b_2)^{-1}$ changes. That the case $\delta<0$ plays a somewhat special role in the situation where $b_2+d_2>0$ is not surprising since for low particle density the evolution is similar to the Branching Random Walk.

Another widespread feature can be seen here, namely that there is a dichotomy between low and high dimensions: low dimensions tend to make a process (ξ_t^{μ}) forming clusters, while high dimensions tend to make the number of particles at different sites less correlated and to spread out more regularly.

If we are in the case where $\sup_t (E^{\mu}(\xi_t^2(x))) < \infty$ we can say something about the dependence structure of the field $\{\xi_{\infty}^{\mu}(x)\}_{x\in S}$:

THEOREM 3: Suppose that we are in the case $\sup_t (E^{\mu}(\xi_t^2(x))) < \infty$. Then

$$\begin{aligned} &E^{\nu_\rho}(\eta(x)\eta(y)) \geq \rho^2 \quad (> \text{ if } b_1+b_2>0 \text{ and } \tfrac{1}{2}(\tilde{q}(x,y)+\tilde{\tilde{q}}(x,y)) \neq \delta(x,y)) \\ &E^{\nu_\rho}(\eta(x)\eta(y)) \underset{|x-y|\to\infty}{\to} \rho^2 \end{aligned} \tag{16}$$

Note that here we need not distinguish between the cases $\delta>0$, $\delta<0$, $\delta=0$. (16) expresses a universal property of our class of processes.

We should make a short remark here how the results described above ((11)-(16)) can be proved. The crucial step is to obtain the asymptotic behaviour of $\{E(\eta_t^{\mu}(x)\eta_t^{\mu}(y)), E(\eta_t^{\mu}(z))\}_{x,y,z\in S}$ for $t\to\infty$, $|x-y|\to\infty$. The point is that the differential equations for these functions form a closed system. This closed system can be analysed by the iterative use of the relation: $V_t = U_t + \int_0^t U_{t-s}(L_V - L_U)V_s\,ds$ where L_V, L_U denote the generators of two semigroups $(V_t),(U_t)$. This relation allows us to express the evolution of the moments in terms of random walks of two particles with fluctuating mass. Details of the somewhat lengthy calculations can be found in [4].

b) Now we come to another very interesting difference in the behaviour of systems with interdependence between particles compared to those without. In the case of no interdependence we will see that the behaviour of the second moments $E^{\mu}(\xi_t^2(x))$ for $t\to\infty$ determines wether, $\nu_\rho = \delta_{\{\xi\equiv 0\}}$ or $\int \xi(x)d\nu_\rho = \rho$. This is not true in the case where there is interdependence between particles. (A similar picture has been observed for the Generalized Potlatch Process [6].)

We demonstrate these features using some particular examples: We compare an infinite system of Random Walks, Branching Random Walks and the Coupled Branching Process. The fact that we have to consider special cases here reflects the fact that in order to show that $L(\xi_t^{\mu}) \underset{t\to\infty}{\to} \delta_{\{\xi\equiv 0\}}$ one has to use essentially the special properties of the evolution mechanism. This is different to the situation in theorem 1-3.

THEOREM 4: Denote by ν_ρ measures on $(\mathbb{R}^+)^S$ with $\int \xi(x)d\nu_\rho = \rho$

a) case: $b_1+d_2+b_2 = 0$

$$L(\xi_t^{\mu}) \underset{t\to\infty}{\overset{w}{\to}} \nu_\rho$$

b) case: $b_1 > 0$ but $b_2+d_2 = 0$

$$L(\xi_t^{\mu}) \underset{t\to\infty}{\overset{w}{\to}} \delta_{\{\xi\equiv 0\}} \quad \text{if } X_t \text{ is recurrent and } \delta = 0 \ (7) \tag{17}$$

$$L(\xi_t^{\mu}) \underset{t\to\infty}{\overset{w}{\to}} \nu_\rho \quad \text{if } X_t \text{ is transient or } \delta \neq 0 \tag{18}$$

c) case: $b_2 = 0$ but $b_1 > 0$, $d_2 > 0$. (Coupled Branching Process)

Assume furthermore that $\delta = 0$. Then the following holds: There exists some $\gamma\in(1,\infty)$ such that:

$$L(\xi_t^{\mu}) \underset{t\to\infty}{\overset{w}{\to}} \delta_{\{\xi\equiv 0\}} \quad \text{if } \frac{1}{2}\frac{d_2}{m+b_1k} > \gamma(G_{\hat{r}}(o,o))^{-1} \tag{19}$$

$$L(\xi_t^{\mu}) \underset{t\to\infty}{\overset{w}{\not\to}} \delta_{\{\xi\equiv 0\}} \quad \text{if } \frac{1}{2}\frac{d_2}{m+b_1k} < \gamma(G_{\hat{r}}(o,o))^{-1} \tag{20}$$

(Remark: It can be shown that there are kernels $r(x,y)$ where γ is such that (19) and (20) hold for feasible values of the parameters).

REMARK: γ can be determined explicitly in terms of a random walk problem. (Compare (27)).

Note again that relevant for phase transition is the strength of the interdependence between particles which is measured by d_2 in this case. Probably the analogue of (19) (20) hold also in the case $\delta < 0$, $\delta > 0$.

In order to shed some more light on the different behaviour in the cases $d_2 = 0$, $d_2 > 0$ we study also the higher moments of (ξ_t^{μ}) which will show how with increasing d_2 the tendency to form clusters build up slowly until the critical point is reached. To do so it is convenient to concentrate on the case $\mu = \delta_{\{\eta\equiv 1\}}$ and $N\leq$const.

Then we obtain the following theorem:

THEOREM 5: Suppose that $\delta = 0$. Then we observe the following three different cases: ($m = 2,3....$)

a) case $b_1 + b_2 + d_2 = 0$

$$\sup_t \left(E^{\mu}(\xi_t^m(x))\right) < \infty \tag{21}$$

b) case $b_1 > 0$ but $b_2 + d_2 = 0$

$$\sup_t \left(E^{\mu}(\xi_t^m(x))\right) < \infty \text{ if } \delta \neq 0 \text{ or } X_t \text{ is transient} \tag{22}$$

$$\lim_{t\to\infty} \left(E^{\mu}(\xi_t^m(x))\right) = +\infty \text{ if } \delta = 0 \text{ and } X_t \text{ is recurrent} \tag{23}$$

c) case: $d_2 > 0$, $\delta = 0$ (but $b_2 = 0$) and X_t transient

There exists a sequence $0 < \ldots < d_m^* \leqq \ldots \leqq d_3^* < d_2^* < d_0^*$ (= $=(\gamma^{-1} \frac{1}{2} \frac{1}{m+b_1k+b_2} G_{\hat{r}})^{-1}$ such that:

$$\sup_t \left(E^{\mu}(\xi_t^m(x))\right) < \infty \quad \text{if} \quad \frac{d_2}{m+b_1k} < d_m^* \tag{24}$$

$$\lim_{t\to\infty} \left(E^{\mu}(\xi_t^m(x))\right) = +\infty \quad \text{if} \quad \frac{d_2}{m+b_1k} > d_m^* \tag{25}$$

It is conjectured that the picture as in c) above is true for all models in our class with interdependence between the particles, unfortunately no rigorous proof exists at the moment.

c) In this section we try to explain how the forming of clusters (which causes the phase transition) arises from the interdependence between the evolution of different particles. We look at the case c) of the Coupled Branching Process for that purpose. How do we understand the clustering in this situation intuitively? If we find a particle of the Coupled Branching Process at site 0 at time t then at certain sites and times no site-killing took place, which gives another particle using a similar path also the chance to survive the site-killing, so that clusters appear. This clustering has to overcome the tendency of particles to be spread out in space by the displacement mechanism. The tendency of spreading out can be described in terms of two independent versions of random walks (with kernels $r(x,y)$).

Let us explain the above reasoning a little bit more precise. For this purpose it is useful to construct the Coupled Branching Process in the following way: We take a Branching Random Walk (with parameters m, b_1, k, d_1 the same as the process we want to construct) and give every particle one of two colours: black or white. We use the rule that colours are inherited. At time $t = 0$ every particle is white. At each site we have a Poisson process with rate d_2 which gives us the times where all particles at that site are coloured black. Then we have for the Coupled Branching Process (η_t^μ) the following representation:

$$\eta_t^\mu(x) = \#\{\text{white particles at site } x \text{ at time } t\} \qquad (26)$$

Now if we observe at time t at site 0 at least one white particle we can ask what is the expected number of white particles under that condition. The Coupled Branching Process will die out if and only if this conditional expectation diverges for $t \to \infty$.

To proceed further one proves the following criterion:

$$\left(E(\eta_t^\mu(o) \mid \eta_t^\mu(o) > 0) \underset{t\to\infty}{\to} \infty \text{ a.s. } \right) \Leftrightarrow$$

$$\left(F_t(\zeta) := E\Big(\exp\big(d_2 \int_0^t 1_{\{\hat{X}_s^o = \zeta_s\}} ds\big)\Big) \underset{t\to\infty}{\to} \infty \qquad \zeta\text{-a.s. } \right) \qquad (27)$$

Here ζ is a path of the random walk with kernel $r(x,y)$ and jump-rate $m+b_1 k$ starting at 0 and $(\hat{X}_t^o)$ is another independent version of that random walk. Of course we have:

$$(E_\zeta F_t(\zeta) \le K < \infty, \forall t \in \mathbb{R}^+) \Rightarrow (F_t(\zeta) \le K(\zeta) < \infty \text{ a.s., } \forall t \in \mathbb{R}^+)$$

This gives us a sufficient condition for $\{L(\eta_t^\mu)\}$ to have nontrivial weak limitpoints. This condition is, as one easily checks, equivalent to $(m+b_1 k)^{-1} d_2 G_{\hat{r}}(o,o) < 1$, which was the condition for having $\sup_t(E^\mu(\eta_t^2(x))) < \infty$. It is not true however that $E_\zeta(F_t(\zeta)) < \infty$ is necessary for $F_t(\zeta) \le K(\zeta) < \infty$! Of course we see also immediately that if X_t is recurrent that then $L(\eta_t^\mu) \underset{t\to\infty}{\to} \delta_{\{\eta\equiv 0\}}$ (Note X_t is up to timescaling distributed as $\hat{X}_t^o - \hat{X}_t'^o$ where $\hat{X}_t^o, \hat{X}_t'^o$ are independent versions of the walk since in that case

$$\int_0^\infty 1_{\{\hat{X}_s^o - \zeta_s = 0\}} ds = +\infty \qquad \zeta\text{-a.s}$$

d) Now we relate the facts described so far to work done in other models. The class discussed here and the dynamic Ising model show the common feature that the interaction strength is the essential parameter for phase transition once the dimension is large enough (respectively $G_{\hat{r}}(o,o) \in (1,2)$).

The Branching Random Walk and Generalised Potlatch (as introduced in [6]), have in common that the particle number (mass) is not preserved but fluctuates randomly. In the potlatch case however we have interdependence (in the sense that η_t^μ with $\mu = \delta_{\{\eta\equiv 1\}}$ is not a sum of independent $(\eta_t^{\nu_x})$ where $\nu_x = \delta_{\{\eta = \delta_x\}}$) so that the ergodic behaviour of the two systems shows different patterns just in the way we would expect from what we found in our class of processes: The phase transition occurs for the Branching Random Walk only on a submanifold of the parameterspace, namely $\delta = 0$, and the dimension is the only relevant parameter, where as in the case of the Potlatch process the variability of the masschanging mechanism is the relevant parameter and the sign of δ is unimportant.

It is interesting to note that the situation described in theorem 5c) seems to hold also in the case of the Generalised Potlatch process.

There is however a model on $(\mathbb{N})^S$ where it seems at the first look that the point of phase transition is determined by the second moments and where we have also interdependence between particles. This is a model occuring in [8]: Particles perform Random Walks in the following way: At rate 1 a clock rings at a site say x. Then a site y is chosen with probability $p(x,y)$. Now every particle flips a coin with success-probability p to decide wether to jump or to stay. If $p = 1$ the system will die out locally, while for $p < 1$ a nontrivial equlibrium with finite second moments is approached.

But this case is comparable to take our class and compare processes with $b_1 > 0$ and $b_1 = 0$. If (X_t) is recurrent then we would observe a phase transition in b_1 and the critical point is determined by the second moments.

What this shows is that one should talk about phase transition only if one compares a one parameter set of models such that all types of transitions are possible for all the models. In the above example the transition for $p < 1$, $p = 1$ are of a different type. So it is still an open problem if there could exist an evolution on $(\mathbb{R}^+)^S$ where we have phase transition along a continuously changing parameter of the evolution mechanism, such that the critical point is also the point where the n-th moments start to diverge for $t \to \infty$.

There is another class of models exhibiting phase transition and where comparisons inside that class are very interesting. These are the so-called nearest neighbour particle systems on $\{0,1\}^{\mathbb{Z}}$ (the contact process is a special case). They are discussed and compared in a variety of papers by F. Spitzer and by T. Liggett. Compare [9] for references and newest results.

4. EVOLUTIONS IN A RANDOM ENVIRONMENT

a) An important question arising now is which of the features of our processes, as described in section 3 are stable against perturbations of the evolution mechanism. Here we do not think so much about a perturbation of the homogeneity in time but in space. Consider for example the case of a Branching Random Walk (i.e. $b_2+d_2=0$ $b_1>0$). It seems more realistic to assume that the offspring distribution $L(N)$ is really site dependent and the evolution mechanism therefore has a nontrivial spatial structure. So we would like to consider the evolution as described earlier in section 1 but with a collection $\{L(N(x))\}_{x\in S}$ assigned at random at time $t=0$ instead of a single distribution $L(N)$. Denote by $P_x(\cdot) = L(N(x))$. Now we want to investigate wether the structure of the weak limit points of $\{L(\xi_t^\mu \mid \text{random environment})\}_{t\in\mathbb{R}^+}$ for $t\to\infty$ is different from the one we observe for an analogue process in a homogeneous medium. D. Dawson and F. Fleischmann considered such a question in a discrete time setting and under the restriction $\sum_{n=0}^{\infty} P_x(n)n \equiv \text{const}$, $b_1 E(N(x))-d_1 = b_1$. In this case it turns out that for an evolution in a random environment, $\sum_n^{\infty} P_x(n)n^2 < \infty$, the situation is the same as described for a homogeneous medium in (12), (13) and we can prove this with different methods for continuous time setting as well. If however the average (over ω) tails of $P_x(\cdot)$ are thick enough (i.e. $E_\omega(\sum_{n=0}^{\infty} P_x(n)\lambda^n) \underset{\lambda\to 0}{\sim} \lambda^\alpha$ for α small), a process with nontrivial equilibrium may suddenly have only trivial equilibria. If we allow even $\sum_{n=0}^{\infty} P_x(n)n$ to be nonconstant in x then another change in the longterm behaviour of (η_t^μ) occurs. If $p(x,y)$, $\tilde{q}(x,y)$ are symetric we have the following dichotomy:

<u>THEOREM 6:</u> Let $\{P_x(\cdot)\}_{x\in S}$ be an i.i.d. collection of probability distributions on $\mathbb{N}$. Assume that $\sum_{n=0}^{\infty} P_x(n)n \not\equiv \text{const}$ (in x) and define $\lambda := b_1 \operatorname{supess}_\omega(\sum_{n=0}^{\infty} P_x(n)n) - d_1 - b_1$. Then

$$L_\eta(\eta_t^\mu(x)) \underset{t\to\infty}{\overset{w}{\to}} \begin{cases} \delta_{\{\eta\equiv+\infty\}} & \text{for } \lambda>0 \\ \delta_{\{\eta\equiv 0\}} & \text{for } \lambda\leq 0 \end{cases} \tag{28}$$

This means there exist no equilibria of the process whatever the values of b_1 and d_1 are (as long as $\sum_n P_x(n)n \not\equiv \text{const}$).

So we observe that there are changes in the patterns of the longterm behaviour of our processes if they evolve in a random environment instead of a homogeneous medium. What we can hope to be true though is that if the random environment does not fluctuate too much then the behaviour of the process (ξ_t^μ) for $t\to\infty$ paralells the one for a homogeneous medium. For example one would like to prove that for Branching Random Walks in a random environment such that: $\sum_{n=0}^{\infty} P_x(n)n^2 \leq K < \infty \ \forall x\in S$, still (12,(13);(17),(18) holds.

Only little is known in that direction so far. We state now two partial results obtained by the author in [3].

THEOREM 7: Let $\{P_x(\cdot)\}_{x\in S}$ be an i.i.d. collection of probability distributions such that $\sum_n P_x(n)n^2 \leq K < \infty$ a.s.. We set $L(N(x)) = P_x(\cdot)$. Assume $b_2=d_2=0$ and $\tilde{q}(x,y)=\delta(x,y)$. Then we have:

$$L(\xi_t^\mu \mid \text{random environment}) \underset{t\to\infty}{\overset{w}{\to}} \delta_{\{\xi\equiv 1\}} \tag{29}$$

THEOREM 8: Let $\{d_1(x)\}_{x\in S}$ be an i.i.d. collection of random variables with values in $\mathbb{R}^+$. Then for $b_2(x)\equiv 0$ and $p(x,y),\tilde{q}(x,y)$ symetric we have in the case $\delta\geq 0$:

$$\lim_{t\to\infty} \log \frac{1}{t} \hat{E}^\mu(\xi_t^2(x)) = \begin{cases} = 0 & \text{for } \frac{1}{2}\frac{\bar{d}_1}{m+b_1k} \leq G_{\hat{r}}(o,o)^{-1} \\ > 0 & \text{for } \frac{1}{2}\frac{\bar{d}_1}{m+b_1k} > G_{\hat{r}}(o,o)^{-1} \end{cases} \tag{30}$$

where $\bar{d}_1 = \operatorname{infess}_\omega (d_1(x))$.

b) If we could prove that the fluctuations around the averages (i.e. ξ_t^μ) are not disturbed by the perturbation of the evolution mechanism the question still remains wether and how the behaviour of the averages itself is changed. Here the changes are quite significant and another curious phenomenon can be observed for evolutions in a random environment and that is the difference between the behaviour of systems with a symetric kernel $r(x,y)$ and those with a $r(x,y)$ (compare (8)) with drift.

Suppose that we have an evolution where $b_1,b_2,d_1,d_2,L(N)$ are replaced by i.i.d. collections $\{b_2(x)\}_{x\in S},\{d_1(x)\}_{x\in S},\{d_2(x)\}_{x\in S}$ and $\{L(N(x))\}_{x\in S}$ $(L(n(x)) = P_x(\cdot))$. We have to redefine $r(x,y)$ now:

$$r(x,y) := \varepsilon^{-1}(x)(mp(y,x)+b_1(\sum_n P_y(n)n)\tilde{q}(y,x)+b_2(y)\tilde{\tilde{q}}(y,x))$$

$$\varepsilon(x) := m+\sum_y (b_1(\sum_n P_y(n)n)\tilde{q}(y,x)+b_2(y)\tilde{\tilde{q}}(y,x))$$

Define furthermore:

$$\delta(x) := b_1\sum P_x(n)n+b_2(x)-d_1(x)-d_2(x) \qquad \lambda := (\sup_x \delta(x))-b_1 \tag{31}$$

$$\hat{E}(x) := E(X \mid \text{random environment}) \tag{32}$$

THEOREM 9: Assume that $\delta(x) \not\equiv \text{const}$, $\delta(x) \leq K < \infty$ a.s..
If $r(x,y)$ is symetric, then:

$$\lim_{t\to\infty} \frac{1}{t} \log(\hat{E}(\eta_t^\mu(x))) = \lambda \quad \text{a.s.} \tag{33}$$

If $r(x,y)$ is self-avoiding i.e. $\{r(x,\cdot)\}_{x\in S}$ are concentrated on a quadrant of $\mathbb{Z}^n$, then:

$$E_\omega(\delta(x)) \leq \overline{\lim_{t\to\infty}}(\frac{1}{t} \log \hat{E}(\eta_t^\mu(x))) < \lambda \quad \text{a.s.} \tag{34}$$

That means, that if we compare the behaviour of the particle density for: the system in a homogeneous medium with $b_2 = E_\omega(b_2(o)),\ldots$ and the system evolving in random environment, then the difference in the growthrate of the particle density is maximal for symetric systems.

REMARK: The assumption for (34) can be weakened to the following condition (which implies the transience of $r(x,y)$).

Denote by (X_n) the random walk with kernel $r(x,y)$.

$$R(n) := \{x \mid \exists i : X_i = x\} \tag{35}$$

We require now that there exists $\delta, \delta' > 0$ such that:

$$\text{Prob}\left(\frac{|R(n)|}{n} \geqq \delta\right) \leqq 1-e^{-\delta' t} \tag{36}$$

An important open problem here is to calculate the exponential growth rate of the particle density in the asymetric situation and to analyse its dependence on the strength of the drift of $r(x,y)$.

REFERENCES

[1] Dawson, D.; Fleischmann, K.: On spatially homogeneous branching processes in a random environment, Math.Nach. 113 (1983).

[2] Durrett, R.: An Introduction to Infinite Particle Systems. Stochastic Processes and their Applications 11 (1981).

[3] Greven, A.: On a class of Infinite Particle Systems evolving in a Random Environment. To appear: Proceedings of the workshop: Stochastic Spatial Processes, Heidelberg 1984 (Springer, Lecture Notes in Biomathematics).

[4] Greven, A.: Critical phenomenon for the Coupled Branching Process. Preprint (last version Dec. 1984).

[5] Griffeath, D.: The binary Contact Path Process; Ann.Prob. II no. 3 (1983).

[6] Holley, R.; Liggett, T.: Generalized Potlatch and Smoothing Process. Zeitschrift f. Wahrscheinlichkeitstheorie verw. Gebiete 55 (2981).

[7] Kallenberg, O.: Stability of critical cluster fields. Math. Nachr. 77 (2977).

[8] Liggett, T.; Spitzer, F.: Ergodic Theorems for Coupled Random Walks and other systems with locally interacting components. Zeitschrift f. Wahrscheinlichkeitstheorie verw. Gebiete 56 (1981).

[9] Liggett, T.: Finite nearest particle systems. Preprint 1984.

[10] Spitzer, F.: Interaction of Markov Processes. Advances in Mathematics 5 (1970).

Andreas Greven
Universität Heidelberg
Insitut für Angewandte Mathematik
Im Neuenheimer Feld 294
D-6900 Heidelberg
West-Germany

Contemporary Mathematics
Volume 41, 1985

LARGE DEVIATIONS IN SUBADDITIVE PROCESSES AND FIRST-PASSAGE PERCOLATION

Geoffrey Grimmett

ABSTRACT. The principal theme of this paper is the study of large deviations in subadditive processes, with particular reference to first-passage percolation. For subadditive processes in general the picture is quite incomplete, although certain results are available for processes which either are independent or satisfy a condition of association. Quite a lot is known on the other hand about large deviations in first-passage percolation.

There is a secondary theme which runs throughout the paper, and this is the relationship between probability theory and operations research. We give two examples of the use of probabilistic ideas in the design of algorithms, both related to the problem of graph-colouring. Our main applications of subadditive ergodic theory are to the bin-packing problem of operations research and to first-passage percolation. The latter process has as dual the famous network flow process of Ford and Fulkerson, and thus subadditive theory has useful applications to problems of random flows through networks, including network flows and electrical flows.

CONTENTS

1. INTRODUCTION. Nearly twenty years have elapsed since Hammersley and Welsh (1965) formulated the idea of a subadditive stochastic process whilst studying first-passage percolation. In the intervening period many subadditive processes have been discovered in a variety of guises, and the general theory of such processes has been developed to some considerable degree. The main landmarks in this development are Kingman's ergodic theorem and the Hammersley-Kesten theory of superconvolutive sequences of distributions. In addition

1980 Mathematics Subject Classification. Primary: 60F10, secondary: 60K35, 90B15.

there is the central limit theorem of Ishitani (1977) and the improved ergodic theorem of Liggett (1984), to mention but two recent advances. Only incomplete progress has been made towards understanding small and large deviations in subadditive processes. For example, we know of no interesting application of Ishitani's theorem, and Hammersley's theory of large deviations for independent processes lacks a condition for the rate function to be *strictly* positive. The principal purpose of this paper is to discuss large-deviation results for subadditive processes, with particular reference to the problem of first-passage percolation.

This paper has a secondary theme also, and this is to say something about the use of probability theory in studying problems of operations research. First-passage percolation is only one of many random processes which arise naturally out of operations research, and we shall see several other such processes in Sections 2 and 3. In Section 2 we discuss two aspects of the relationship between probability and the design of algorithms. The first of these is the use of "randomized" algorithms to solve deterministic problems, and our example of such an algorithm is Welsh's "antivoter" method for colouring graphs; the second aspect is the study of the typical behaviour of algorithms, and we use as an example the asymptotic behaviour of the greedy graph-colouring algorithm applied to large random graphs.

Section 3 contains an account of the large-deviation theory of subadditive processes, together with several important open problems. As an example of subadditive theory in action, we describe a simple application to the bin-packing problem of operations research.

Finally we return to first-passage percolation, for which many of the open questions about subadditive processes may be answered. For this process there is a complete large-deviation theorem but currently no central limit theorem. This large-deviation theorem has various consequences for processes involving flows through random media, such as network flows and electrical flows, and the final section of the paper contains sketches of some of these consequences.

2. PROBABILITY AND OPERATIONS RESEARCH. The relationship between probability and operations research is rich and fruitful. We mention two aspects of this relationship here, and hope that our examples indicate something of the ways in which the two subjects are bound together. First we give an example of the use of so-called "randomized algorithms" to solve deterministic problems of optimization, and secondly we discuss the behaviour of algorithms when assessed according to "typical" rather than "worst-case" instances of the underlying problems. Rather than discuss these matters in abstraction, we present two particular examples involving random processes on graphical networks; both of

these examples are concerned with the problem of graph-colouring. For more general discussions of such questions the reader may refer to Jaffe (1984), Welsh (1983) and Karp (1976).

Let G be a finite graph without loops. Let k be a positive integer and suppose that $c_1, c_2, \ldots, c_k$ are distinct colours. A k-*colouring* of G is an assignment of colours to the vertices of G in such a way that no pair of adjacent vertices have the same colour. The *chromatic number* $\chi(G)$ is the smallest value of k for which G has a k-colouring. The problem of determining whether G has a k-colouring is NP-complete for $k \geq 3$, and so is the problem of finding $\chi(G)$ (see Garey and Johnson (1979) for an account of NP-completeness).

Randomized algorithms are recognized as powerful tools for solving difficult optimization problems. Welsh (1984) has found such an algorithm for determining whether a graph G has a k-colouring, and his algorithm seems to work fairly quickly and efficiently in practice. We sketch this briefly here for the case when k = 3; it will be clear how to adapt the method for larger values of k. The approach is based upon observing the evolution of a type of *antivoter process* on G, and waiting for it to visit a particular low-energy state. Suppose that the three available colours are blue (b), red (r) and green (g). The process in question is a random process $\xi = \{\xi_t : t \geq 0\}$ indexed by time, in which $\xi_t(i)$ equals either r, b or g for each time t and vertex i of G, and we suppose that the process evolves in the following way. For each $t \geq 0$, let X_t be the set of vertices i of G with the property that the colour of i is the same as the colour of at least one of its neighbours. We are provided with an exponential alarm clock, and ξ changes its value whenever this clock rings. If the clock rings at time t then a vertex is selected at random from X_t and its colour flips to colour c with probability

$$\pi(c) = \frac{\theta^{\kappa}}{\theta^{\beta}+\theta^{\rho}+\theta^{\gamma}} \qquad \text{for} \quad c = b,r,g,$$

where κ (respectively β, ρ, γ) is the number of neighbours with colour c (respectively b, r, g) and θ is a constant of the process satisfying $0 < \theta < 1$. The flip rates here have been chosen in such a way that the colour of each vertex is repelled by similarly coloured neighbours. 3-colourings of G correspond to absorbing states of ξ. Thus Welsh's algorithm for deciding whether G is 3-colourable proceeds as follows. Set the process ξ running. If, before a pre-determined time T, the process ξ is absorbed in a state in which no pair of neighbours has the same colour, then we decide that G is 3-colourable; if this does not occur, then we decide that G is not 3-colourable. There is a certain probability $\pi(T)$ that this decision rule reaches the false conclusion that G is not 3-colourable, and clearly $\pi(T) \to 0$ as $T \to \infty$. We may control $\pi(T)$ by choosing T large enough. Unfortunately, little is currently known about the random time which elapses before the antivoter process passes through a prescribed state of low energy, and so it is not yet known how large T should be taken to guarantee that $\pi(T)$ is sufficiently small. For a further discussion of this algorithm and for some related conjectures, we refer the reader to Welsh (1984).

Next we present an example of the average behaviour of algorithms, and this also arises out of the problem of graph-colouring. The best algorithm currently known for finding the chromatic number of an n-vertex graph has a very long worst-case run-time. If we are prepared to accept a colouring which uses more than the minimal possible number of colours then we may be able to find this rather faster. However, it turns out to be difficult to find a k-colouring of G for any value of k satisfying $\chi(G) \leq k < \rho\chi(G)$ where $\rho < 2$, in the sense that this problem is NP-hard for any given value of ρ (see Garey and Johnson (1979, p.144)). A simple example of a non-optimal graph-colouring algorithm is the so-called "greedy algorithm", described as follows. Let the vertices of G be labelled in some arbitrary order as $v_1, v_2, \ldots, v_n$ and let $c_1, c_2, \ldots$ be the colours available for colouring these vertices. We colour the vertices in increasing lexicographic order, colouring v_1 with colour c_1, and more generally colouring v_j with c_i where i is the least integer such that v_j is adjacent to no vertex already coloured c_i. We write $\Gamma(G)$ for the number of colours used by this greedy algorithm to colour the vertices of G, and we note that $\chi(G) \leq \Gamma(G)$. Certainly the greedy algorithm is fast, running in $O(n^2)$ time, but it generally uses more colours than necessary; indeed the ratio $\Gamma(G)/\chi(G)$ has order n in the worst case (see Johnson (1974)). It turns out that the "typical" discrepancy between $\chi(G)$ and $\Gamma(G)$ is not nearly so bad as the worst-case discrepancy. Assessment of "typical" behaviour involves putting a probability measure on the set of n-vertex graphs, and it may not be easy to decide which measure is appropriate. One simple measure is constructed as follows. Fix p satisfying $0 < p < 1$, and let

$$P(G) = p^e(1-p)^{N-e}$$

for all n-vertex graphs G, where e is the number of edges of G and $N = \binom{n}{2}$; this is the random graph model of Erdös and Rényi (see Grimmett (1983) for a recent review of random graphs). Writing χ_n and Γ_n for $\chi(G)$ and $\Gamma(G)$ when G is chosen according to the above probability measure, it turns out that

$$\frac{\log n}{n} X_n \to \tfrac{1}{2} \quad \text{in probability,}$$

and

$$\frac{\log n}{n} \Gamma_n \to 1 \quad \text{in probability,}$$

as $n \to \infty$, where the logarithms are calculated to the base $(1-p)^{-1}$. Thus, on average, the greedy algorithm uses approximately twice as many colours as are necessary. See Grimmett and McDiarmid (1975), Bollobás and Erdös (1976) and Korshunov (1980) for proofs of these facts. More information about graph-colouring algorithms and their behaviour may be found in Johnson (1974) and McDiarmid (1979, 1984).

3. SUBADDITIVE PROCESSES. A *subadditive process* is a family $X = \{X_{mn} : 0 \le m < n, m,n = 0,1,2,\ldots\}$ of random variables satisfying the following three conditions:

(3.1) *subadditivity*: $X_{rt} \le X_{rs} + X_{st}$ whenever $r < s < t$,

(3.2) *stationarity*: the joint distributions of $\{X_{mn} : 0 \le m < n\}$ are the same as the joint distributions of $\{X_{m+1,n+1} : 0 \le m < n\}$,

(3.3) *moment condition*: $E|X_{0n}| < \infty$ for each n, and $E(X_{0n}) > -\alpha n$ for some $\alpha \in \mathbb{R}$ and all n.

A subadditive process X is called *independent* if the random variables $X_{ab}, X_{cd}, \ldots, X_{ef}$ are independent whenever $a<b\le c<d\le\ldots\le e<f$.

We denote by $\mathcal{T}$ the σ-field of events defined in terms of X and invariant under the shift $X_{mn} \to X_{m+1,n+1}$. The following ergodic theorem is the principal result about the asymptotic behaviour of a subadditive process.

THEOREM 1. *If* X *is a subadditive process then*

$$\xi = \lim_{n\to\infty} \frac{1}{n} X_{0n} \tag{3.4}$$

exists almost surely and in L^1. *If* $\mathcal{T}$ *is trivial then* $\xi = \gamma$ *almost surely, where*

$$\gamma = \lim_{n\to\infty} \frac{1}{n} E(X_{0n}). \tag{3.5}$$

The constant γ, given by (3.5), is called the *time constant* of the subadditive process X.

In particular, if X is an independent subadditive process then $\mathcal{T}$ is trivial, so that the limit ξ in (3.4) is equal to the time constant γ almost surely.

An important step in the original proof of Theorem 1, due to Kingman (1968, 1976) is to show the existence of a subadditive process A which lies beneath X in the sense that

$$A_{mn} \leq X_{mn} \qquad \text{for all } m < n \tag{3.6}$$

and with the further properties that $E(A_{01}) = \gamma$, the time constant of X, and that A is *additive*, so that

$$A_{rt} = A_{rs} + A_{st} \qquad \text{for } r < s < t.$$

Other proofs of Theorem 1 avoid this step, thereby shortening Kingman's proof but presenting less information about the structure of the subadditive process. The existence of such an additive process is of course interesting and useful in its own right. Perhaps the most recent version of the ergodic theorem is due to Liggett (1984), who has shown that the conclusions of Theorem 1 are valid subject to a weaker set of conditions than (3.1) - (3.3); there are several interesting examples of processes which satisfy Liggett's conditions but not those of Kingman's theorem.

We present one example of the subadditive ergodic theorem in action, and this example is taken from the field of operations research. Consider the so-called *bin-packing problem*. Suppose that $y_1, y_2, \ldots, y_n$ are real numbers, each lying in the interval $[0,1]$, and that we are provided with a sequence $\beta_1, \beta_2, \ldots$ of bins, each of which has capacity 1. For $m = 1,2,\ldots$, we say that there is a *packing* of $\beta_1, \beta_2, \ldots, \beta_m$ if each y_i may be assigned to one of these bins in such a way that the sum of the y's in each bin does not exceed the capacity of that bin. We write $b = b(y_1, y_2, \ldots, y_n)$ for the smallest value of m for which there exists a packing of $\beta_1, \beta_2, \ldots, \beta_m$. It is a well-known problem of operations research to determine b for a given set of y's; see Garey and Johnson (1979, pp.124,226) for an account of the computational complexity of the bin-packing problem. It is interesting to ask how large b will be if the y's are in some sense "typical", and to this end we shall suppose that we are provided with an infinite sequence $Y_1, Y_2, \ldots$ of random variables which are independent and identically distributed, taking values in $[0,1]$. Let B_{mn} be the minimum number of bins with unit capacity into which the numbers $Y_{m+1}, Y_{m+2}, \ldots, Y_n$ may be packed. It is clear that $B = \{B_{mn} : 0 \leq m < n\}$ is an independent subadditive process, and hence

$$\frac{1}{n} B_{0n} \to \nu \qquad \text{a.s.},$$

where ν is the time constant of B. It is easy to see that $\nu \geq \mu$ where $\mu = E(Y_1)$, and we say that the Y's are *well-packed* if $\nu = \mu$, since the amount of empty bin-space after packing $Y_1, Y_2, \ldots, Y_n$ is $o(n)$ almost surely in this case. It may be shown that $\nu = \mu = \frac{1}{2}$ if Y_1 and $1-Y_1$ have the same distribution, and it is easy to construct examples for which $\nu \neq \mu$. It is an open problem to exhibit a general formula for ν in terms of the underlying distribution of the Y's. See Lueker (1983) and Karp et al. (1984) for recent results in this direction.

It is important that we be able to say something useful about small and large deviations in subadditive processes, but this seems to be very difficult in general. Even for stationary sequences of random variables the picture is far from complete with regard to large-deviation results, although the corresponding central limit theory is now standard (see Ibragimov and Linnik (1971)). For a subadditive process X with time constant γ, we formulate two general questions:

(a) under what conditions do there exist sequences $\alpha(n)$ and $\beta(n)$ such that $(X_{0n} - \alpha(n))/\beta(n)$ converges in distribution as $n \to \infty$ to a non-trivial limit, and

(b) under what conditions can we estimate the rate of decay of $P(X_{0n} > n(\gamma+\varepsilon))$ and $P(X_{0n} < n(\gamma-\varepsilon))$ for $\varepsilon > 0$; in particular, when do these probabilities decay geometrically?

Ishitani (1977) has provided a partial answer to the first of these questions. He has shown that $(X_{0n} - \gamma n)n^{-\frac{1}{2}}$ is asymptotically normally distributed with some limiting variance so long as X satisfies a certain collection of four conditions. Three of these conditions relate to mixing properties of X and the fourth requires that $E(X_{0n}) - \gamma n = o(n^{\frac{1}{2}})$; this last condition ensures that X_{0n} is not much larger than the additive process A_{0n} which lies beneath it (see (3.6)), so that known results for stationary sequences may be applied to A and X simultaneously. Unfortunately, we know of no interesting application of this central limit theorem, and the principal reason for this seems to be that the last condition may be verified only in cases when we know so much about X that we need not appeal to subadditive ergodic theory in the first place. We note that this condition does not hold for the process B arising in the bin-packing problem.

We know of no useful large-deviation result for general subadditive processes; indeed it is apparently an open problem to find an attractive mixing condition on a stationary ergodic sequence of random variables which guarantees the geometric decay of the probabilities of large deviations of the partial sums of the sequence. That is to say, if $Y_1, Y_2, \ldots$ is a stationary ergodic sequence with mean μ, under what conditions do $P(S_n > n(\mu+\varepsilon))$ and $P(S_n < n(\mu-\varepsilon))$ decay geometrically as $n \to \infty$ where $S_n = Y_1+Y_2+\ldots+Y_n$ and $\varepsilon > 0$?

Any useful answer to this question has an implication for subadditive processes since if X is subadditive then, for n,r = 0,1,2,...,

$$A_{0r} + A_{r,2r} + \cdots + A_{(n-1)r,nr} \le X_{0,nr} \le X_{0,r} + X_{r,2r} + \cdots + X_{(n-1)r,nr}$$

where A is the additive part of X (see (3.6)); that is to say, $X_{0,nr}$ may be sandwiched between two summations whose means differ by at most $\delta(r)rn$ where $\delta(r) \to 0$ as $r \to \infty$.

Certain large-deviation results are available for subadditive processes X which satisfy extra conditions. For deviations of $n^{-1}X_{0n}$ in *excess* of the time constant, independence of X is sufficient.

THEOREM 2. *Suppose that* X *is an independent subadditive process with time constant* γ, *and that for* $r = 1,2,\ldots$ *it is the case that*

$$E(\exp(tX_{0r})) < \infty \tag{3.7}$$

for $|t| \le \tau(r)$ *and some* $\tau(r) > 0$. *For* $\varepsilon > 0$ *there exists* $R(\varepsilon) > 0$ *such that*

$$P(X_{0n} > n(\gamma+\varepsilon)) \le \exp(-nR(\varepsilon)) \tag{3.8}$$

for all n.

That is to say, if the X's have finite moment generating functions then the probability that X_{0n} exceeds $n(\gamma+\varepsilon)$ decays at least geometrically for $\varepsilon > 0$. The proof of this theorem follows closely that of Theorem 3.2 in Grimmett and Kesten (1984a).

Even for the case of an *independent* subadditive process X, the problem of deviations of $n^{-1}X_{0n}$ *below* γ is rather more subtle; in fact there can be no general result asserting geometric decay for the associated probabilities, and we shall see why this is so later. However, it is the case that Chernoff's theorem about large deviations of sums of independent random variables (see Bahadur (1971)) may be generalized to deal with subadditive processes which satisfy certain ancillary conditions; we shall see that this generalization is incomplete and can never be complete. In advance of stating this result, we introduce some definitions.

Two random variables X and Y are called *positively quadrant dependent* if

$$P(X \le x, Y \le y) \ge P(X \le x)P(Y \le y) \qquad \text{for all } x,y \in \mathbb{R}.$$

Lehmann (1966) has shown that X and Y are positively quadrant dependent if and only if they satisfy the FKG inequality:

$$E(f(X)g(Y)) \ge E(f(X))E(g(Y))$$

for all non-decreasing functions f and g for which these expectations exist. We call the process $X = \{X_{mn} : 0 \le m < n\}$ *pairwise weakly associated* if X_{0m} and X_{mn} are positively quadrant dependent whenever $0 < m < n$; this property is related to the "weak positive association" of Kipnis (1984) and the "weak association" of Rüschendorf (1981). There are many subadditive processes which are pairwise weakly associated without being independent, and the best example of this is perhaps the set of point-to-point times of first-passage percolation, to be discussed in the next section.

We call a sequence $F_1, F_2, \ldots$ of distribution functions *superconvolutive* if, for $m,n = 1,2,\ldots$,

$$F_{m+n}(x) \ge \int_{-\infty}^{\infty} F_m(x-u)dF_n(u) \qquad \text{for all } x \in \mathbb{R}.$$

We call the stationary process $X = \{X_{mn} : 0 \le m < n\}$ *superconvolutive* if the sequence of distribution functions of $X_{01}, X_{02}, \ldots$ is superconvolutive. For a subadditive process X, it is easy to check that X is superconvolutive if it is independent; on the other hand there are superconvolutive processes which are not independent. Also, the property of superconvolutivity neither implies nor is implied by the property of pairwise weak association; see the end of this section for examples which substantiate this assertion.

THEOREM 3. *Suppose that* X *is a subadditive process such that*

$$\phi_r(t) := E(\exp(tX_{0r})) < \infty \tag{3.9}$$

if $0 \ge t \ge \tau$ *for some* $\tau < 0$ *and all* r. *If* X *is either pairwise weakly associated or superconvolutive then the limits*

$$\psi(x) = \lim_{n\to\infty} \frac{1}{n} \log P(X_{0n} \le nx) \tag{3.10}$$

and

$$\kappa(t) = \lim_{n\to\infty} \frac{1}{n} \log \phi_n(t) \tag{3.11}$$

exist for all x *and all* $t \le 0$ *and satisfy*

$$\psi(x) = \inf_t\{\kappa(t) - tx\}, \tag{3.12}$$

$$\kappa(t) = \sup_x\{\psi(x) + tx\}, \tag{3.13}$$

for x *in the interior of* $\{y : \psi(y) > -\infty\}$ *and* t *in the interior of* $\{u : \kappa(u) < \infty\}$.

For superconvolutive processes this theorem was proved by Hammersley; see Hammersley (1974) or Kingman (1976) for the proof. It is easy to see how to adapt this proof if X is pairwise weakly associated. We note that Theorem 3 is basically a result about distributions rather than random variables; corresponding conclusions are valid for any superconvolutive sequence of distribution functions.

Theorem 3 leaves unresolved the most important question: under what conditions on X and for what values of x is it the case that $\psi(x) < 0$ in (3.10)? Under the hypotheses of the theorem, $P(X_{0n} < n(\gamma-\varepsilon))$ decays in the manner of $\exp(n\psi(\gamma-\varepsilon))$, but is it necessarily the case that $\psi(\gamma-\varepsilon) < 0$ if $\varepsilon > 0$? The answer to this question is *no*, despite the assertion of Kingman (1976, p.217) to the contrary. As an example of an independent subadditive process for which $\gamma > 0$ and $\psi(\gamma-\varepsilon) = 0$ for $0 < \varepsilon < \gamma$, we consider the volume swept out by the Wiener sausage in three or more dimensions. Let $\{\beta(t) : t \geq 0\}$ be Brownian motion in d (≥ 3) dimensions with $\beta(0) = \underline{0}$, and let

$$W_{mn} = \bigcup_{x \in I(m,n)} S_\rho(x)$$

where $I(m,n)$ is the image set of $\{\beta(s) : m < s \leq n\}$ and $S_\rho(x)$ is the d-dimensional sphere with radius ρ (> 0) and centre x. It is clear that W is a positive independent subadditive process, and the time constant γ of W satisfies $\gamma > 0$. Donsker and Varadhan (1975) have shown that there exists $k(t,d) > 0$ such that, as $n \to \infty$,

$$n^{-d/(d+2)} \log E(\exp(tW_{0n})) \to -k(t,d)$$

for all $t < 0$. It follows that the corresponding functions ψ and κ in Theorem 3 satisfy $\kappa(t) = 0$ if $t \leq 0$ and hence $\psi(x) = 0$ if $x > 0$. Thus $\psi(\gamma-\varepsilon) = 0$ if $0 < \varepsilon < \gamma$ in this case.

It is thus an important and non-trivial question to determine the range of values of x for which the function $\psi(x)$ in (3.10) is *strictly* negative, and we know of no general way of finding this range. It is easy to see from (3.6) that $\psi(x) < 0$ for $x < \gamma$ so long as the additive part A of the independent subadditive process X is itself an independent process with finite moment generating functions, but we know of no attractive sufficient conditions on X which ensure that this holds.

For certain subadditive processes the answers to some of these questions are known. For example, Le Page (1982) has studied large deviations in products of random matrices, and Grimmett and Kesten (1984a) have shown the geometric decay of the appropriate probabilities for those processes at the root of subadditive ergodic theory, being those which arise in first-passage percolation. The next section contains a short account of this.

We may apply Theorems 2 and 3 to the bin-packing process B with the following conclusions. The process B is independent and $1 \le B_{0n} \le n$ for all n, so that Theorem 2 implies

$$P(B_{0n} > n(\nu+\varepsilon)) \le \exp(-nR(\varepsilon))$$

for all n and some $R(\varepsilon) > 0$, where ν is the time constant of B and $\varepsilon > 0$. From Theorem 3,

$$\psi(x) = \lim_{n\to\infty} \frac{1}{n} \log P(B_{0n} < nx)$$

exists for all x. Moreover,

$$B_{0n} \ge Y_1 + Y_2 + \ldots + Y_n$$

where $Y_1, Y_2, \ldots$ are the sizes of the objects to be packed; exploiting the independence of the Y's, we find that $\psi(x) < 0$ if $x < \mu$, where $\mu = E(Y_1)$. Thus $P(B_{0n} < n(\nu-\varepsilon))$ decays geometrically as $n \to \infty$ for $\varepsilon > 0$, whenever $\nu = \mu$; that is to say, the probabilities of large deviations of $n^{-1}B_{0n}$ below ν decay geometrically whenever the Y's may be well-packed.

We close this section with an elementary demonstration that the property of pairwise weak association neither implies nor is implied by the property of superconvolutivity. First, let Y be a random variable with the normal distribution and let $X_{mn} = (n-m)Y$ for all m and n satisfying $0 \le m < n$. It is easy to check that X is subadditive and pairwise weakly associated but not superconvolutive. Secondly, let A and B be a pair of random variables having as joint density function

$$f(a,b) = \begin{cases} \frac{1}{4} & \text{if } (a,b) \in S_1 \cup S_2, \\ \frac{1}{2} & \text{if } (a,b) \in T_1 \cup T_2, \end{cases}$$

where S_1 is the square $[0,1] \times [0,1]$, S_2 is the square $[1,2] \times [1,2]$, T_1 is the union of the open quadrilateral with vertices $(0,1)$, $(0,1\frac{1}{2})$, $(\frac{1}{2},2)$, $(1,2)$ and the open triangle with vertices $(\frac{1}{2},1)$, $(1,1)$, $(1,1\frac{1}{2})$, and T_2 is the mirror image of T_1 reflected in the line $y = x$. It is easy to see from a diagram that A and B are random variables such that

(a) (A,B) and (B,A) have the same joint distributions,

(b) the marginal distributions of A and B are both uniform on [0,2],

(c) the distribution of A+B is the same as the distribution of the sum of two *independent* random variables with this uniform distribution,

(d) $P(A \le \frac{1}{2}, B \le \frac{5}{4}) = \frac{9}{64} < \frac{5}{32} = P(A \le \frac{1}{2})P(B \le \frac{5}{4})$.

Now, let X be the subadditive process defined as follows:

$$X_{m,m+1} = \begin{cases} A & \text{if } m \text{ is even,} \\ B & \text{if } m \text{ is odd,} \end{cases}$$

$$X_{m,m+2} = A+B \quad \text{for } m = 0,1,2,\ldots,$$

$$X_{mn} = 0 \quad \text{if } n-m \ge 3.$$

It follows from (c) and (d) above that X is superconvolutive but not pairwise weakly associated. This example is based on an example of H.E. Robbins, quoted in Feller (1971, p.99).

4. FIRST-PASSAGE PERCOLATION. One of the most important problems in operations research is to find the shortest path between two given vertices in a prescribed network. If the network is a crystalline lattice (such as the square lattice) and the lengths of individual edges are independent and identically distributed random variables then this is the problem of *first-passage percolation*, introduced by Hammersley and Welsh (1965) as a mathematical description of the flow of such quantities as fluid or disease through a random environment. We formulate this problem next, using the d-dimensional cubic lattice as the underlying network.

Let $\mathbb{Z}^d$ denote the d-dimensional cubic lattice with edges joining pairs of vertices which are unit distance apart in the euclidean metric. To each edge e of $\mathbb{Z}^d$ we assign a non-negative random variable T(e), where the T's are independent with the common distribution function F; we call T(e) the *time co-ordinate* of e. If $\pi = (e_1, e_2, \ldots, e_m)$ is a path using the edges $e_1, e_2, \ldots, e_m$, we define the *passage time* of π by

$$T(\pi) = \sum_{i=1}^{m} T(e_i),$$

and we think of $T(\pi)$ as the time required for the quantity in question (fluid or disease, say) to pass along π from one endpoint to the other. If A and B are disjoint sets of vertices then the passage time from A to B is defined to be

$$T(A,B) = \inf\{T(\pi)\},$$

where the infimum is taken over all paths π which have one endpoint in A and one endpoint in B. First-passage percolation studies the random variables T(A,B) for certain sets A and B, and is particularly concerned with the way in which the distribution of T(A,B) depends on F when A and B are a long way apart from each other. Passage times of particular interest are the *point-to-point times* a_{mn} given by

$$a_{mn} = T(v_m, v_n) \quad \text{for } m < n,$$

where $v_r = (r,0,0,\ldots,0)$, and the *point-to-hyperplane times* b_{mn} given by

$$b_{mn} = T(v_m, H_n) \quad \text{for } m < n,$$

where H_n is the hyperplane $\{\underline{x} \in \mathbb{Z}^d : x_1 = n\}$. Thus a_{mn} is the passage time between the two vertices v_m and v_n on the x-axis, and b_{mn} is the passage time from v_m to the hyperplane containing v_n perpendicular to the x-axis.

It is easy to see that the process a satisfies (3.1) and (3.2), and therefore a is a subadditive process whenever the moment condition (3.3) is valid; it turns out that a is a subadditive process if and only if

$$E(\min\{T_1,\ldots,T_{2d}\}) = \int_0^\infty (1-F(x))^{2d} dx < \infty, \tag{4.1}$$

where the T's are independent time co-ordinates. On the other hand the process b fails to satisfy the subadditive inequality (3.1), although it may be shown that b is stationary and pairwise weakly associated. The two processes a and b have very much the same asymptotic behaviour.

THEOREM 4. *Consider first-passage percolation on* $\mathbb{Z}^d$ *and suppose that (4.1) holds. There exists a constant* $\mu = \mu(F)$ *such that, as* $n \to \infty$,

$$\frac{1}{n}\theta_{0n} \to \mu \text{ almost surely} \tag{4.2}$$

where θ *may represent* a *or* b.

Rather more is known for the planar case when d = 2; for example, Cox and Durrett (1981) have shown that $n^{-1}b_{0n} \to \mu$ almost surely and $n^{-1}a_{0n} \to \mu$ in probability even if (4.1) fails to hold. Similar results should be valid for higher dimensions.

The constant μ is called the *time constant* of the process. See Smythe and Wierman (1978, Chapter 5), Cox and Durrett (1981) and Cox and Kesten (1981) for proofs of Theorem 4 and related facts.

It was through studying the point-to-point process on $\mathbb{Z}^2$ that Hammersley and Welsh came to the idea of studying subadditive processes in their generality, and only by applying the subsequent subadditive ergodic theorem (Theorem 1) did it come to be known that $n^{-1}a_{0n}$ converges almost surely.

Certain large-deviation results are valid for the processes a and b, and we state these next.

THEOREM 5. *Consider first-passage percolation on* $\mathbb{Z}^d$ *with time constant* μ, *and suppose that*

$$\int_0^\infty e^{tx}dF(x) < \infty \qquad \textit{for some} \quad t > 0. \tag{4.3}$$

For $\varepsilon > 0$, *there exists* $A(\varepsilon) > 0$ *such that*

$$P(\theta_{0n} > n(\mu+\varepsilon)) \le \exp(-nA(\varepsilon)) \tag{4.4}$$

for all n, *where* θ *may represent* a *or* b.

Condition (4.3) is both necessary and sufficient for (4.4) to hold for all $\varepsilon > 0$ and n. It may easily be seen that Theorem 5 is a corollary of Theorem 2, and to do this we introduce the *cylinder point-to-point process* $t = \{t_{mn} : m < n\}$, where t_{mn} is the infimum of the passage times of all paths joining v_m to v_n and contained in the *cylinder* $\{\underline{x} \in \mathbb{Z}^d : m < x_1 < n\}$ except for their endpoints. The process t is an independent subadditive process if (4.3) holds, and $b_{mn} \le a_{mn} \le t_{mn}$; furthermore, it may be shown that a and t have the same time constant. We apply Theorem 2 to the process t to deduce (4.4).

THEOREM 6. *Consider first-passage percolation on* $\mathbb{Z}^2$ *with time constant* μ. *For* $\varepsilon > 0$, *there exists* $B(\varepsilon) > 0$ *such that*

$$P(\theta_{0n} < n(\mu-\varepsilon)) \le \exp(-nB(\varepsilon)) \tag{4.5}$$

for all n, *where* θ *may represent* a *or* b. *Furthermore, for* $\varepsilon > 0$, *there exists* $C(\varepsilon)$ *satisfying* $0 < C(\varepsilon) \le \infty$ *such that, as* $n \to \infty$,

$$\frac{1}{n}\log P(a_{0n} < n(\mu-\varepsilon)) \to -C(\varepsilon). \tag{4.6}$$

This result is rather deeper than Theorem 5. The same conclusion is valid for $\mathbb{Z}^d$ if the time co-ordinates have finite variance; probably a much weaker assumption would suffice here. In Section 3 we posed the question of determining whether or not the probabilities of large deviations in subadditive processes *below* the time constant decay geometrically; Theorem 6 answers this question affirmatively for first-passage percolation. Conclusion (4.5) is the important part of this theorem, and with this in the background Theorem 3 may be brought into action to yield both (4.6) and a representation for $C(\varepsilon)$ in terms of the moment generating functions of the a's. See Grimmett and Kesten (1984a) and Kesten (1985) for the proofs of the above results; relation (4.5) is proved by using a "block" argument.

There remains the most interesting question of finding a central limit theorem for a and b. A first step in this direction might be to explore the behaviour of the exponents $A(\varepsilon)$ and $C(\varepsilon)$ in (4.4) and (4.6) for small positive values of ε. It is not too difficult to show that C is convex, continuous and strictly increasing on $(0,\beta)$, where $\beta = \sup\{\varepsilon : F(\mu-\varepsilon) > 0\}$, and $C(\varepsilon) = \infty$ if $\varepsilon \geq \beta$. Kesten has made some progress (unpublished) towards estimating $C(\varepsilon)$ for small ε. On the other hand he has shown that

$$\limsup_{n \to \infty} \frac{1}{n} \log P(t_{0n} > n(\mu+\varepsilon)) = -\infty$$

for $\varepsilon > 0$, whenever $\int e^{tx}dF(x) < \infty$ for all t; that is to say, we may take $A(\varepsilon)$ to be as large as we like and (4.4) remains true for all large n in this case.

5. RANDOM ELECTRICAL NETWORKS. It is an important problem of physics to find the effective resistance of a random electrical network. If a conducting medium contains impurities then its effective conductivity depends on the degree of impurity and on the relative conductivities of the constituent materials. In many cases the impurities may appear to occur at random, and then the composite material is best modelled by a random process. In the simplest such mathematical model we assume that the network takes the form of some specified graph G, with nominated *source* and *sink* labelled s and t respectively, and we assume that the edges of G are assigned electrical resistances which are independent and identically distributed random variables; in many instances such assumptions may be simplistic. Writing F for the common distribution function of the edge-resistances, and R(G) for the (random) effective resistance of G between the pair s and t of vertices, we are interested in the manner in which R(G) depends on F. Only in a few exceptional cases do we know explicit formulae for the distribution of R(G). However, it is sometimes possible to approximate to R(G) by taking a limit as the size of

G tends to infinity; that is to say, we may be interested in the asymptotic behaviour of $R(G_n)$ as G_n passes through an increasing sequence $G_1, G_2, \ldots$ of graphs. Such results are available for sequences of trees and complete graphs (see Stinchcombe (1974) and Grimmett and Kesten (1984b)), but the most interesting and challenging case is when G_n is a subsection of a crystalline lattice such as the square lattice $\mathbb{Z}^2$. For certain types of distribution function F this problem has a close link with first-passage percolation, and the large-deviation results of the last section thus have interesting physical applications.

We consider the plane square lattice only, but note that similar results should be valid in higher dimensions also (see Kesten (1982, Ch. 11) and Grimmett and Kesten (1984a)). We think of the edges of $\mathbb{Z}^2$ as being wires in an electrical network, where the edge e has resistance R(e) and the family $\{R(e) : e \in \mathbb{Z}^2\}$ is a collection of independent non-negative random variables with the common distribution function F. For $n,m = 0,1,2,\ldots$ we write B_{nm} for the box $\{(i,j) \in \mathbb{Z}^2 : 0 \le i \le n,\ 0 \le j \le m\}$ together with all the edges inherited from $\mathbb{Z}^2$. We turn B_{nm} into a finite electrical network by introducing two new vertices, labelled s and t, in such a way that s (respectively t) is joined to each vertex on the left-hand (respectively right-hand) side of B_{nm} by wires with resistance 0. We denote by R_{nm} the effective resistance of B_{nm} between s and t, and we wish to explore the asymptotic behaviour of R_{nm} as $n,m \to \infty$. If F has an atom at 0 or ∞ then percolation theory enters the problem, since "$R(e) = 0$" is equivalent to identifying the two endvertices of the edge e, and "$R(e) = \infty$" is equivalent to deleting e from the network. Here we assume for simplicity that $F(0) = 0$, so that all edges have non-zero resistances almost surely, but we refer the reader to Kesten (1982) for a discussion of the more general case. Thus we shall assume that there exists a constant p satisfying $0 \le p \le 1$ such that

$$F(x) = \begin{cases} pG(x) & \text{if } x < \infty, \\ 1 & \text{if } x = \infty, \end{cases}$$

where G is a probability distribution function with mass concentrated on $(0,\infty)$. If $p < \frac{1}{2}$ then the density of conducting edges is below the percolation threshold, so that $P(R_{nm} = \infty)$ is near 1 unless n is very small compared with m (more precisely, unless $n < \beta(p)\log m$ for a certain function β). If $\frac{1}{2} < p < 1$ then the asymptotic behaviour of R_{nm} is more interesting. It is not difficult to convince oneself that R_{nm} should be approximately proportional to n and inversely proportional to m, since the distance between s and t equals n and the maximum number of edge-disjoint paths from s to t equals m+1. It is currently an open problem to show that $mn^{-1}R_{nm}$ converges as $n,m \to \infty$, even in the physically appealing case when $m = n$, but it is possible to show non-trivial upper and lower bounds for the sequence so long as $n = n(m)$ does not grow too fast as $m \to \infty$.

THEOREM 7. *Suppose that* $\frac{1}{2} < p \leq 1$. *If for all* $c > 0$ *it is the case that* $n = n(m)$ *satisfies*

$$n(m) \leq e^{cm} \qquad \textit{for all large } m,$$

then there exists $\nu(p)$ $(< \infty)$ *such that*

$$\frac{1}{p} \; \|G\|_{-1} \leq \liminf_{m \to \infty} \left(\frac{m}{n} R_{nm}\right) \leq \limsup_{m \to \infty} \left(\frac{m}{n} R_{nm}\right) \leq \nu(p) \; \|G\|_{1}$$

almost surely, where $\|G\|_{\alpha} = \left(\int_0^{\infty} x^{\alpha} dG(x)\right)^{1/\alpha}$ *for* $\alpha = \pm 1$.

We interpret $\|G\|_{-1}$ as 0 if $\int x^{-1} dG(x) = \infty$. Thus $mn^{-1}R_{nm}$ is uniformly bounded away from 0 and ∞ if G has finite mean and non-zero harmonic mean and if n is a sub-exponential function of m. The principal step in the proof of this theorem is to estimate the maximum number of edge-disjoint conducting paths in B_{nm} which join s to t. The max-flow min-cut theorem may be applied to deduce that this number is essentially the length of a shortest path crossing the dual rectangle of B_{nm} from top to bottom. In terms of first-passage percolation, this is a *line-to-line* passage time, and its asymptotic and large-deviation behaviour follow from the results in the previous section.

If $n = n(m)$ grows as an exponential function of m then a critical phenomenon takes place. Suppose that $n = n(m) = \lfloor e^{cm} \rfloor$ for some $c > 0$ and $\frac{1}{2} < p < 1$. Then there exist numbers σ and τ satisfying $0 < \sigma \leq \tau < \infty$ such that

$$\limsup_{m \to \infty} \left(\frac{n}{m} R_{nm}\right) < \infty \text{ a.s. if } c < \sigma$$

and

$$\liminf_{m \to \infty} \left(\frac{n}{m} R_{nm}\right) = \infty \text{ a.s. if } c > \tau.$$

The exact nature of this phenomenon is not yet understood. The values of σ and τ may be estimated in terms of certain rate functions of first-passage percolation, including the function C given in (4.6).

Electrical flow is a particular physical process involving the flow of indestructible uncreatable matter through a graphical network; other such flow processes are the classical *network flows* of operations research. There are as many differences as there are similarities between these two types of flow, but it may be possible to treat some of their properties in a unified approach (see Iri (1969)). Just as *random* electrical networks pose interesting probabilistic questions, so do network flows through *randomly capacitated* networks; see Grimmett (1985) for a survey of some of this area, and Kesten (1985) for some results in three dimensions.

ACKNOWLEDGEMENT. I am grateful to Harry Kesten for discussing the asymptotics of the Wiener sausage with me.

BIBLIOGRAPHY

Bahadur, R.R. (1971), Some limit theorems in statistics, *Conference Board of the Mathematical Sciences Regional Conference Series in Applied Mathematics* no. 4, SIAM, Philadelphia.

Bollobás, B. and Erdös, P. (1976), Cliques in random graphs, *Mathematical Proceedings of the Cambridge Philosophical Society* 80, 419-427.

Cox, J.T. and Durrett, R. (1981), Some limit theorems for percolation processes with necessary and sufficient conditions, *Annals of Probability* 9, 583-603.

Cox, J.T. and Kesten, H. (1981), On the continuity of the time constant of first-passage percolation, *Journal of Applied Probability* 18, 809-819.

Donsker, M.D. and Varadhan, S.R.S. (1975), Asymptotics for the Wiener sausage, *Communications of Pure and Applied Mathematics* 28, 525-565.

Feller, W. (1971), *An introduction to probability theory and its applications*, vol. 2 (2nd edn.), John Wiley, New York.

Garey, M.R. and Johnson, D.S. (1979), *Computers and intractability, a guide to the theory of NP-completeness*, Freeman, San Francisco.

Grimmett, G.R. (1983), Random graphs, in *Selected topics in graph theory 2*, ed. L. Beineke and R. Wilson, Academic Press, London, 201-235.

Grimmett, G.R. (1985), Random flows: networks flows and electrical flows through random media, in *Surveys in Combinatorics 1985*, ed. I. Anderson, Cambridge University Press, Cambridge, to appear.

Grimmett, G.R. and Kesten, H. (1984a), First-passage percolation, network flows and electrical resistances, *Zeitschrift für Wahrscheinlichkeitstheorie und verwandte Gebiete* 66, 335-366.

Grimmett, G.R. and Kesten, H. (1984b), Random electrical networks on complete graphs, *Journal of the London Mathematical Society*, to appear.

Grimmett, G.R. and McDiarmid, C.J.H. (1975), On colouring random graphs, *Mathematical Proceedings of the Cambridge Philosophical Society* 77, 313-324.

Hammersley, J.M. (1974), Postulates for subadditive processes, *Annals of Probability* 2, 652-680.

Hammersley, J.M. and Welsh, D.J.A. (1965), First-passage percolation, subadditive processes, stochastic networks and generalized renewal theory, in *Bernoulli-Bayes-Laplace Anniversary Volume*, ed. J. Neyman and L. Le Cam, Springer, Berlin, 61-110.

Ibragimov, I.A. and Linnik, Yu.V. (1971), *Independent and stationary sequences of random variables*, Walters-Noordhoff, Groningen.

Iri, M. (1969), *Network flow, transportation and scheduling*, Academic Press, New York.

Ishitani, H. (1977), A central limit theorem for the subadditive process and its application to products of random matrices, *Publications of Research Institute for Mathematical Sciences, Kyoto University*, 12, 565-575.

Jaffe, A.M. (1984), Ordering the universe: the role of mathematics, *Notices of the American Mathematical Society* 31, 589-608.

Johnson, D.S. (1974), Worst case behavior of graph coloring algorithms, in *Proceedings of Fifth Southeastern Conference on Combinatorics, Graph Theory and Computing*, Utilitas Mathematica Publishing, Winnipeg, 513-528.

Karp, R.M. (1976), The probabilistic analysis of some combinatorial search algorithms, in *Algorithms and Complexity*, ed. J.F. Traub, Academic Press, New York, 409-429.

Karp, R.M., Luby, M. and Marchetti-Spaccamela, A. (1984), A probabilistic analysis of multi-dimensional bin-packing problems, *Proceedings of the 16th Association of Computing Machinery Symposium on the Theory of Computing*, 289-298.

Kesten, H. (1982), *Percolation theory for mathematicians*, Birkhäuser, Boston.

Kesten, H. (1985), Aspects of first-passage percolation, in *Ecole d'Eté de Probabilités de Saint Flour XIV - 1984*, Lecture Notes in Mathematics, Springer, Berlin, to appear.

Kingman, J.F.C. (1968), The ergodic theory of subadditive stochastic processes, *Journal of the Royal Statistical Society, Series B* 30, 499-510.

Kingman, J.F.C. (1976), Subadditive processes, in *Ecole d'Eté de Probabilités de Saint Flour V-1975*, ed. P.-L. Hennequin, Lecture Notes in Mathematics no. 539, Springer, Berlin, 167-223.

Kipnis, C. (1984), Central limit theorems for infinite series of queues and applications to simple exclusion, Internal report no. 105, Centre of Applied Mathematics, Ecole Polytechnique, Palaiseau, France.

Korshunov, A.D. (1980), The chromatic number of n-vertex graphs, *Metody Diskretnogo Analiza* (in Russian) 35, 15-44, 104.

Lehmann, E.L. (1966), Some concepts of dependence, *Annals of Mathematical Statistics* 37, 1137-1153.

Le Page, E. (1982), Théorèmes limites pour les produits de matrices aléatoires, in *Probability Measures on Groups*, ed. H. Heyer, Lecture Notes in Mathematics no. 928, Springer, Berlin, 258-303.

Liggett, T.M. (1984), An improved subadditive ergodic theorem, preprint.

Lueker, G.S. (1983), Bin packing with items uniformly distributed over intervals [a,b], *Proceedings of the 24th IEEE Symposium on the Foundations of Computer Science*, 289-297.

McDiarmid, C.J.H. (1979), Determining the chromatic number of a graph, *SIAM Journal of Computing* 8, 1-14, and *Stanford University Computing Science Preprint* STAN-CS-76-576.

McDiarmid, C.J.H. (1984), Colouring random graphs, *Annals of Operations Research* 1, 183-200.

Rüschendorf, L. (1981), Weak association of random variables, *Journal of Multivariate Analysis* 11, 448-451.

Smythe, R.T. and Wierman, J.C. (1978), *First-passage percolation on the square lattice*, Lecture Notes in Mathematics no. 671, Springer, Berlin.

Stinchcombe, R.B. (1974), Conductivity and spin-wave stiffness in disordered systems - an exactly soluble model, *Journal of Physics C* 7, 179-203.

Welsh, D.J.A. (1983), Randomized algorithms, *Discrete Applied Mathematics* 5, 133-145.

Welsh, D.J.A. (1984), A randomized algorithm for an NP-complete problem, to appear.

SCHOOL OF MATHEMATICS
UNIVERSITY OF BRISTOL
UNIVERSITY WALK
BRISTOL BS8 1TW
UNITED KINGDOM.

Contemporary Mathematics
Volume 41, 1985

RANDOM MEDIA AND SELFDIFFUSION OF INTERACTING PARTICLES

M. Z. Guo and G. Papanicolaou*
Peking University and Courant Institute, New York University

ABSTRACT. This is a review of some recent work on diffusion in random media and related problems.

1. INTRODUCTION.

Let $(\Omega, \mathcal{F}, P)$ be a probability space and let $\Phi(x,\omega)$ be a stationary and ergodic random function on Ω with values in $\mathbb{R}$, with $x \in \mathbb{R}$. Assume that Φ and its derivative Φ' are bounded.

$$(1.1)\qquad \begin{aligned} |\Phi(x,\omega)| &\le C \\ |\Phi'(x,\omega)| &\le C \qquad \text{for all } x \in \mathbb{R} \text{ and } \omega \in \Omega . \end{aligned}$$

Consider the one-dimensional stochastic differential equation

$$(1.2)\qquad dx(t) = \Phi'(x(t),\omega)dt + dw(t) , \qquad x(0) = x$$

where $w(t)$ is the standard Brownian motion process. In integral form (1.2) is

$$(1.3)\qquad x(t) = x + \int_0^t \Phi'(x(s),\omega)ds + w(t) .$$

Since we are in one space dimension the process $x(t)$ is well defined for each $\omega \in \Omega$ in the sense that there is a probability measure $Q_{x,\omega}$ on continuous paths $C([0,\infty),\mathbb{R}^1)$ for which

$$(1.4)\qquad f(x(t)) - \int_0^t \left[\frac{1}{2} f''(x(s)) + \Phi'(x(s),\omega)f'(x(s))\right] ds$$

*Work supported by the National Science Foundation.

is a martingale for each f twice differentiable and bounded and

$$(1.5) \qquad Q_{x,\omega}\{x(0) = x\} = 1 .$$

The question we are interested in is this. How does $x(t)$ behave as $t \to \infty$? Of course we are interested in multidimensional problems as well and in problems for which the random medium is not a gradient force field as in (1.2).

Let us first prove the following

THEOREM.

The process $\frac{x(t) - x(0)}{\sqrt{t}}$ converges as $t \to \infty$ to a Gaussian random variable, in probability with respect to $\omega \in \Omega$.

Note that only stationarity and ergodicity of $\Phi(x,\omega)$ are needed, along with (1.1). Ergodicity means that the only functionals of $\Phi(\cdot,\omega)$ measurable with respect to the σ algebra generated by $\Phi(x,\omega)$ for $|x| \geq M$ for all M are constants. Note also that (1.2) is a gradient system.

2. PROOF OF THEOREM

Because the method is general and instructive as well as easy we will give the proof of this theorem. One can easily generalize it to an invariance principle along standard lines. This generalization is not always automatic but here (1.1) and the gradient form of (1.2) make it so.

Let L_ω be the generator of the process $x(t)$ in the medium labeled by ω.

$$(2.1) \qquad L_\omega = \frac{1}{2}\frac{d^2}{dx^2} + \Phi'(x,\omega)\frac{d}{dx} = \frac{1}{2}\frac{1}{e^{2\Phi(x,\omega)}}\frac{d}{dx}\left(e^{2\Phi(x,\omega)}\frac{d}{dx}\right) .$$

In analyzing $t^{-\frac{1}{2}}(x(t) - x)$ by (1.3) one wants to replace the integral by a term that is lower order in t as $t \to \infty$. This is perturbation theory that is somewhat clearer when expressed via a small parameter ε. We will claculate here without such a parameter, directly with t for t large.

Introduce $\chi(x,\omega)$ by solving

$$(2.2) \qquad L_\omega \chi = -\Phi'(x,\omega)$$

or

$$\frac{1}{2}\chi'' + \Phi'\chi' = -\Phi'$$

This can be written as

$$\text{(2.3)} \qquad \frac{d}{dx}\left[e^{2\Phi(x,\omega)}(\chi'+1)\right] = 0$$

so that

$$\text{(2.4)} \qquad \chi' = e^{-2\Phi}c - 1$$

where c is a constant. Demanding that χ' be (stationary) with mean zero we determine c and obtain

$$\text{(2.5)} \qquad \chi(x,\omega) = \int_0^x \left[\frac{e^{-2\Phi(s,\omega)}}{E(e^{-2\Phi})} - 1\right] ds \quad .$$

Note that with Φ bounded and ergodic, the ergodic theorem tells us that

$$\text{(2.6)} \qquad \frac{1}{|x|}\,\chi(x,\omega) \to 0 \quad \text{as} \quad |x| \to \infty$$

with P probability one. Note also that from (2.2) $x + \chi(x,\omega)$ is an L_ω harmonic function

$$\text{(2.7)} \qquad L_\omega(x + \chi(x,\omega)) = 0 \quad .$$

Note finally that χ is not stationary and will be so only under special circumstances. For example when $\Phi(x,\omega)$ is periodic almost surely or almost periodic with a frequency gap at zero.

Now apply Ito's lemma to $f(x) = x + \chi(x,\omega)$ to find that

$$\text{(2.8)} \qquad x(t) + \chi(x(t),\omega) = x + \chi(x,\omega) = \int_0^t (1+\chi'(x(s),\omega))dw(s) \; .$$

Since χ' is stationary and bounded, the Brownian stochastic integral is well defined. Moreover it behaves like t for large t in mean square with respect to $Q_{x,\omega}$, for $\omega \in \Omega$. Thus $t^{-\frac{1}{2}}(x(t)-x)$ is of order one in mean square as $t \to \infty$ and hence $t^{-\frac{1}{2}}\chi(x(t),\omega)$ tends to zero on $t \to \infty$ in mean square, for almost all ω by (2.6).

It follows that

$$\text{(2.9)} \qquad \frac{x(t)-x}{\sqrt{t}} \quad \text{and} \quad \frac{1}{\sqrt{t}}\int_0^t (1+\chi'(x(s),\omega))dw(s)$$

have the same limit distributions as $t \to \infty$, with probability one with respect to P.

3. PROOF OF THEOREM, CONTINUATION

At this point we can invoke a general central limit theorem for the martingale

$$\frac{1}{\sqrt{t}} \int_0^t (1+\chi'(x(s),\omega))dw(s) \tag{3.1}$$

to conclude that it has a gaussian limit distribution. This is not very difficult to actually show in detail and the proof is actually an extension of the argument below given for the calculation of the variance σ^2.

Let in fact σ^2 be the limit variance of (3.1)

$$\sigma^2 = \lim_{t \uparrow \infty} \frac{1}{t} \int_0^t (1+\chi'(x(s),\omega))^2 ds \quad . \tag{3.2}$$

We want to show that the limit exists with probability one in ω and that

$$\sigma^2 = E(e^{2\Phi})^{-1} E(e^{-2\Phi})^{-1} \quad , \tag{3.3}$$

i.e., the limit variance is the harmonic mean of the diffusion coefficient in the selfadjoint form of (2.1), suitably normalized. The calculation of the limit (3.2) should involve the ergodic theorem somehow but that is not immediately clear.

We shall now explain in detail how σ^2 is computed. We need for that purpose some definitions which we provide.

We may assume that the stationary process $\Phi(x,\omega)$ in the space (Ω,F,P) is the translate of $\Phi(0,\omega)$ with respect to a group of shift transformations $\tau_x : \Omega \to \Omega$, $x \in \mathbb{R}$ (τ_x is one-one and $\tau_x \tau_y = \tau_{x+y}$) that leave P invariant. We shall write

$$\Phi(x,\omega) = \tilde{\Phi}(\tau_{-x}\omega) \tag{3.4}$$

with $\tilde{\Phi}(\omega) = \Phi(0,\omega)$. In fact for any square integrable function $\tilde{f}(\omega)$ we may define a unitary group of transformations

$$(T_x \tilde{f})(\omega) = \tilde{f}(\tau_{-x}\omega) \tag{3.5}$$

in $H = L^2(\Omega,F,P)$. The unitarity is a consequence of the invariance of P.

Since T_x is a unitary group it has a densely defined generator D in H which is skew adjoint

$$\int_\Omega \tilde{f}(\omega)D\tilde{g}(\omega)dP(\omega) = -\int_\Omega D\tilde{f}(\omega)\tilde{g}(\omega)dP(\omega)$$

for every $\tilde{f}$ and $\tilde{g}$ that belong to the domain of D.

Let $x(t)$ be a continuous path with $x(0) = 0$. For each $\omega \in \Omega$ we may associate a path in Ω by

$$(x(t),\omega) \to \omega(t) = \tau_{-x(t)}\omega \quad .$$

We may assume that Ω is a complete separable metric space and that $\tau_x\omega$ is continuous in x without loss in generality. Then $\omega(t)$ will be a continuous trajectory in Ω and we have

LEMMA

Let $\tilde{Q}_{\omega,x}$ be the image of $Q_{x,\omega}$ induced by (3.7). Then $\tilde{Q}_{\omega,x}$ is the measure of a diffusion process on Ω , i.e., a measure on $C([0,\infty),\Omega)$ such that for any $\tilde{f}(\omega)$ in the domain of D^2

$$\tilde{f}(\omega(t)) - \int_0^t \left(\frac{1}{2} D^2 + D\tilde{\Phi}(\omega(s))D\right)\tilde{f}(\omega(s))ds \tag{3.8}$$

is a martingale and

$$\tilde{Q}_{\omega,x}\left\{\omega(0) = \tau_{-x}\omega\right\} = 1 \quad . \tag{3.9}$$

Let us give the proof of this. Consider

$$E^{\tilde{Q}_{\omega,x}}\left\{\tilde{f}(\omega(t))\right\} = E^{\tilde{Q}_\omega}\left\{\tilde{f}(\tau_{-x(t)}\omega)\right\}$$
$$= E^{Q_{x,\omega}}\left\{f(x(t),\omega)\right\}$$

where $f(x,\omega) = \tilde{f}(\tau_{-x}\omega)$. The statement of the lemma is seen thus to be nothing but a restatement of the Markov property for $Q_{x,\omega}$. In fact, more explicitly

$$E^{\tilde{Q}_{\omega,x}}\left\{\tilde{f}(\omega(t)) \mid \omega(\sigma),\ \sigma \le s\right\}$$

$$(3.11) \qquad = E\left\{\tilde{f}(\tau_{-x(t)}\omega) \mid \tau_{-x(\sigma)}\omega,\ \sigma \le s\right\}$$

$$= E\left\{f(x(t),\omega) \mid \omega,\ x(\sigma) \text{ for } \sigma \le s\right\}$$

$$= E\left\{f(x(t),\omega) \mid \omega,\ x(s)\right\}$$

$$= E^{Q_{x(s),\omega}}\left\{f(x(t),\omega)\right\}$$

$$= E^{\tilde{Q}_{\omega(s)}}\left\{\tilde{f}(\omega(t))\right\}$$

which is the Markov property for $\tilde{Q}_{\omega,x}$.

A simple observation is now that in fact this diffusion process $\omega(t)$ has

$$(3.12) \qquad \frac{e^{2\tilde{\Phi}(\omega)}}{E\left\{e^{2\tilde{\Phi}}\right\}} P(d\omega)$$

as an invariant measure which is ergodic.

Let us now return to (3.2). We have that

$$(3.13) \qquad \frac{1}{t}\int_0^t (1+\chi'(x(s),\omega))^2 ds = \frac{1}{t}\int_0^t (1+\tilde{\psi}(\tau_{-x(s)}\omega))^2 ds$$

$$= \frac{1}{t}\int_0^t \left[1+\tilde{\psi}(\omega(s))\right]^2 ds$$

where, since $\chi'(x,\omega)$ is stationary (cf. (2.5))

$$(3.14) \qquad \tilde{\psi}(\omega) = \frac{e^{-2\tilde{\Phi}(\omega)}}{E(e^{-2\Phi})} - 1$$

so that $\tilde{\psi}(\tau_{-x}\omega) = \chi'(x,\omega)$. We can now apply the ergodic theorem in (3.13) to obtain that

$$(3.15)\qquad \lim_{t\uparrow\infty} \frac{1}{t}\int_0^t (1+\chi'(x(s),\omega))^2 ds = \int_\Omega (1+\tilde{\psi}(\omega))^2 \frac{e^{+2\tilde{\Phi}(\omega)}}{E\left\{e^{2\Phi}\right\}} P(d\omega)$$

$$= E(e^{-2\Phi})^{-1} E\left\{e^{2\Phi}\right\}^{-1} ,$$

almost everywhere with respect to (3.12) and hence with respect to P.

This completes the proof of the theorem.

4. OTHER ONE DIMENSIONAL PROBLEMS

In one dimension one can carry out computations explicitly to an extent that is impossible in more dimensions. Not surprisingly, the theorem of section 1 can be generalized considerably and one can compute corrections to it in the sense that the next term in

$$(4.1)\qquad E\left\{\frac{x(t)-x(0))^2}{t}\right\} = \sigma^2 + o(1) , \qquad t \to \infty$$

can be calculated. This will depend now on further assumptions about $\Phi(x,\omega)$ and not just on stationarity.

The most interesting aspect of one dimensional problems is that in a case considered by Sinai [1] the behavior of $x(t)$ is not diffusive (i.e., linear in t mean square displacement for large t) but $x(t) \sim (\log t)^2$ in a suitable sense. In Sinai's case $\Phi'(x,\omega)$ is not the derivative of a bounded stationary function as in (1.2) but white noise, i.e., the derivative of Brownian motion. (Sinai actually did the equivalent random walk problem.) And of course Brownian motion is neither stationary nor bounded. Sinai's result is a striking instance of localization in a random medium which is not felt by gradient systems like (1.3).

5. MULTIDIMENSIONAL PROBLEMS

The theorem of section 1 is proved in the general d-dimensional case in [2], in [3] and in [4]. In [2] the operator L_ω (corresponding to (2.1)) has the form

$$(5.1)\qquad L_\omega = \frac{1}{2}\sum_{i,j=1}^{d} \frac{\partial}{\partial x_i}\left(a_{ij}(x,\omega)\frac{\partial}{\partial x_j}\right)$$

with $(a_{ij}(x,\omega))$ stationary processes that satisfy a uniform ellipticity condition. The operator

$$(5.2)\qquad L_\omega = \frac{1}{2}\Delta + \sum_{j=1}^{d} \frac{\partial\Phi(x,\omega)}{\partial x_j}\frac{\partial}{\partial x_j}$$

reduces to (5.1) if we note that (cf. (2.1)) we can write it as

$$L_\omega = \frac{1}{2}\frac{1}{e^{2\Phi(x,\omega)}}\sum_{j=1}^{d}\frac{\partial}{\partial x_j}\left(e^{2\Phi(x,\omega)}\frac{\partial}{\partial x_j}\right) .$$

Random walk versions of these results are proven in [5] and [6]. In [3] the nondivergence form of (5.1) is considered

$$L_\omega = \frac{1}{2}\sum_{i,j=1}^{d} a_{ij}(x,\omega)\frac{\partial^2}{\partial x_i \partial x_j} .$$

An interesting case contained in (5.1) is when the $(a_{ij}(x,\omega))$ do not form a symmetric matrix. Operators of the form

$$\text{(5.3)} \qquad L_\omega = \frac{1}{2}\Delta + \sum_{j=1}^{d} b_j(x,\omega)\frac{\partial}{\partial x_j}$$

can be reduced to (5.1) with nonsymmetric (a_{ij}) when for example

$$\text{(5.4)} \qquad 0 = \sum_{j=1}^{d}\frac{\partial}{\partial x_j} b_j(x,\omega) .$$

In three or more dimensions there will exist then a skew symmetric matrix $(a_{ij}^{(1)}(x,\omega))$ of stationary processes such that (5.3) transforms to (5.1) with $(a_{ij}(x,\omega)) = (\delta_{ij} + 2a_{ij}^{(1)}(x,\omega))$. In 3 dimensions this means that the vector field b can be written as the curl of a stationary field when it is divergence free. This observation was made in [7]. It was developed extensively by Ossada [8]. S. Kozlov pointed out that it also works in two dimensions (private communication).

The fully general case (5.3) with no assumptions on the vector field $b(x,\omega)$ except stationarity, ergodicity and mean zero (with say conditions on smoothness and boundedness) is a major open problem. From Sinai's one-dimensional problem [1] we know that we do not always have diffusive behavior. It is widely anticipated that when $b(x,\omega)$ is a small perturbation of a constant field and the dimension is larger than or equal to two then again the asymptotic behavior is diffusive. Additional hypotheses are probably needed as can be seen from a theorem of Kozlov [9] that treats this case when $b(x,\omega)$ is a quasi-periodic vector field whose periods satisfy a strong nonresonance condition (a KAM condition).

Another interesting question arises in connection with diffusion in a random medium with

(5.5) $$L_\omega = \frac{\kappa}{2}\Delta + \sum_{j=1}^{d} b_j(x,\omega)\frac{\partial}{\partial x_j}$$

and under condition (5.4) again, with $b(x,\omega)$ stationary. There is a numerical result of Kraichnan [10] that indicates the limiting diffusion constant $\sigma^2(\kappa)$ for (5.5), which we know is well defined by the theory of [2] (of also [7]), tends to a positive constant as $\kappa \to 0$. Mathematically this should translate to a conjecture that if $b(x,\omega)$ is a divergence free field that is sufficiently mixing and the dimension is bigger than or equal to 3, then $\sigma(\kappa)$ has a positive limit as $\kappa \to 0$. This seems to be an extremely difficult analytical problem.

In one special case something can be done however. If $b(x,\omega)$ has the form

(5.6) $$b = b_0 + \varepsilon b_1(x,\omega)$$

then presumably the limit

$$\lim_{\kappa \downarrow 0} \sigma^2(x,\varepsilon) = \sigma^2(0,\varepsilon)$$

exists and

(5.7) $$\sigma^2(0,\varepsilon) = \varepsilon^2 \int_0^\infty E\left\{b_1(x+b_0 s)b_1(x)\right\} ds + o(\varepsilon^2)$$

as $\varepsilon \to 0$. The coefficient of ε^2 in (5.7) is the so-called Kubo formula for the diffusion coefficient. The coefficient of ε^2 can be recovered in a simpler way via a weak coupling limit as in [11]. Various asymptotic formulas for diffusion coefficients with respect to κ and ε and other parameters are given in [12].

6. SELFDIFFUSION OF INTERACTING BROWNIAN MOTIONS

The main point of the theorem in section 1 is that we have found a fairly general way to get the central limit theorem for $x(t)$ given by

$$x(t) = x + \int_0^t \Phi'(x(s),\omega)ds + w(t) .$$

In view of the discussion in section 3, the whole analysis reduces to showing the central limit theorem for

$$(6.1)\qquad y(t) = \int_0^t D\tilde{\Phi}(\omega(s))ds$$

where $\omega(t)$ is the process described in the Lemma of section 3. It is an infinite dimensional, reversible diffusion process with invariant measure given by (3.12). Moreover the integrand in (6.1) is a gradient. Varadhan noted that these are the only essential hypotheses required to prove the central limit theorem. A surprisingly large number of apparently diverse problems can be handled this way [13]. Selfdiffusion of interacting Brownian particles is an example. In [14] this problem was analyzed without taking advantage of this observation, by exploiting the analytical techniques of [2]. The analytical techniques are of course not much different from the ones in [13]. Occasionally they can be used in problems that can not be formulated via martingales [15]. Let us briefly describe the selfdiffusion problem [14].

A system of infinitely many interacting Brownian particles in $\mathbb{R}^d$ is described by the system of stochastic differential equations

$$(6.2)\qquad dx_i(t) = \sum_{\substack{j=1\\ j\neq i}}^{\infty} \nabla\Phi(x_i(t)-x_j(t))dt + \sqrt{2}\,dw_i(t)\ , \qquad i=1,2,\ldots$$

where $x_i(t)$ is the position vector in $\mathbb{R}^d$ of the particle labelled by i and $\Phi(x)$ is the pair interaction potential assumed smooth, positive, even and with compact support. The $\{w_i(t)\}$ $i = 1,2,\ldots$ are independent standard Brownian Motions in $\mathbb{R}^d$. Lang [16] has studied the existence theory of (6.2) and the ergodicity of the Gibbs measure [17] which is an invariant measure with respect to which (6.2) is reversible.

Following a suggestion of Kipnis (private communication) the selfdiffusion coefficient for (6.2) can be turned into a problem similar to the one of section 1 for random media. What is the selfdiffusion problem? Give the particle labeled by subscript 1 black color and all the rest white color. Set

$$(6.3)\qquad \begin{cases} \xi(t) = x_1(t) \\ y_i(t) = x_{i+1}(t) - x_1(t)\ , \qquad i=1,2,\ldots\ . \end{cases}$$

The system of stochastic differential equations for $\xi(t),(y_1(t),y_2(t),\ldots)$ is then

$$(6.4)\qquad d\xi(t) = \sum_{i=1}^{\infty} \nabla\Phi(y_i(t))dt + \sqrt{2}\, dw_1(t)$$

$$(6.5)\qquad dy_i(t) = -\sum_{\substack{j=1 \\ j\neq i}}^{\infty} \nabla\Phi(y_i(t)-y_j(t))dt - \nabla\Phi(y_i(t))dt$$

$$-\sum_{j=1}^{\infty} \nabla\Phi(y_j(t))dt + \sqrt{2}\, dw_{i+1}(t) - \sqrt{2}\, dw_1(t)\ , \quad i=1,2,\dots\ .$$

The important thing to notice here is that the black particle satisfies an equation like (1.2) (in $\mathbb{R}^d$) or (6.1) with $\omega(t) = (y_1(t), y_2(t), \dots)$. That is the environment looked at from the relative frame is just the position of the white particles relative to the position of the black. Moreover, except for the $w_1(t)$ on the right side of (6.5), the environment evolves independently. So it is not surprising that $t^{-\frac{1}{2}}(\xi(t)-\xi(0))$ obeys the central limit theorem as $t\to\infty$. The variance of the resulting gaussian is the selfdiffusion coefficient.

The detailed treatment of this problem by extending the methods in [2] is carried out in [14].

REFERENCES

[1] Yu. Sinai, Limit behavior of one-dimensional random walks in random environments, Theory Prob. Appl. 27 (1982) pp. 247-258.

[2] G.C. Papanicolaou and S.R.S. Varadhan, Boundary value problems with rapidly oscillating random coefficients, Seria Coll. Math. Soc. J. Bolyai, Vol. 27, pp. 835-873, North Holland, 1981.

[3] G.C. Papanicolaou and S.R.S. Varadhan, Diffusions with random coefficients, in Essays in Statistics and Probability, edited by G. Kallianpur, P.R. Krishnaiah and J.K. Ghosh, North Holland, 1982, pp. 253-262.

[4] S. Kozlov, Differential equations with almost periodic and random coefficients, DAN, 236 (1977) pp. 1068-1071

[5] R. Kunnemann, The diffusion limit for reversible jump processes on Z^d with ergodic random bond conductivities, Comm. Math. Phys. 90 (1983) pp. 27-68.

[6] G. Lawler, A discrete stochastic integral inequality and balanced random walk in a random environment, Duke Univ. Math. J. 50 (1983) pp. 1261-1274.

[7] G. Papanicolaou and O. Pironneau, On the asymptotic behavior of motion in random flows, in Stochastic Nonlinear Systems, edited by L. Arnold and R. Lefever, Springer, New York, 1981, pp. 36-41.

[8] H. Ossada, Homogenization of diffusion processes with random coefficients, preprint, Kyoto University.

[9] S. Kozlov, Ground states of stochastic quasiperiodic operators, DAN 271 1983, pp. 532-538.

[10] R. Kraichnan, Diffusion by a random velocity field, Phys. Fluids 13 (1970) pp. 22-31.

[11] H. Kesten and G. Papanicolaou, A Limit theorem for turbulent diffusion, Comm. Math. Phys. 65 (1979) pp. 97-128.

[12] C.L. Winter, C.M. Newman and S.P. Neuman, A perturbation expansion for diffusion in a random velocity field, SIAM J. Appl. Math. 44 (1984) pp. 411-424.

[13] C. Kipnis and S.R.S. Varadhan, Central limit theorem for additive functionals of reversible Markov processes and application to simple exclusions, to appear.

[14] M.Z. Guo, Limit theorems for interacting particle systems, Ph.D. thesis, Courant Institute, September 1984.

[15] K. Golden and G. Papanicolaou, Bounds for effective parameters of heterogeneous media by analytic continuation, Comm. Math. Phys., 90 (1983) pp. 473-491.

[16] R. Lang, Unendlichdimensionale Wiener prozesse in Wechselwirkung I, II, Z. Wahr. Verw. Geb. 38 (1977) pp. 55-72, 39 (1977) pp. 277-299.

[17] D. Ruelle, Statistical Mechanics, New York, Benjamin, 1969.

G. C. Papanicolaou
Courant Institute, New York University
251 Mercer Street
New York, New York 10012

M. Z. Guo
Department of Mathematics
Beijing University
Beijing, Peoples Republic of China

Contemporary Mathematics
Volume 41, 1985

A WEAK VERSION OF RSW THEOREM FOR THE TWO-DIMENSIONAL ISING MODEL

Yasunari Higuchi

ABSTRACT. We present a very weak version of Russo-Seymour-Welsh's famous result in classical percolation problem, for the two-dimensional Ising model with attractive interaction in the case of zero external magnetic field. By using this result, we give a simple proof of the fact that there exist no infinite clusters above the critical temperature, which was proved by Coniglio et al. ([1]).

1. INTRODUCTION.

In [3] and [5], Russo and Seymour & Welsh discovered an important relation between classical percolation problem and the so-called sponge percolation problem, independently. The key statement for connecting these two percolation problems appears both in [3] and in [5] just in the same manner. Therefore this statement is now called RSW Theorem (see [2]).

To state their result, we need some notations and terminologies. The terminologies we adopt are mainly due to Russo, [4].

Let Z^2 be the square lattice, whose element is expressed by a pair of integers; $x = (x^1, x^2)$. Two points $x, y \in Z^2$ are said to be *nearest neighbors* (denote it by $\langle x, y\rangle$) if and only if

$$|x^1 - y^1| + |x^2 - y^2| = 1,$$

and **nearest neighbors* (denote it by $\langle x, y\rangle^*$) if and only if

$$\max\{|x^1 - y^1|, |x^2 - y^2|\} = 1.$$

A *chain* (**chain*) is a sequence of points $x_1, x_2, \ldots, x_n \in Z^2$, such that $\langle x_i, x_j\rangle$ ($\langle x_i, x_j\rangle^*$) is equivalent to $|i - j| = 1$. A subset $V \subset Z^2$ is said to be *connected* (**connected*) if for any pair of points $x, y \in V$, there exists a chain γ (a *chain δ) in V such that $x, y \in \gamma$ ($x, y \in \delta$). A *cluster* (**cluster*) of a subset $V \subset Z^2$ is a maximal connected

---

1980 Mathematics Subject Classification. 60K35.

(*connected) component of V. A *circuit* (**circuit*) is a sequence of points $\{x_1, x_2, \dots, x_n\}$ such that both $\{x_2, x_3, \dots, x_n\}$, $\{x_n, x_1, \cdots, x_{n-2}\}$ are chains (*chains). Let π be a half space of Z^2, i.e. for some integer k, either one of the following i) ~ iv) holds;

i) $\pi = \{x \in Z^2, x^1 \geq k\}$,

ii) $\pi = \{x \in Z^2; x^1 < k\}$,

iii) $\pi = \{x \in Z^2; x^2 \geq k\}$,

iv) $\pi = \{x \in Z^2, x^2 < k\}$.

A *half circuit* (**half circuit*) of π is a chain $\gamma = \{x_1, x_2, \dots, x_n\}$ (a chain $\delta = \{x_1, x_2, \dots, x_n\}$) such that

a) $\gamma \subset \pi$

b) $\gamma \cap \partial^-\pi = \{x_1, x_n\}$,

where $\partial^-\pi$ is the inner boundary of π, i.e. corresponding to above i) ~ iv) cases, $\partial^-\pi$ is given by

$$\partial^-\pi = \{x^1 = k\} \qquad \text{for i), ii),}$$

$$\partial^-\pi = \{x^2 = k\} \qquad \text{for iii), iv).}$$

Let $\Omega \equiv \{-1, +1\}^{Z^2}$, with product topology, and F be its topological Borel field. Then (Ω, F) is the basic measurable space we are going to consider. The above notions are now carried into Ω-space. For $\omega \in \Omega$, *+cluster of* ω is a cluster of $\omega^{-1}(+1) \equiv \{x \in Z^2; \omega(x) = 1\}$, a *+chain of* ω is a chain in $\omega^{-1}(+1)$, a *+circuit*, a *+half circuit* are a circuit in $\omega^{-1}(+1)$, and a half circuit in $\omega^{-1}(+1)$ respectively. *-cluster*, *+*cluster*, *-*cluster*, *-chain*, *+*chain*, *-*chain*, *-circuit*, *+*circuit*, *-*circuit*, *-half circuit*, *+*half circuit*, *-*half circuit* are defined in the same way.

For positive integers m, n, let $\Lambda(m, n)$ be the following rectangle;

$$\Lambda(m, n) = \{x \in Z^2; |x^1| \leq m, |x^2| \leq n\}$$

and $A^+(m, n)$ ($A^{+*}(m, n)$) be the event that there exists a +chain γ (a $+^*$chain δ) in $\Lambda(m, n)$, connecting two vertical sides of $\Lambda(m, n)$, i.e.

$$\gamma \subset \Lambda(m, n), \ |\gamma \cap \{x^1 = -m\}| = 1, \ |\gamma \cap \{x^1 = m\}| = 1.$$

$$(\delta \subset \Lambda(m, n), \ |\delta \cap \{x^1 = -m\}| = 1, \ |\delta \cap \{x^1 = m\}| = 1.)$$

RSW THEOREM ([3], [5])

Let ν_ρ, $0 \leq \rho \leq 1$, *be the Bernoulli probability measure on* (Ω, F) *with density* ρ: $\nu_\rho(\omega(0) = +1) = \rho$. *Then there exist functions* f, g *such that*

$$\nu_\rho(A^{+(*)}(2n, n)) \geq f(\nu_\rho(A^{+(*)}(n, n))),$$

$$\nu_\rho(A^{+(*)}(3n, n)) \geq g(\nu_\rho(A^{+(*)}(n, n))),$$

where f, g *are continuous, increasing functions such that*

$$f(0) = g(0) = 0, \quad f(1) = g(1) = 1.$$

REMARK. In [4], f, g are taken as in the following:

$$f(x) = x(1 - \sqrt{1 - x})^6, \quad g(x) = x^3(1 - \sqrt{1 - x})^{12}.$$

2. A VERY WEAK VERSION OF RSW THEOREM.

Now let us turn to our problem. We are going to discuss a similar problem as above, but with an extremal Gibbs state of the two-dimensional Ising model (i.e. μ_+ and μ_-). The Hamiltonian of the system we consider is given by

$$(1) \qquad H_\Lambda^\omega(\sigma) = - \sum_{\substack{x\cdot y\in\Lambda \\ \langle x,y\rangle}} \sigma(x)\sigma(y) - \sum_{\substack{x\in\Lambda, \\ y\in\Lambda^c \\ \langle x,y\rangle}} \sigma(x)\omega(y) \qquad \sigma \in \{-1, +1\}^\Lambda$$

for any $\omega \in \Omega$, and any finite $\Lambda \subset Z^2$. To discuss the properties of an extremal Gibbs state corresponding to (1), it is convenient to introduce some transformations on Ω:

(a) *Translation;* $\{\tau_x; \ x \in Z^2\}$.

$$(\tau_x\omega)(y) = \omega(x + y), \qquad y \in Z^2.$$

(b-1) *Reflection with respect to vertical lines;*

$$\{V_k; \ k = \pm 1, \pm 2, \ldots\}.$$

$$(V_k\omega)(x) = \omega(k - x^1, x^2).$$

(b-2) *Reflection with respect to horizontal lines;*

$$\{W_k; \ k = \pm 1, \pm 2, \ldots\}.$$

$$(W_k\omega)(x) = \omega(x^1, k-x^2).$$

(c) *Rotation;* R.

$$(R\omega)(x) = \omega(-x^2, x^1).$$

(d) *Spin reversal;* T.

$$(T\omega)(x) = -\omega(x).$$

As an extremal Gibbs state corresponding to (1), we consider a probability measure μ on (Ω, F) having the following properties;

(i) μ is trivial on $F_\infty \equiv \bigcap_{\Lambda\subset Z^2:\text{finite}} F_{\Lambda^c}$,

where for any $\Lambda \subset Z^2$, $F_\Lambda \equiv \sigma\{\omega(x);\ x \in \Lambda\}$

(ii) for any finite $\Lambda \subset Z^2$,

$$\mu(\cdot|F_{\Lambda^c})(\omega) = P_\Lambda^\omega(\cdot) \qquad \mu\text{-a.s.},$$

where $P_\Lambda^\omega(\cdot)$ is a probability measure on $\{-1, +1\}^\Lambda$ such that

$$P_\Lambda^\omega(\sigma) \propto \exp\{-\beta H_\Lambda^\omega(\sigma)\}, \qquad \sigma \in \{-1, +1\}^\Lambda.$$

(iii) μ is invariant under translations, reflections, and the rotation.

(iv) μ satisfies FKG inequality.

THEOREM 1.

If μ *is an extremal Gibbs state for* (1) *in two dimension, then we have functions* $\hat{f}$, $\hat{g}$ *such that*

$$\mu(A^{+*}(2n, n)) \geq \hat{f}(\mu(A^{+*}(n, n))),$$

$$\mu(A^{+*}(3n, n)) \geq \hat{g}(\mu(A^{+*}(n, n))),$$

where we can thake $\hat{f}$, $\hat{g}$ *as*

$$(2)\qquad \begin{cases} \hat{f}(x) = \{x(1 - \sqrt{1-x})^2/2\}^2, \\ \hat{g}(x) = \hat{f}(x)\{x(1 - \sqrt{1-x})^4\}/4, \end{cases}$$

or, simply

$$(3)\qquad \begin{cases} \hat{f}(x) = (x^3/8)^2, \\ \hat{g}(x) = \hat{f}(x)\cdot x^5/2^6. \end{cases}$$

REMARK. As is easily seen, Theorem 1 is not a satisfactory result. We can control only *clusters, but not clusters. Also functions $\hat{f}$, $\hat{g}$ are very bad because $\hat{f}(1)$, $\hat{g}(1)$ are much smaller than 1, so that we can not expect having a statement that

"If $\lim_{n\to\infty} \mu(A^{+*}(n, n)) = 1$, then $\lim_{n\to\infty} \mu(A^{+*}(3n, n)) = 1$."

which is a crucial point for applications. Still, we have an application of our theorem, which we will give in section 3.

PROOF OF THEOREM 1.

Let $B^{+*}(n, n)$ be the event that there exists a $+^*$chain δ in $\Lambda(n, n)$

connecting $\{x^1 = -n\}$ with $\{x^1 = n,\ x^2 \geq 0\}$, and $W_0[B^{+*}(n, n)]$ be the reflection of the event $B^{+*}(n, n)$ with respect to x^1- axis. Obviously,

$$B^{+*}(n, n) \cup W_0[B^{+*}(n, n)] = A^{+*}(n, n),$$

and since μ is invarinat under W_0, we get

$$\mu(B^{+*}(n, n)) = \mu(W_0[B^{+*}(n, n)]) \geq \mu(A^{+*}(n, n))/2. \tag{4}$$

Or, as usual we can argue in the following; Since $B^{+*}(n, n)$ and $W_0[B^{+*}(n, n)]$ are positive events, and since

$$\mu((B^{+*}(n, n))^c \cap (W_0[B^{+*}(n, n)])) = 1 - \mu(A^{+*}(n, n)),$$

from FKG inequality we obtain

$$\begin{aligned} \mu(B^{+*}(n, n)) &= \mu(W_0[B^{+*}(n, n)]) \\ &\geq 1 - \sqrt{1 - \mu(A^{+*}(n, n))}. \end{aligned} \tag{5}$$

By FKG inequality and rotation invariance of μ, we have

$$\begin{aligned} &\mu(B^{+*}(n, n) \cap W_0[B^{+*}(n, n)] \cap R[A^{+*}(n, n)]) \\ &\qquad \geq \mu(A^{+*}(n, n))\cdot\mu(B^{+*}(n, n))^2. \end{aligned} \tag{6}$$

If $\omega \in \{B^{+*}(n, n)\} \cap \{W_0[B^{+*}(n, n)]\} \cap \{R[A^{+*}(n, n)]\}$, then there exists a $+^*$half circuit S of $\{x^1 \leq n\}$, such that

(#) S surrounds $(n, 0)$, and

(##) $S \subset \Lambda(n, n)$.

For any *half circuit S of $\{x^1 \leq n\}$, satisfying (#) and (##), let

$$E_S = \left\{ \omega \in B^{+*}(n,n) \cap W_0[B^{+*}(n,n)] \cap R[A^{+*}(n,n)] \;\middle|\; \begin{array}{l} S \text{ is the maximal} \\ +^*\text{half circuit of} \\ \{x^1 \leq n\} \text{ satisfying} \\ (\#) \text{ and } (\#\#) \end{array} \right\}$$

and S' be the reflection of S with respect to $\{x^1 = n\}$. Let $\Lambda(S)$ be the region enclosed by $S \cup S'$, and F_S^{+*} be the event that either $(0, n)$ is surrounded by a $+^*$circuit in $\Lambda(S)$ which is $+^*$connected to S in $\Lambda(S)$, or $(0, n)$ is $+^*$connected to S in $\Lambda(S)$. Then, by FKG inequality we have

$$\mu(F_S^{+*} | E_S) \geq \mu(F_S^{+*} | E_S \cap \{\omega(x) = -1,\ x \in S' \backslash S\}).$$

The right hand side is equal to

$$P_{(S)}^{\omega^{\pm}, n}(F_S^{+*}),$$

where $\omega^{\pm,n}$ is defined by

$$\omega^{\pm,n}(x) = \begin{cases} +1 & \text{if } x^1 \le n \\ -1 & \text{if } x^1 > n. \end{cases}$$

Reversing all spin in $\Lambda(S) \cup S \cup S'$ and taking reflection with respect to $\{x^1 = n\}$, we get

$$P^{\omega^{\pm,n}}_{\Lambda(S)}(F^{+*}_S) = P^{\omega^{\pm,n-1}}_{\Lambda(S)}(T\circ V_{2n}[F^{+*}_S]) \ge P^{\omega^{\pm,n}}_{\Lambda(S)}(T\circ V_{2n}[F^{+*}_S]).$$

The last inequality follows from FKG inequality. Noting that $T\circ V_{2n}[F^{+*}_S]$ is the event that $(n, 0)$ is either surrounded by a +*circuit in $\Lambda(S)$ which is connected by a +*chain in $\Lambda(S)$ with S, or +*connected directly with S in $\Lambda(S)$, it is trivial that

$$P^{\omega^{\pm,n}}_{\Lambda(S)}(F^{+*}_S \cup T\circ V_{2n}[F^{+*}_S]) = 1,$$

and hence we have

$$P^{\omega^{\pm,n}}_{\Lambda(S)}(F^{+*}_S) \ge 1/2. \tag{7}$$

Denoting by $F^{+*}(n)$ the event that $(n, 0)$ is either surrounded by a +*circuit in $\Lambda(2n, n) + (n, 0) = \{x + (n, 0);\ x \in \Lambda(2n, n)\}$, which is +*connected with $\{x^1 = -n\}$ in $\Lambda(2n, n) + (n, 0)$, or $(n, 0)$ is +*connected with $\{x^1 = -n\}$ in $\Lambda(2n, n) + (n, 0)$, we have from (6) and (7),

$$\begin{aligned} \mu(F^{+*}(n)) &\ge \mu(B^{+*}(n,n) \cap V_0[B^{+*}(n,n)] \cap R[A^{+*}(n,n)] \cap F^{+*}(n)) \\ &= \sum_S \mu(E_S \cap F^{+*}(n)) \\ &\ge \sum_S \mu(E_S \cap F^{+*}_S) \\ &\ge \sum_S \mu(E_S)/2 \\ &\ge \mu(A^{+*}(n, n))\cdot\mu(B^{+*}(n, n))^2/2. \end{aligned}$$

Thus, by FKG inequality, we get

$$\begin{aligned} \mu(F^{+*}(n) \cap V_{2n}[F^{+*}(n)]) \\ \ge \{\mu(A^{+*}(n, n))\cdot\mu(B^{+*}(n, n)^2/2\}^2. \end{aligned} \tag{8}$$

But if $\omega \in F^{+*}(n) \cap V_{2n}[F^{+*}(n)]$, then we can find a +*chain δ in ω, connecting two vertical sides of $\Lambda(n, n) \cup \{\Lambda(n, n) + (2n, 0)\}$.

From the translation invariance of μ and from (8), we get

$$\mu(A^{+*}(2n, n)) \ge \{2^{-1}\cdot\mu(A^{+*}(n, n))\mu(B^{+*}(n, n))^2\}^2,$$

which, together with (4) and (5), implies the first part of (2) and (3).

The second part of (2) and (3) will be easy applications of this argument. (Q.E.D.)

3. AN APPLICATION OF THEOREM 1.

In this section, we give a simple alternative proof of Theorem 2, [1].

THEOREM 2 (CONIGLIO-NAPPI-PERUGGI-GUSSO [1]).

If μ *is an extremal Gibbs state for* (1) *in two dimension, then there is no coexistence of infinite +cluster and infinite -cluster* μ*-a.s.. Especially, if* $\beta \le \beta_c$, *then there is neither infinite +cluster nor infinite -cluster* μ*-a.s..*

PROOF.

It is enough to show that

$$\mu_+(\text{there exists an infinite -cluster}) = 0,$$

since $\mu_+ \circ T = \mu_-$, where T: $\Omega \to \Omega$ is the spin reversal transformation defined by (d). Note that

$$\mu_+(A^{+*}(n, n) \cup A^-(n, n)) = 1, \tag{9-a}$$

and

$$\mu_+(A^{+*}(n, n)) \ge \mu_-(A^{+*}(n, n)) = \mu_+(A^{-*}(n, n)) \ge \mu_+(A^-(n, n)). \tag{9-b}$$

The first inequality follows from FKG inequality. Thus, we obtain for any $n \ge 1$,

$$\mu_+(A^{+*}(n, n)) \ge 1/2.$$

Then, from Theorem 1, we have

$$\mu_+(A^{+*}(3n,n)) \ge \hat{g}(1/2).$$

From this, and from FKG inequality, we obtain

$$\begin{aligned}
&\mu_+(\text{there exists a +*circuit surrounding } \Lambda(N, N)) \\
&\ge \mu_+(\tau_{(-3N,0)} \circ R[A^{+*}(3N, N)] \cap \tau_{(0,-3N)}[A^{+*}(3N, N)] \cap \\
&\qquad \cap \tau_{(3N,N)} \circ R[A^{+*}(3N, N)] \cap \tau_{(0,3N)}[A^{+*}(3N, N)]) \\
&\ge \{\mu_+(A^{+*}(3N, N))\}^4 \\
&\ge \hat{g}(1/2)^4 > 0.
\end{aligned}$$

By tail triviality (i) of μ_+, this implies

μ_+(for infinitely many N, there exist + circuits surrounding $\Lambda(N, N)$)

= 1,

which proves that

μ_+(there exists an infinite -cluster) = 0.

(Q.E.D.)

BIBLIOGRAPHY.

1. Coniglio, A., Nappi, C. R., Peruggi, F., Russo, L.: Comm. Math. Phys., 51, 315-323 (1976).

2. Kesten, H.: Percolation theory for mathematicians, Birkhauser, Boston-Basel-Stuttgart, 1982.

3. Russo, L.: Z. Wahrscheinlichkeitstheorie verw. Geb., 43, 39-48 (1978).

4. Russo, L.: i. b. i. d., 56, 229-238 (1981).

5. Seymour, P. D., Welsh, D. J. A.: Ann. Discrete Math., 3, 227-245 (1978).

DEPARTMENT OF MATHEMATICS
KOBE UNIVERSITY
ROKKO, KOBE 657
JAPAN

Contemporary Mathematics
Volume 41, 1985

POSSIBLE RATES OF CONVERGENCE IN FINITE RANGE, ATTRACTIVE SPIN SYSTEMS

Richard Holley[1]

ABSTRACT. We prove that for finite range, attractive spin systems the rate of convergence to equilibrium is either exponentially fast or it is no faster than the reciprocal of a polynomial of degree d, where d is the dimension of the underlying space. Applications are then given to contact processes and the stochastic Ising model. For the stochastic Ising model new sufficient conditions for exponentially fast convergence are given, and different necessary conditions for exponentially fast convergence are also given. These conditions apply in any dimension.

0. Introduction

As an illustration of the type of phenomenon described by our main theorem consider a finite state space continuous time Markov process with state space $S = \{0,1,\ldots,N\}$. Let $P_t(x,y)$ be the transition function for this Markov process. We are concerned with the rate of convergence of $P_t(x,y)$ to the stationary measure. With finite S there are only two cases. Either there is only one recurrence class in which case there is an $\varepsilon > 0$ such that for all $x_1,x_2,y \in S$

$$|P_t(x_1,y) - P_t(x_2,y)| < e^{-\varepsilon t}$$

or there are more than one recurrence class in which case for some $x_1,x_2,y \in S$ $|P_t(x_1,y) - P_t(x_2,y)| \not\to 0$.

This paper resulted from an attempt to find a similar dichotomy for infinitely many interacting stochastic processes with finite state space S. We were motivated to do this by the results in [5], which when applied to a d-dimensional ferromagnetic (attractive) stochastic Ising model of sufficiently high temperatures yield convergence to equilibrium at a rate which is at least as fast as $\exp(-\varepsilon t^{1/d})$ for some $\varepsilon > 0$. Theorem (0.1) below, which is proved in section 2, is more than adequate to prove that if the convergence is as fast as $\exp(-\varepsilon t^{1/d})$ then it must be exponentially fast.

1980 Mathematics Subject Classification. Primary 60K35; secondary 82A05
[1]Research supported in part by NSF Grant MCS-8310542

Before giving a precise statement of our main result we establish some notation. Let S be as above and set $E = S^{Z^d}$. E is the state space for our interacting stochastic processes, and elements of E will be referred to as configurations and denoted by such letters as η,σ,ω. Elements of Z^d will be referred to as sites. $\eta(k) \in S$ is the value of configuration η at site k. The configuration which is identically $0(N)$ will be denoted by $\vec{0}(\vec{N})$.

The evolution of the interacting system is determined by its jump rates $\{c(k,\eta,a): k \in Z^d, \eta \in E, a \in S\}$. Intuitively the probability of a jump into the state a at site k between times t and $t+h$ is $c(k,\eta t,a)h + o(h)$ if the configuration at time t is η_t. We assume that the jump rates have finite range, R (i.e., if $\eta(j) = \sigma(j)$ for all $|k-j| \leq R$, then $c(k,\eta,a) = c(k,\sigma,a)$ for all $a \in S$). (Throughout this paper the norm $|\cdot|$ on Z^d is given by $|(k_1,\cdots,k_d)| = \max\{|k_1|,\ldots,|k_d|\}$.) We need some sort of uniform control over the jump rates and the one which we choose is to assume that the jump rates are translation invariant. That is, if $\sigma(j) = \eta(k+j)$ for all $j \in Z^d$, then $c(k,\sigma,a) = c(0,\eta,a)$ for all $a \in S$. We need one more assumption on the jump rates but we postpone describing it for a moment.

Let C be the set of continuous functions from E (with the product topology) into R. If $\Lambda \subset Z^d$ we let

$$D(\Lambda) = \{f \in C : f(\eta) = f(\sigma) \text{ if } \eta(j) = \sigma(j) \text{ for all } j \in \Lambda\}$$

and we set $D = \cup D(\Lambda)$, where the union is over all finite subsets Λ of Z^d.

The infinitesimal generator of the interacting system when restricted to D is given by

$$Lf(\eta) = \sum_{k\in Z^d} \sum_{a\in S} c(k,\eta,a)[f(\eta^{k,a}) - f(\eta)],$$

where $\eta^{k,a}$ is the configuration obtained from η by changing its value at site k to be a. It is known (see [8] or [9]) that L generates a unique positive contraction semigroup on C. This semi-group is the semi-group of the interacting stochastic processes with jump rates $\{c(k,\cdot,\cdot): k \in Z^d\}$, and we denote it by $\{T_t: t \geq 0\}$.

The final assumption which we make concerning the jump rates is that they are attractive. Formally this means that if $\sigma(j) \geq \eta(j)$ for all $j \in Z^d$, then

i) if $a < \eta(j)$, then $\sum_{b\leq a} c(0,\sigma,b) \leq \sum_{b\leq a} c(0,\eta,b)$ and

ii) if $a > \sigma(j)$, then $\sum_{b\geq a} c(0,\sigma,b) \geq \sum_{b\geq a} c(0,\eta,b)$.

This is equivalent to the following condition on $\{T_t: t \geq 0\}$ (see [4]): If $f \in C$ has the property that $f(\sigma) \geq f(\eta)$ for any pair $\sigma,\eta \in E$ with $\sigma(j) \geq \eta(j)$ for all $j \in Z^d$, then $T_t f(\cdot)$ also has this property. It is this latter condition which we will take as the operational definition of attractive jump rates.

Finally we define $f_k \in C$ by $f_k(\eta) = \eta(k)$.

(0.1) THEOREM. Let $\{c(k,\cdot,\cdot): k \in Z^d\}$ be finite range, translation invariant, attractive jump rates and let $\{T_t: t \geq 0\}$ be the corresponding semi-group. Let

$$c = \max\{\sup_{\eta} \sum_{a\in S} c(0,\eta,a) \quad \sum_{|j|\leq R} \sup_{\eta\in E} \sup_{b\in S} \sum_{a\in S} |c(0,\eta,a) - c(0,\eta^{j,h},a)|\}.$$

Then either

i) There is an $\varepsilon > 0$ such that for all $f \in D$ there is a constant $A_f < \infty$ and for all $\eta,\sigma \in E$

$$|T_t f(\eta) - T_t f(\sigma)| \leq A_f e^{-\varepsilon t}$$

or

ii) There are constants $\alpha_0 = \alpha_0(R,d)$ and $\alpha_1 = \alpha_1(R,d)$ such that

$$t^d |T_t f_0(\vec{N}) - T_t f_0(\vec{0})| > \alpha_0/c^d$$

for all $t > \alpha_1/c$.

This theorem shows that for finite range, translation invariant, attractive interacting stochastic processes in d-dimensions either the convergence to equilibrium takes place exponentially fast or it takes place at a rate at most t^{-d}. Of course it may not converge at all, a situation which we include in the latter case.

Examples of finite range, translation invariant, attractive interacting stochastic processes which have only one stationary measure and fall under case ii) of Theorem (0.1) are rare. In fact the author knows of just one such example. The two-dimensional nearest neighbor ferromagnetic stochastic Ising model at the critical temperature and with no external field is the example (see [6]). According to Theorem (0.1) the convergence to equilibrium in this model must be at most quadratically fast.

The applications of Theorem (0.1) which we have in mind involve showing that ii) fails and hence i) must hold. Whenever we are able to show that ii) fails we can show that

$$\lim_{t\to\infty} t^d (T_t f_0(\vec{N}) - T_t f_0(\vec{0})) = 0.$$

We have chosen to phrase ii) of Theorem (0.1) as it is stated in order to facilitate the proof of the following theorem.

(0.2) THEOREM. Let $\{c(k,\cdot,\cdot): k \in Z^d\}$ be the range $R < \infty$, translation invariant, attractive jump rates for an infinite system of interacting processes and suppose that the corresponding semi-group converges exponentially fast to its equilibrium on D. Then there is a $\delta > 0$ such that if $\{\bar{c}(k,\cdot,\cdot): k \in Z^d\}$ is a similar collection of jump rates and

$$\sup_{\eta \in E} \sum_{a \in S} |c(0,\eta,a) - \bar{c}(0,\eta,a)| < \delta$$

then the semi-group generated by the $\{\bar{c}(k,\cdot,\cdot): k \in Z^d\}$ also converges to equilibrium exponentially fast on D.

Theorem (0.2) says that in the appropriate topology the set of jump rates for which there is exponentially fast convergence is open. As an immediate application of this theorem note that for a contact process (see [3]) at the critical value of the parameter either the system is not ergodic or convergence to equilibrium takes place at most at rate $1/t^d$. For $d = 1$ this was already known (see [2]), however for $d \geq 2$ it appears to be new. Theorem (0.2) also has implications for the regularity of the stationary measures for the infinite system as a function of some parameter of the process. We investigate these in section 4.

Section three is devoted to an application of Theorem (0.1) to the stochastic Ising model. It takes quite a bit of notation to describe this application precisely, so we postpone that until section three. However the general theme of section 3 is that exponentially fast convergence of the semi-group of the stochastic Ising model and exponentially fast decay of the correlations of the corresponding Gibbs state are very closely related. Of course the boundary conditions must be taken into account to get a correct statement.

Finally section one contains the preparatory material and statements of known results which we need.

1. Preparatory Results

Throughout the rest of this paper we let FTAS stand for "finite range, translation invariant, attractive system(s) of infinitely many interacting stochastic processes".

This section contains some notation and known results which are needed to prove the theorems of section two. We begin with a lemma which says that for any FTAS, information about what is going on in some part of the system flows at a rate which is essentially constant. In what follows $||\cdot||$ denotes the uniform norm on C.

(1.1) LEMMA. Consider a FTAS whose jump rates $\{c(k,\cdot,\cdot): k \in Z^d\}$ have range R. Let $\{T_t: t \geq 0\}$ be the corresponding semi-group. Then if $0 \in \Lambda \subset Z^d$ and $\eta,\sigma \in E$ satisfy $\eta(k) = \sigma(k)$ for all $k \in \Lambda$ we have

$$|T_t f_0(\eta) - T_t f_0(\sigma)| \leq 4e^{ct} \frac{(ct)^{M+2}}{(M+2)!}, \tag{1.2}$$

where $M = \text{dist}(0,\Lambda^c)/R$, and c is as defined in Theorem (0.1).

The assumption of attractiveness plays no role here. The important assumption is the finite range. A proof of this lemma for the case $N = 1$ may be found in Lemma (4.15) of [7]. The changes in that proof needed to prove the present statement are merely notational.

As noted in the introduction the assumption of attractiveness means that the semi-group preserves monotone functions. Thus for all increasing functions f and all $t \geq 0$

$$E_{\vec{N}}[f(\eta_t)] \geq E_{\vec{0}}[f(\eta_t)] \tag{1.3}$$

The inequality in (1.3) means that the distribution at time t when starting from $\vec{N}$ is stochastically larger than when the initial configuration is $\vec{0}$. Thus (see [9]) there is a measure ν_t in $E \times E$ such that

$$\begin{aligned} &\nu_t(\{(\eta,\sigma): \eta(k) \geq \sigma(k) \quad \text{for all} \quad k\}) = 1 \\ &\nu_t(A \times E) = P_{\vec{N}}(\eta_t \in A) \\ &\nu_t(E \times A) = P_{\vec{0}}(\eta_t \in A). \end{aligned} \tag{1.4}$$

Thus if $f \in D(\Lambda)$ for some finite $\Lambda \subset Z^d$ and if f is increasing, then

$$\begin{aligned} &|T_t f(\eta) - T_t f(\sigma)| \leq T_t f(\vec{N}) - T_t f(\vec{0}) \\ &= \int (f(\eta) - f(\sigma))\nu_t(d\eta,d\sigma) \leq 2||f|| \sum_{k\in A} \nu_t(\eta(k) > \sigma(k)) \\ &\leq 2||f|| \; |\Lambda|(T_t f_0(\vec{N}) - T_t f_0(\vec{0})). \end{aligned} \tag{1.5}$$

For the last inequality we have used the inequality

$$\nu_t(\eta(k) > \sigma(k)) \leq \int (f_k(\eta) - f_k(\sigma))\nu_t(d\eta,d\sigma) = T_t f_k(\vec{N}) - T_t f_k(\vec{0}),$$

which follows from (1.4), and then we used translation invariance.

Since linear combinations of increasing functions in $D(\Lambda)$ make up all of $D(\Lambda)$, exponentially fast convergence of the semi-group on D is equivalent to:

$$\text{for some } \varepsilon > 0, \; T_t f_0(\vec{N}) - T_t f_0(\vec{0}) \leq Ne^{-\varepsilon t} \quad \text{for all} \quad t \geq 0. \tag{1.6}$$

2. Main Results

We begin with two calculus lemmas.

(2.1) LEMMA. If $\{a_n\}_{n=0}^{\infty}$ is a sequence of nonnegative numbers and there are numbers $0 < A, b < \infty$ such that

$$a_{n+1} \leq A_n^2 + b^{2^n}, \quad n \geq 0 \tag{2.2}$$

then

$$a_n \leq \frac{C^{2^n}}{A+1}\text{, where } C = (A+1)(a_0 \vee b^{1/2}). \tag{2.3}$$

PROOF: Clearly $a_0 \leq C^{2^0}/(A+1)$.

We now proceed by induction on n. Assume that the lemma is true for n. Then

$$a_{n+1} \leq Aa_n^2 + b^{2^n}\text{, and if } a_n^2 > b^{2^n} \text{ then}$$

$$a_{n+1} \leq (A+1)a_n^2 \leq (A+1)(C^{2^n}/(A+1))^2 = C^{2^{n+1}}/(A+1).$$

If $a_n^2 \leq b^{2^n}$ then

$$a_{n+1} \leq (A+1)b^{2^n} \leq (A+1)(C^2/(A+1)^2)^{2^n} \leq C^{2^{n+1}}/(A+1).$$

(2.4) LEMMA. If $\rho(t) \geq 0$ is a decreasing function which satisfies

$$\rho(2t) \leq A_0 t^d(\rho(t))^2 + e^{-\gamma t} \quad \text{for all } t \geq t_0, \tag{2.5}$$

where $A_0 < \infty$ and $\gamma > 0$, and if in addition there is some $t_1 \geq t_0 \vee t_0' \vee 1$ (where t_0' is chosen so that $t \geq t_0'$ implies that $2^d t^d \leq e^{(\gamma/2)t}$) such that

$$(2^d A_0+1)[t_1^d \rho(t_1) \vee e^{-\frac{\gamma}{4}t_1}] < 1\text{, then} \tag{2.6}$$

$$\rho(t) < e^{-\varepsilon t} \quad \text{for all } t \geq t_1, \tag{2.7}$$

where $\varepsilon = -\frac{1}{2t_1}\ln[(2^d A_0+1)(t_1^d \rho(t_1) \vee e^{-(\gamma/4)t_1})] > 0$.

PROOF: Let $\phi(t) = t^d\rho(t)$. Then for all $t \geq t_0 \vee t_0'$

$$\phi(2t) = 2^d t^d \rho(2t) \leq 2^d A_0(t^d\rho(t))^2 + 2^d t^d e^{-\gamma t}$$

$$\leq A(\phi(t))^2 + e^{-\frac{\gamma}{2}t},$$

where $A = 2^d A_0$. Now set $a_n = \phi(2^n t_1)$ and $b = e^{-(\gamma/2)t_1}$. Then $a_{n+1} \leq Aa_n^2 + b^{2^n}$. Therefore by Lemma (2.1), $a_n \leq (e^{-2\varepsilon t_1})^{2^n}/(A+1)$, i.e., $\phi(2^n t_1) \leq e^{-2\varepsilon 2^n t_1}/(A+1)$. Therefore

$$\rho(2^n t_1) \leq (2^n t_1)^{-d} e^{-\varepsilon 2^{n+1} t_1}/(A+1) \leq e^{-2^{n+1}\varepsilon t_1}.$$

Given $t \geq t_1$ take n such that $2^n t_1 \leq t < 2^{n+1}t_1$. Then $\rho(t) \leq \rho(2^n t_1) \leq e^{-\varepsilon 2^{n+1} t_1} < e^{-\varepsilon t}$.

(2.8) LEMMA. Let η_t be a FTAS and set

$$\rho(t) = E_{\vec{N}}[\eta_t(0)] - E_{\vec{0}}[\eta_t(0)] = T_t f_0(\vec{N}) - T_t f_0(\vec{0}).$$

Let $\{c(k,\cdot,\cdot): k \in Z^d\}$ be the jump rates of η_t and let R be the range of the $\{c(k,\cdot,\cdot): k \in Z^d\}$. Let c be as in Theorem (0.1). Then if $c' \geq c$, and we set $A = (1 + 8Rc')^d$ and $\gamma = c'(4 \ln 4 - 5 \ln 3)$, we have

$$\rho(2t) \leq t^d A(\rho(t))^2 + e^{-\gamma t} \quad \text{for all} \quad t \geq [5(\ln 3-1)c']^{-1}. \tag{2.9}$$

PROOF: Let γ_t be as in section 1. Then

$$\rho(t) = E_{\vec{N}}[\eta_t(0)] - E_{\vec{0}}[\eta_t(0)] = \iint(\eta(0) - \sigma(0))\nu_t(d\eta,d\sigma).$$

Similarly

$$\begin{aligned}\rho(2t) &= E_{\vec{N}}[T_t f_0(\eta_t)] - E_{\vec{0}}[T_t f_0(\eta_t)] \\ &= \iint(T_t f_0(\eta) - (T_t f_0(\sigma))\nu_t(d\eta,d\sigma).\end{aligned}$$

Let $U = \{(\eta,\sigma) \in E \times E: \eta(k) = \sigma(k) \text{ for all } |k| \leq 4Rc't\}$. Then by Lemma (1.1) and the monotonicity of $T_t f_0$,

$$\begin{aligned}\rho(2t) &= \int_U\int(T_t f_0(\eta) - T_t f_0(\sigma))\nu_t(d\eta,d\sigma) \\ &\quad + \int_{U^c}\int (T_t f_0(\eta) - T_t f_0(\sigma))\nu_t(d\eta,d\sigma) \\ &\leq \int_U\int 4e^{c't}\frac{(c't)^{[4c't+2]}}{[4c't+2]!}\nu_t(d\eta,d\sigma) + \rho(t)\nu_t(U^c).\end{aligned}$$

From Stirling's formula we have

$$4e^{c't}\frac{(c't)^{[4c't+2]}}{[4c't+2]!} < 4e^{c't}(\tfrac{e}{4})^{[4c't+2]} < e(e^5 4^{-4})^{c't} < e^{-\gamma t}$$

if $t \geq [5(\ln 3-1)c']^{-1}$. Also

$$\begin{aligned}\nu_t(U^c) &\leq \sum_{|k|\leq 4Rc't} \nu_t((\eta,\sigma): \eta(k) > \sigma(k)) \\ &\leq \sum_{|k|\leq Rc't} \iint (\eta(k) - \sigma(k))\nu_t(d\eta,d\sigma) \\ &= \sum_{|k|\leq 4Rc't}(T_t f_k(\vec{N}) - T_t f_k(\vec{0})) \leq (1+8Rc')^d t^d \rho(t).\end{aligned}$$

Substituting our last two inequalities into the right side of (2.10) we obtain the desired result.

We are now in a position to prove the theorems mentioned in the introduction.

PROOF OF THEOREM (0.1). We begin with the case when $c = 1$. Let $\rho(t)$ be defined as in Lemma (2.8). If we set $A = (1+8R)^d$ and $\gamma = 4 \ln 4 - 5 \ln 3$, then from Lemma (2.8) we have

$$\rho(2t) \leq At^d \rho^2(t) + e^{-\gamma t} \quad \text{for all} \quad t \geq 1/(5(\ln 3-1)). \tag{2.11}$$

From Lemma (2.4), if

$$(2^d A+1)[t^d \rho(t) \vee e^{-\gamma t/4}] < 1 \quad \text{for any} \quad t > 1000e^d \tag{2.12}$$

(Note $1000e^d > (5(\ln 3-1))^{-1} \vee 1 \vee t_0'$ where t_0' is such that $t \geq t_0'$ implies that $(2t)^d \leq e^{\gamma t/2}$) then there is an $\varepsilon_0 > 0$ such that $\rho(t) \leq e^{-\varepsilon_0 t}$ for all large t. Since $\rho(t)$ is decreasing and $\frac{d}{dt}\rho(t)\big|_{t=0} < 0$ unless both $\vec{0}$ and $\vec{N}$ are traps for the process, it follows that if (2.12) holds, then there is an $\varepsilon > 0$ such that $\rho(t) \leq Ne^{-\varepsilon t}$ for all $t \geq 0$. Thus by the remark at the end of section one, if (2.12) holds, then the process falls into case i).

If (2.12) fails, then we must have $\alpha_0^{-1}(R,d)t^d\rho(t) \geq 1$ for all $t \geq 1000e^d$ such that $\alpha_0(R,d)e^{-\gamma t/4} < 1$, i.e., for all $t > \frac{4 \ln \alpha_0^{-1}(R,D)}{\gamma} \vee 1000e^d \equiv \alpha_1(R,d)$ where $\alpha_0(R,d) = (2^d(1+8R)^d+1)^{-1}$.

When $c = 1$ this is case ii) of Theorem (0.1). For general c make the change of time scale $t \to t/c$ to reduce the problem to a process with $c = 1$.

PROOF OF THEOREM (0.2). Let $\{T_t : t \geq 0\}$ be the semi-group determined by the $\{c(k,\cdot,\cdot) : k \in Z^d\}$ and let $\varepsilon > 0$ be such that

$$T_t f_0(\vec{N}) - T_t f_0(\vec{0}) < Ne^{-\varepsilon t}.$$

Then let c be as in Theorem (0.1) and choose $t_1 > 2\alpha_1(R,d)/c$ so large that

$$t_1^d 4Ne^{-\varepsilon t_1} < \alpha_0(R,d)/c^d. \tag{2.13}$$

Now let $\{\{c_n(k,\cdot,\cdot) : k \in Z^d\} : n = 1,2,\ldots\}$ be a sequence of translation invariant jump rates such that

$$\sum_a \sup_\eta |c(0,\eta,a) - c_n(0,\eta,a)| \to 0 \quad \text{as} \quad n \to \infty.$$

Then if $L_n f(\eta) = \sum_{k \in Z^d} \sum_{a \in S} c_n(k,\eta,a)[f(\eta^{k,a}) - f(\eta)]$, it follows that $||L_n f - Lf|| \to 0$ as $n \to \infty$ for all $f \in D$, and hence that $e^{L_n t_1} f_0 - e^{L t_1} f_0 \to 0$ as $n \to 0$. (See [8] or [9].) Thus there is a $\delta_0 > 0$ such that if

$$\sum_a \sup_\eta |c(0,\eta,a) - \bar{c}(0,\eta,a)| < \delta_0 \tag{2.14}$$

for some translation invariant jump rates $\{\overline{c}(k,\cdot,\cdot): k \in Z^d\}$, then $||e^{Lt_1}f_0 - e^{\overline{L}t_1}f_0|| < \alpha_0(R,d)/(8N(ct_1)^d)$. Thus if (2.14) holds, then $t_1^d(\overline{T}_t f_0(\vec{N}) - \overline{T}_t f_0(\vec{0})) < \alpha_0(R,d)/(2c^d)$. Here $\overline{T}_t = e^{\overline{L}t}$.

Finally choose $0 < \delta \leq \delta_0$ such that

$$\sum_a \sup|c(0,\eta,a) - \overline{c}(0,\eta,a)| < \delta \tag{2.15}$$

implies that $c/2 < \overline{c} \equiv \max\{\sup_\eta \sum_{a\in S} \overline{c}(0,\eta,a),$

$$\sum_{|j|\leq R} \sup_\eta \sup_{b\in S} \sum_{a\in S} |\overline{c}(0,\eta,a) - \overline{c}(0,\eta^{j,b},a)|\} < 2^{1/d}c.$$

Then if (2.15) holds for some jump rates $\{\overline{c}(k,\cdot,\cdot): k \in Z^d\}$, the corresponding semigroup, $\{\overline{T}_t: t \geq 0\}$, does not fall into case ii) of Theorem (0.1) and hence must converge exponentially fast on D.

3. The Stochastic Ising Model

In this section we apply the results of section two to ferromagnetic finite range stochastic Ising models. The interaction is given via a pair potential as follows. For each pair $\{x,y\} \subset Z^d$ with $0 < |x-y| \leq R$ let $J_{\{x,y\}} \geq 0$. We assume that $J_{\{x,y\}} = J_{\{0,y-x\}}$ for all x,y. For the state space we take $E = \{-1,1\}^{Z^d}$ and refer to the value of the configuration η at site k as the spin at k. The jump rates will be called flip rates and since $c(k,\eta,\eta(k))$ is irrelevant, we need only define $c(k,\eta,-\eta(k))$, which we denote by $c_k(\eta)$. The relation between the flip rates and the pair potential is given by the detailed balance equation

$$\begin{aligned} &c_k(\eta)\exp\left[\eta(k)\Big(\sum_j J_{\{k,j\}}\eta(j) + H\Big)\right] \\ &\quad = c_k(\eta^k)\exp\left[-\eta(k)\Big(\sum_j J_{\{k,j\}}\eta(j) + H\Big)\right], \end{aligned} \tag{3.1}$$

where η^k is the configuration obtained from η by flipping the spin at k, and H is a real number which represents the external magnetic field. In addition to (3.1) we assume that the flip rates are attractive. Possible choices for c_k are

$$c_k(\eta) = \exp\left[-\eta(k)\Big(\sum_j J_{\{k,j\}}\eta(j) + H\Big)\right] \tag{3.2}$$

or more commonly

$$c_k(\eta) = 1/(1+\exp\left[2\eta(k)\Big(\sum_j J_{\{k,j\}}\eta(j) + H\Big)\right]. \tag{3.3}$$

It is easily checked that if all $J_{\{k,j\}} \geq 0$ then both of these choices of $\{c_k : k \in Z^d\}$ are attractive. Note that there are infinitely many possible choices of attractive flip rates which satisfy (3.1); however, if c_k and c'_k are two such choices, then $c_k(\eta)/c'_k(\eta)$ is independent of $\eta(k)$. One of the points of this section is that whether or not the semi-group T_t, with attractive flip rates $\{c_k : k \in Z^d\}$ satisfying (3.1) converges exponentially fast on D is a function only of $\{J_{\{k,j\}} : k,j \in Z^d\}$ and not of the particular choice of c_k used. As just stated this is not entirely proved. What we do in this section is the following. The $\{J_{\{k,j\}} : k,j \in Z^d\}$ determine the Gibbs states and the Gibbs states have certain mixing properties. We derive sufficient conditions for exponentially fast convergence of the semi-group T_t in terms of mixing properties of the Gibbs states, and we also derive necessary conditions for exponentially fast convergence of the semi-group T_t in terms of (apparently different) mixing properties of the Gibbs states. While the author believes that these various mixing properties are equivalent, he has not been able to prove it.

The mixing conditions are given in terms of Gibbs states on finite sets with arbitrary boundary conditions. These Gibbs states are defined as follows. If $\Lambda \subset Z^d$ let $E(\Lambda) = \{-1,1\}^\Lambda$. If Λ is finite and $\eta \in E(Z^d \setminus \Lambda)$, let μ_Λ^η be the measure on $E(\Lambda)$ given by

$$\mu_\Lambda^\eta(\{\sigma\}) = \exp\Big[\sum_{\{j,k\}\subset\Lambda} J_{\{j,k\}}\sigma(j)\sigma(k) + H\sum_{k\in\Lambda}\sigma(k) + \sum_{j\in\Lambda}\sum_{k\in\Lambda^c} J_{\{j,k\}}\sigma(j)\eta(k)\Big]/Z(\Lambda,\eta) \tag{3.4}$$

where $Z(\Lambda,\eta)$ is the normalization constant which makes μ_Λ^η a probability measure. We think of μ_Λ^η as a measure on the Borel sets of E which is concentrated on $\{\eta\} \times E(\Lambda)$. I.e., configurations drawn from μ_Λ^η are equal to η off of Λ with probability one.

$\Lambda \subset Z^d$ is called a box if Λ is a product of d finite intervals in Z. For any box, Λ, $\partial\Lambda = \{k \in Z^d \setminus \Lambda : \text{dist}(k,\Lambda) \leq R\}$, where R is the range of the interaction. If $\eta \in E(Z^d \setminus \Lambda)$ and $\sigma \in E(\Lambda)$ and $k \in Z^d$, then $\tau^k(\eta\sigma)$ is the configuration given by

$$\tau^k(\eta\sigma)(j) = \begin{cases} \eta(k+j) & \text{if } k+j \notin \Lambda \\ \sigma(k+j) & \text{if } k+j \in \Lambda \end{cases}$$

and $\tau^k g(\eta\sigma) = g(\tau^{-k}(\eta\sigma))$ for functions g on E. Finally if ν is a measure on the Borel sets of E and $f \in D$ we write $\nu(f)$ for $\int f(\eta)\nu(d\eta)$.

The three mixing conditions on Gibbs states which we consider are:

(GS1) There exists $\varepsilon > 0$ such that for all $f,g \in D$ there exists $A_{f,g} < \infty$ such that for all boxes Λ and all $\eta \in E(Z^d \setminus \Lambda)$ and all $j,k \in Z^d$

$$|\mu_\Lambda^\eta(\tau^j f \tau^k g) - \mu_\Lambda^\eta(\tau^j f)\mu_\Lambda^\eta(\tau^k g)| \leq A_{f,g} e^{-\varepsilon|f-k|}.$$

(GS2) There exists a sufficiently large cube Λ_0 and a $\delta < 1$ such that

$$\sum_{k \in \partial\Lambda_0} \sum_{j \in \Lambda_0} \sup_\eta \sup_{\Lambda_1 \ni j} |\mu_{\Lambda_1}^\eta(\sigma(j)=1) - \mu_{\Lambda_1}^{\eta^k}(\sigma(j)=1)| = \delta|\Lambda_0|.$$

Here the sumpremum over $\Lambda_1 \ni j$ is over all boxes containing j and contained in Λ_0.

(GS3) There exists $\varepsilon > 0$ such that for all boxes Λ

$$|\mu_\Lambda^1(f_k) - \mu_\Lambda^{-1}(f_k)| \leq \exp[-\varepsilon \operatorname{dist}(k,\Lambda^c)].$$

It is easy to show that (GS1) implies (GS2) and (GS1) implies (GS3). We will show below that (GS2) implies (GS1), and although we have not been able to prove it we believe (GS3) also implies (GS1). (GS2) is the condition which implies rapid convergence in the stochastic Ising model.

We also have three different statements of exponentially fast convergence for the stochastic Ising model. They are:

(SI1) There exists $\varepsilon > 0$ such that for all boxes Λ and all boundary conditions η

$$||T_t^{\Lambda,\eta} f - \mu_\Lambda^\eta(f)||_{L^2(\mu_\Lambda^\eta)} \leq e^{-\varepsilon t}||f - \mu_\Lambda^\eta(f)||_{L^2(\mu_\Lambda^\eta)},$$

where $\{T_t^{\Lambda,\eta}: t \geq 0\}$ is the semi-group generated by $L_{\Lambda,\eta}$, which is given by

$$L_{\Lambda,\eta} f(\sigma) = \sum_{k \in \Lambda} c_k(\sigma,\eta)(f(\sigma^k) - f(\sigma)),$$

and $c_k(\sigma,\eta) = c_k(\omega)$, where

$$\omega(j) = \begin{cases} \sigma(j) & \text{if } j \in \Lambda \\ \eta(j) & \text{if } j \notin \Lambda. \end{cases}$$

(SI2) There exists $\varepsilon > 0$ such that for all $f \in D$ there exists $A_f < \infty$ for which

$$||T_t f - \mu(f)|| \leq A_f e^{-\varepsilon t}.$$

Here μ is the (assumed unique) Gibbs state for the pair potential $\{J_{\{j,k\}}: j,k \in Z^d\}$ with external field H.

(SI3) The pair potential $\{J_{\{j,k\}}: j,k \in Z^d\}$ with external field H admits a unique Gibbs state μ and there is an $\varepsilon > 0$ such that

$$||T_t f - \mu(f)||_{L^2(\mu)} \leq e^{-\varepsilon t}||f - \mu(f)||_{L^2(\mu)}.$$

We will prove the following implications among the statements (GS1)-(GS3) and (SI1)-(SI3).

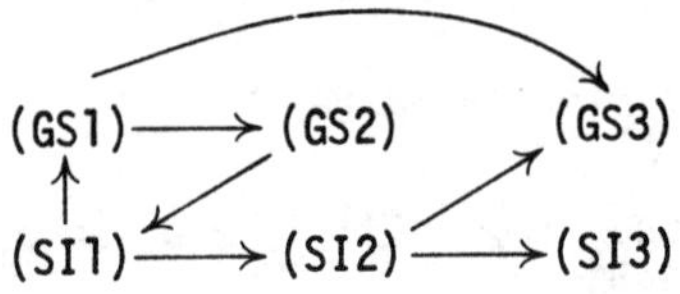

Note that an additional implication from (GS3) to (GS1) would make all of the mixing conditions on the Gibbs states equivalent to (SI2) (exponentially fast convergence of the semi-group of the stochastic Ising model).

(3.5) THEOREM. (SI2) implies (GS3).

PROOF: This is just Theorem (4.7) in [7].

(3.6) THEOREM. (SI1) implies (GS1).

PROOF: This is just Theorem (2.16) in [6] applied to the finite stochastic Ising model with generator $L_{\Lambda,\eta}$.

(3.7) THEOREM. (GS1) implies (GS2).

PROOF: Let $g(\eta) = \exp[-2\eta(0)(\sum_i J_{\{0,i\}}\eta(i) + H)]$ and note that

$$\frac{d\mu_{\Lambda_1}^{\eta^k}}{d\mu_{\Lambda_1}^{\eta}}(\sigma) = \frac{\tau^k g(\sigma\eta)}{\mu_{\Lambda_1}^{\eta}(\tau^k g(\eta))}, \tag{3.8}$$

where $\sigma\eta$ equals σ on Λ_1 and η on Λ_1^c. Thus

$$\begin{aligned}
&|\mu_{\Lambda_1}^{\eta}(\sigma(j) = 1) - \mu_{\Lambda_1}^{\eta^k}(\sigma(j) = 1)| \\
&= |\mu_{\Lambda_1}^{\eta}(\tau^j(f_0+1)/2) - \frac{\mu_{\Lambda_1}^{\eta}((\tau^j(f_0+1)/2)\tau^k g)}{\mu_{\Lambda_1}^{\eta}(\tau^k g)} \\
&\leq \left(\exp\left[2\left(\sum_i J_{\{0,i\}} + |H|\right)\right/2|\mu_{\Lambda_1}^{\eta}(\tau^j f_0\tau^k g) - \mu_{\Lambda_1}^{\eta}(\tau^j f_0)\mu_{\Lambda_1}^{\eta}(\tau^k g)|.
\end{aligned}$$

Let $\exp[2(\sum_i J_{\{0,i\}} + |H|]A_{f_0,g}/2 = K$. Then by (GS1)

$$|\mu_{\Lambda_1}^{\eta}(\sigma(j) = 1) - \mu_{\Lambda_1}^{\eta^k}(\sigma(j) = 1)| \leq K\exp[-\varepsilon|j-k|] \tag{3.9}$$

uniformly in Λ_1 and η. (GS2) follows easily from this.

(3.10) THEOREM. (SI2) implies (SI3).

PROOF: This is clear since D is dense in $L^2(\mu)$.

(3.11) THEOREM. (GS1) implies (GS3).

PROOF: Let $\eta_0 \equiv -1$ and $\eta_1, \eta_2, \ldots, \eta_{|\partial\Lambda|}$ be a sequence such that $\eta_j = \eta_{j-1}^{i_j}$

for some $i_j \in \partial\Lambda$ and $\eta_{|\partial\Lambda|} \equiv 1$ for $j \in \partial\Lambda$. Then

$$\begin{aligned}|\mu_\Lambda^1(f_k) - \mu_\Lambda^{-1}(f_k)| &= |\mu_\Lambda^{\eta_{|\partial\Lambda|}}(f_k) - \mu_\Lambda^{\eta_0}(f_k)| \\ &\leq \sum_{j=1}^{|\partial\Lambda|} |\mu_\Lambda^{\eta_j}(f_k) - \mu_\Lambda^{\eta_{j-1}}(f_k)|.\end{aligned} \tag{3.12}$$

Let g be as in (3.8). Then by (3.9)

$$|\mu_\Lambda^{\eta_j}(f_k) - \mu_\Lambda^{\eta_{j-1}}(f_k) \leq 2K \exp[-\varepsilon|k - i_j|].$$

Therefore

$$|\mu_\Lambda^1(f_k) - \mu_\Lambda^{-1}(f_k)| \leq 2K \sum_{j\in\Lambda^c} \exp[-\varepsilon|k - j|].$$

(GS3) follows easily from this.

(3.13) THEOREM. (SI1) implies (SI2).

PROOF: By Theorems (3.6) and (3.11), (SI1) implies (GS3) which implies that there is a unique Gibbs state for the interaction $\{J_{\{i,j\}}: i,j \in Z^d\}$ with external field H. In [5] it was noted that (GS3) and (SI1) imply that there is a $\delta > 0$ such that for all $f \in D$ there is an $A_f < \infty$ for which

$$||T_t f - \mu(f)|| \leq A_f \exp[-\delta t^{1/d}]. \tag{3.14}$$

(See Remark (2.14) in [5]). The result now follows from Theorem (0.1).

All that remains is to prove that (GS2) implies (SI1). Before doing that we first construct an auxiliary process. Let Λ_0 be a cube satisfying (GS3) and centered at the origin. If Λ is any box and η is any boundary configuration let A_Λ^η be the bounded generator on $D(\Lambda)$ defined by

$$A_\Lambda^\eta f(\sigma) = \frac{1}{|\Lambda_0|} \sum_{k:(\Lambda_0+k)\cap\Lambda\neq\emptyset} \int (f(\sigma\omega) - f(\sigma))\mu_{(\Lambda_0+k)\cap\Lambda}^\eta(d\omega), \tag{3.15}$$

where $\sigma\omega$ is the configuration given by

$$\sigma\omega(j) = \begin{cases}\sigma(j) & \text{if } j \in \Lambda\smallsetminus(\Lambda_0+k) \\ \omega(j) & \text{if } j \in \Lambda \cap (\Lambda_0+k).\end{cases}$$

A_Λ^η generates a semi-group on $D(\Lambda)$ which we denote by $S_t^{\Lambda,\eta}$.

(3.16) LEMMA. If there is an $\varepsilon_0 > 0$ such that for all cubes Λ, all $\eta \in E(Z^d\smallsetminus\Lambda)$ and all $f \in D(\Lambda)$ there is a $B_f < \infty$ (possibly depending on Λ and η) such that

$$||S_t^{\Lambda,\eta} f - \mu_\Lambda^\eta(f)|| \leq B_f \exp[-\varepsilon_0 t], \tag{3.17}$$

then there is an $\varepsilon > 0$ depending on ε_0 but not on Λ or η such that

$$||T_t^{\Lambda,\lambda}f - \mu_\Lambda^\eta(f)||_{L^2(\mu_\Lambda^\eta)} \le ||f - \mu_\Lambda^\eta(f)||_{L^2(\mu_\Lambda^\eta)} \exp[-\varepsilon t]. \tag{3.18}$$

PROOF: This is essentially the content of Lemmas(1.1)-(1.17) in [5]. In those lemmas the state μ_Λ^η is replaced by a Gibbs state μ on all of E and d is assumed to be one. However, this latter assumption plays no role at this stage, and the argument with μ replaced by μ_Λ^η is unchanged.

(3.19) THEOREM. (GS2) implies (SI1).

PROOF: According to Lemma (3.16) we must show that (3.17) holds for all boxes Λ with the same $\varepsilon_0 > 0$. (Conceivably B_f may depend on the box Λ or η; however, we will see that we can take B_f independently of Λ and η.)

We use the following notation. If Λ is the box under consideration we set $\Lambda_k = (\Lambda_0 + k) \cap \Lambda$.

We first construct a process which corresponds to coupling together two processes with semi-group $S_t^{\Lambda,\eta}$ starting from two different configurations. We do this in such a way that if the spins in the initial configuration of the first process are all at least as large as the corresponding spins in the initial configuration of the second process, then this situation is maintained for all time. To do this first note that by the FKG inequalities (since all $J_{\{i,j\}} \ge 0$) if σ and ω satisfy $\sigma(j) \ge \omega(j)$ for all $j \in \Lambda$ then there is a measure $r_{\Lambda_k}^{\eta\sigma,\eta\omega}$ on $E(\Lambda_k) \times E(\Lambda_k)$ such that

$$r_{\Lambda_k}^{\eta\sigma,\eta\omega}((\tau^1,\tau^2) \in E(\Lambda_k)^2: \tau^1(j) \ge \tau^2(j) \text{ for all } j \in \Lambda_k) = 1, \tag{3.20}$$

and

$$r_{\Lambda_k}^{\eta\sigma,\eta\omega}(A \times E(\Lambda_k)) = \mu_{\Lambda_k}^{\eta\sigma}(A) \text{ for all } A \subset E(\Lambda_k), \tag{3.21}$$

and

$$r_{\Lambda_k}^{\eta\sigma,\eta\omega}(E(\Lambda_k) \times A) = \mu_{\Lambda_k}^{\eta\omega}(A) \text{ for all } A \subset E(\Lambda_k). \tag{3.22}$$

(See Theorem (9.8) in [10]). We use these measures to define the generator, $\overline{A}_\Lambda^\eta$, of the coupled process. Thus if f is a real valued function on $E(\Lambda)^2$, then we define

$$\overline{A}_\Lambda^\eta f(\sigma,\omega) = \frac{1}{|\Lambda_0|} \sum_{k:\Lambda^k \ne \emptyset} \int (f(\tau^1,\tau^2) - f(\sigma,\omega)) r_{\Lambda_k}^{\eta\sigma,\eta\omega}(d\tau^1,d\tau^2).$$

set $\overline{S}_t^{\Lambda,\eta} = e^{t\overline{A}_\Lambda^\eta}$, and let $\overline{P}_{(\sigma,\omega)}^{\Lambda,\eta}$ be the distribution of the corresponding Markov process with initial configuration (σ,ω). Then for all $f \in D(\Lambda)$,

$$S_t^{\Lambda,\eta}f(\sigma) = \overline{S}_t^{\Lambda,\eta}f^{(1)}(\sigma,\omega) \text{ for all } \omega \le \sigma, \text{ and}$$

$$S_t^{\Lambda,\eta}f(\omega) = \overline{S}_t^{\Lambda,\eta}f^{(2)}(\sigma,\omega) \text{ for all } \omega \le \sigma,$$

where $f^{(1)}(\sigma,\omega) = f(\sigma)$ and $f^{(2)}(\sigma,\omega) = f(\omega)$. (See [8] or [9]). Thus if f is an increasing function on $D(\Lambda)$, then $S_t^{\Lambda,\eta}f$ is again increasing and for all σ

$$S_t^{\Lambda,\eta}f(-1) \leq S_t^{\Lambda,\eta}f(\sigma) \leq S_t^{\Lambda,\eta}f(1).$$

Also it is easily checked from the definition of A_Λ^η that $\int A_\Lambda^\eta f d\mu_\Lambda^\eta = 0$ for all $f \in D(\Lambda)$ (see [5]). Thus if f is increasing and depends only on the coordinates in $\Delta \subset \Lambda$, then

$$\begin{aligned} ||S_t^{\Lambda,\eta}f - \mu_\Lambda^\eta(f)|| &= ||S_t^{\Lambda,\eta}f - \int S_t^{\Lambda,\eta}f(\sigma)\mu_\Lambda^\eta(d\sigma)|| \\ &\leq S_t^{\Lambda,\eta}f(1) - S_t^{\Lambda,\eta}f(-1) = E_{P^{\Lambda,\eta}_{(1,-1)}}[f(\tau_t^{(1)}) - f(\tau_t^{(2)})] \\ &\leq 2||f|| \sum_{j\in\Delta} P^{\Lambda,\eta}_{(1,-1)}(\tau_t^{(1)}(j) > \tau_t^{(2)}(j)). \end{aligned}$$

Since every $f \in D(\Lambda)$ is the difference of the increasing functions in $D(\Lambda)$, the proof will be complete if we show that there is an $\varepsilon_0 > 0$ such that

$$\sup_{j\in\Lambda} P^{\Lambda,\eta}_{(1,-1)}(\tau_t^{(1)}(j) > \tau_t^{(2)}(j)) \leq e^{-\varepsilon_0 t}. \tag{3.23}$$

The verification of (3.23) is where we use the assumption (GS2). From the definition of $\overline{A_\Lambda^\eta}$ we have

$$\begin{aligned} &\frac{d}{dt} P^{\Lambda,\eta}_{(1,-1)}(\tau_t^{(1)}(j) > \tau_t^{(2)}(j)) \\ &= \frac{1}{|\Lambda_0|} \sum_{k:\Lambda_k\ni j} E^{\Lambda,\eta}_{(1,-1)}\Big[r_{\Lambda_k}^{\eta\tau_t^{(1)},\eta\tau_t^{(2)}}(\sigma^{(1)}(j) > \sigma^{(2)}(j))\Big] \\ &\qquad - \frac{1}{|\Lambda_0|} \sum_{k:\Lambda_k\ni j} P^{\Lambda,\eta}_{(1,-1)}(\tau_t^{(1)}(j) > \tau_t^{(2)}(j)) \\ &\leq \frac{1}{|\Lambda_0|} \sum_{k:\Lambda_k\ni j} E^{\Lambda,\eta}_{(1,-1)}\Big[\sum_{t\in(\partial\Lambda_k)\cap\Lambda}\Big\{\sup_w|\mu^w_{\Lambda_k}(\sigma(j)=1) \\ &\qquad\qquad - \mu^{w^t}_{\Lambda_k}(\sigma(j)=1)|\chi_{[\tau_t^{(1)}(t)>\tau_t^{(2)}(t)]}\Big\}\Big] \\ &\qquad - P^{\Lambda,\eta}_{(1,-1)}(\tau_t^{(1)}(j) > \tau_t^{(2)}(j)). \end{aligned}$$

Here we have used (3.20) to conclude that

$$r_{\Lambda_t}^{\eta\tau_t^{(1)},\eta\tau_t^{(2)}}(\sigma^{(1)}(j) > \sigma^{(2)}(j)) = \mu_{\Lambda_k}^{\eta\tau_t^{(1)}}(\sigma(j) = 1) - \mu_{\Lambda_t}^{\eta\tau_t^{(2)}}(\sigma(j) = 1)$$

$$\leq \sum_{t\in(\partial\Lambda_k)\cap\Lambda} \sup_w |\mu_{\Lambda_k}^{w}(\sigma(j) = 1) - \mu_{\Lambda_k}^{w^t}(\sigma(j) = 1)|\chi_{[\tau_t^{(1)}(t)>\tau_t^{(2)}(t)]}.$$

The last inequality is obtained by moving from $\tau_t^{(1)}$ to $\tau_t^{(2)}$ by flipping the necessary spins in $(\partial\Lambda_k) \cap \Lambda$ one at a time. Continuing (3.24) we have

$$\frac{d}{dt} P_{(1,-1)}^{\Lambda,\eta}(\tau_t^{(1)}(j) > \tau_t^{(2)}(j))$$

$$\leq \frac{1}{|\Lambda_0|} \sum_{k:\Lambda_k\ni j} \sum_{\ell\in(\partial\Lambda_k)\cap\Lambda} \sup_w |\mu_{\Lambda_k}^{w}(\sigma(j) = 1) - \mu_{\Lambda_k}^{w^\ell}(\sigma(j) = 1)|$$

$$\cdot P_{(1,-1)}^{\Lambda,\eta}(\tau_t^{(1)}(t) > \tau_t^{(2)}(t)) - P_{(1,-1)}^{\Lambda,\eta}(\tau_t^{(1)}(j) > \tau_t^{(2)}(j))$$

$$= \frac{1}{|\Lambda_0|} \sum_{k:\Lambda_k\ni j} \sum_{\ell\in\partial\Lambda_0} \sup_w |\mu_{\Lambda_0\cap(\Lambda-k)}^{w}(\sigma(j-k) = 1) - \mu_{\Lambda_0\cap(\Lambda-k)}^{w^\ell}(\sigma(j) = 1)|$$

$$\cdot P_{(1,-1)}^{\Lambda,\eta}(\tau_t^{(1)}(\ell-k) > \tau_t^{(2)}(\ell-k)) - P_{(1,-1)}^{\Lambda,\eta}(\tau_t^{(1)}(j) > \tau_t^{(2)}(j)) \qquad (3.25)$$

$$\leq \frac{1}{|\Lambda_0|} \sum_{\Lambda_0+k\ni j} \sum_{\ell\in\partial\Lambda_0} \sup_w \sup_{\Lambda_1\subset\Lambda_0} |\mu_{\Lambda_1}^{w}(\sigma(j-k) = 1) - \mu_{\Lambda_1}^{w^\ell}(\sigma(j-k) = 1)|$$

$$\cdot P_{(1,-1)}^{\Lambda,\eta}(\tau_t^{(1)}(\ell-k) > \tau_t^{(2)}(\ell-k)) - P_{(1,-1)}^{\Lambda,\eta}(\tau_t^{(1)}(j) > \tau_t^{(2)}(j))$$

$$\leq \delta \sup_{t\in\Lambda} P_{(1,-1)}^{\Lambda,\eta}(\tau_t^{(1)}(t) > \tau_t^{(2)}(t)) - P_{(1,-1)}^{\Lambda,\eta}(\tau_t^{(1)}(j) > \tau_t^{(2)}(j)).$$

Thus

$$P_{(1,-1)}^{\Lambda,\eta}(\tau_t^{(1)}(j) > \tau_t^{(2)}(j)) \leq e^{-t} + \delta \int_0^t e^{-(t-s)} \sup_{t\in\Lambda} P_{(1,-1)}^{\Lambda,\eta}(\tau_s^{(1)}(t) > \tau_s^{(2)}(t))ds. \qquad (3.26)$$

Since the right side of (3.26) does not depend on j, if we take the supremum over $j \in \Lambda$ on the left side of (3.26), we obtain the same bound. From this it is easy to conclude that

$$\sup_{j\in\Lambda} P_{(-1,1)}^{\Lambda,\eta}(\tau_t^{(1)}(j) > \tau_t^{(2)}(j)) \leq e^{-(1-\delta)t}. \qquad (3.27)$$

4. Smoothness of the Stationary Measures as a Function of a Parameter

Consider a family, $\{c^{(\lambda)}(k,\cdot,\cdot): k \in Z^d\}_{\lambda\in I}$, of FTAS's indexed by a parameter λ contained in some interval, I, of the reals and suppose that $R < \infty$ is an upper bound on the range of interaction of all of them. In addition assume that $c^{(\lambda)}(0,\eta,a)$ is continuous in λ for all $\eta \in E$ and $a \in S$. Let $L^{(\lambda)}$ denote the generator determined by the $c^{(\lambda)}$'s and let $\{T_t^{(\lambda)}: t \geq 0\}$ be the semi-group generated by $L^{(\lambda)}$. By Theorem (0.2), if for some $\lambda_0 \in I$,

(4.1) there is an $\varepsilon > 0$ such that for all $f \in D$ there is an $A_f < \infty$ such that

$$|T_t^{(\lambda_0)} f(\eta) - T_t^{(\lambda_0)} f(\sigma)| < A_f e^{-\varepsilon t}$$

for all $\eta,\sigma \in E$, then (4.1) holds for all λ in a neighborhood, I_0, of λ_0. If (4.1) holds for some $\lambda \in I$, then the corresponding semi-group has a unique stationary measure, which we denote by $\mu^{(\lambda)}$. Our goal in this section is to study the behavior of $\mu^{(\lambda)}(f)$ as a function of λ, for $f \in D$. The following lemma is crucial.

(4.2) LEMMA. If (4.1) holds, then

$$\phi(\lambda_0,f,t) \equiv \sum_{k\in Z^d} \sum_{a\in S} \sup_\eta |T_t^{(\lambda_0)} f(\eta^{k,a}) - T_t^{(\lambda_0)} f(\eta)| \tag{4.3}$$

satisfies

$$\int_0^\infty \phi(\lambda_0,f,t)dt < \infty. \tag{4.4}$$

PROOF: By Lemma (4.15) of [6], if Λ is such that $f \in D(\Lambda)$ and $\Lambda(M) = \{k: \mathrm{dist}(k,\Lambda) > MR\}$ then

$$\sum_{k\in\Lambda(M)} \sum_{a\in S} \sup_\eta |T_t^{(\lambda_0)} f(\eta^{k,a})-T_t^{(\lambda_0)} f(\eta)| \leq 2||f|| \; |\Lambda| \; Ne^{ct} \frac{(ct)^{M+1}}{(M+1)!} . \tag{4.5}$$

Here c is as in Theorem (0.1). Thus if we set $M(t) = [4ct]$ and apply Stirling's formula, we obtain

$$\sum_{k\in\Lambda(M(t))} \sum_{a\in S} \sup_\eta |T_t^{(\lambda_0)} f(\eta^{k,a})-T_t^{(\lambda_0)} f(\eta)| \leq \frac{e}{2}|\Lambda|N||f||(\frac{e^6}{4^4})^{ct}, \tag{4.6}$$

which goes to zero exponentially fast as $t \to \infty$. Also

$$\begin{aligned} \sum_{k\in\Lambda(M(T))^c} \sum_{a\in S} \sup_\eta |T_t^{(\lambda_0)} f(\eta^{k,a})-T_t^{(\lambda_0)} f(\eta)| \\ \leq |\Lambda(M(t))^c| NA_f e^{-\varepsilon t}. \end{aligned} \tag{4.7}$$

Since $|\Lambda(M(t))^c| \sim (8Rct)^d$, it follows that the expression on the left side of (4.7) is integrable, and hence $\phi(\lambda_0,f,\cdot)$ is integrable.

(4.8) THEOREM. If $\{c^{(\lambda)}(k,\cdot,\cdot): k \in Z^d\}_{\lambda \in I}$ is as above and if (4.1) holds for λ_0, then for all $f \in D$, $\mu^{(\lambda)}(f)$ is continuous at λ_0.

PROOF: This is an easy consequence of Proposition (2.14) in [9]. In fact one only needs the ergodicity of $\{T_t^{(\lambda_0)}: t \geq 0\}$ not the rapid convergence to equilibrium implied by (4.1).

If we require more smoothness on the jump rates, we may conclude more smoothness on the stationary measures as the following theorem shows.

(4.9) THEOREM. Suppose that (4.1) holds at λ_0 and that $\frac{d}{d\lambda} c^{(\lambda)}(k,\eta,a)|_{\lambda_0} = \gamma(k,\eta,a)$ exists for each k, η, and a.

Then for each $f \in D$, $\mu^{(\lambda)}(f)$ is differentiable at λ_0.

PROOF: Let

$$e^{(\lambda)}(k,\eta,a) = (c^{(\lambda)}(k,\eta,a) - c^{(\lambda_0)}(k,\eta,a))/(\lambda-\lambda_0)-\gamma(k,\eta,a).$$

Then $e^{(\lambda)}(k,\eta,a) \to 0$ uniformly in k,η, and a as $\lambda \to \lambda_0$. Set

$$\Gamma g(\eta) = \sum_k \sum_a \gamma(k,\eta,a)(g(\eta^{k,a}) - g(\eta)) \tag{4.10}$$

and

$$\xi^{(\lambda)}g(\eta) = \sum_k \sum_a e^{(\eta)}(k,\eta,a)(g(\eta^{k,a}) - g(\eta)). \tag{4.11}$$

Then

$$\begin{aligned}(L^{(\lambda_0)}-L^{(\lambda)})T_s^{(\lambda_0)}f(\eta) &= (\lambda_0-\lambda)\Gamma T_s^{(\lambda_0)}f(\eta) \\ &\quad + (\lambda_0-\lambda)\xi^{(\lambda)}T_s^{(\lambda_0)}f(\eta).\end{aligned} \tag{4.12}$$

By a standard formula from perturbation theory we have

$$\begin{aligned}\mu^{(\lambda_0)}(f) - \mu^{(\lambda)}(f) &= \lim_{t\to\infty} (T_t^{(\lambda_0)}f(\vec{0}) - T_t^{(\lambda)}f(\vec{0})) \\ &= \lim_{t\to\infty} \int_0^t T_{t-s}^{(\lambda)}(L^{(\lambda_0)} - L^{(\lambda)})T_s^{(\lambda_0)}f(\vec{0})ds.\end{aligned} \tag{4.13}$$

Thus

$$\begin{aligned}&(\mu^{(\lambda)}(f) - \mu^{(\lambda_0)}(f))/(\lambda-\lambda_0) \\ &= \lim_{t\to\infty} \left[\int_0^t T_{t-s}^{(\lambda)}\, T_s^{(\lambda_0)}f(\vec{0})ds + \int_0^t T_{t-s}^{(\lambda)}\xi^{(\lambda)}T_s^{(\lambda_0)}f(\vec{0})ds\right]\end{aligned} \tag{4.14}$$

If $G = \sup_{\eta,a}|\gamma(0,\eta,a)|$ and $e(\lambda) = \sup_{\eta,a}|e^{(\lambda)}(0,\eta,a)$, then by Lemma (4.1)

$||\Gamma T_s^{(\lambda_0)} f|| \leq G\phi(\lambda_0,f,s)$ and $||\xi^{(\lambda)} T_s^{(\lambda_0)} f|| \leq e(\lambda)\phi(\lambda_0,f,s)$. Also since $\Gamma T_s^{(\lambda_0)} \in C$ (the range is finite so $\gamma(k,\cdot,a)$ is continuous and recall (4.5)), $T_{t-s}^{(\lambda)}\Gamma T_s^{(\lambda_0)} f(\vec{0}) \to \mu^{(\lambda)}(\Gamma T_s^{(\lambda_0)} f)$ as $t \to \infty$. Thus by dominated convergence

$$\left| \frac{\mu^{(\lambda)}(f)-\mu^{(\lambda_0)}(f)}{\lambda - \lambda_0} - \iint_0^\infty \Gamma T_s^{(\lambda_0)} f(\eta)ds\, \mu^{(\lambda)}(d\eta)\right| \leq e(\lambda)\int_0^\infty \phi(\lambda_0,f,s)ds. \tag{4.15}$$

As $\lambda \to \lambda_0$, $e(\lambda) \to 0$, and thus to complete the proof we need only note that $||\Gamma|T_s^{(\lambda_0)} f||$ is integrable on $[0,\infty)$ and hence, since D is dense in C and $\Gamma T_s^{(\lambda_0)} f \in C$, Theorem (4.8) and Fubini's Theorem yield

$$\begin{aligned}\iint_0^\infty \Gamma T_s^{(\lambda_0)} f(\eta)ds\mu^{(\lambda)}(d\eta) &= \int_0^\infty\!\!\int \Gamma T_s^{(\lambda_0)} f(\eta)\mu^{(\lambda)}(d\mu)ds \\ &\to \int_0^\infty\!\!\int \Gamma T_s^{(\lambda_0)} f(\eta)\mu^{(\lambda_0)}(d\eta)ds.\end{aligned} \tag{4.16}$$

Hence $\frac{d}{d\lambda}\mu^{(\lambda)}(f)|_{\lambda_0}$ exists and is given by the right side of (4.16).

We conclude with one final remark. When applying the results of this section to the Ising model, one needs to check a condition such as (GS2). In the case that $\Lambda_0 = \{0\}$ this comes down to checking

$$\sum_k \sup_\eta |\mu_{\{0\}}^\eta(\sigma(0) = 1) - \mu_{\{0\}}^{\eta^k}(\sigma(0) = 1)| < 1. \tag{4.17}$$

The inequality in (4.17) is exactly Debrushins condition for uniqueness [1]. The condition (GS2) is a generalization of (4.17) and by the results of this section implies not only uniqueness, but differentiability of the expected value of the local observables as a function of the interaction. As a practical matter this is not a very useful generalization because (GS2) is very difficult to check unless $\Lambda_0 = \{0\}$. In some cases it can be done with a lot of tedious computation. For example if $d = 2$ and $J_{\{0,k\}} = \beta$ for $|k| = 1$ 0 otherwise, then for $\Lambda_0 = \{0\}$, (4.17) holds if $\beta < (\ln 3)/4 = .275$. If Λ_0 is a square with side length one (i.e. two lattice sites per side), then (GS2) holds for $\beta < .294$, and if Λ_0 has side length two (i.e., three lattice sites per side), then (GS2) holds for $\beta < .318$. For this model it is known that the conclusions of the theorems in this section hold for $\beta < \frac{1}{2}\ln(1 + \sqrt{2}) \sim .44$.

BIBLIOGRAPHY

[1] Dobrushin, R. L., "The description of a random field by means of conditional probabilities and conditions of its regularity", Theory Prob. Appl., (1968) 197-224.

[2] Griffeath, D., "The basic contact process", Stochastic Processes and their Applications, 11 (1981) 151-185.

[3] Harris, T. E., "Contact interactions on a lattice", Annals of Prob., 2 (1974) 969-988.

[4] Holley, R., "An ergodic theorem for interacting systems with attractive interactions", Z. Wahr. verw. Geb., 24 (1972) 325-334.

[5] Holley, R., "Rapid convergence to equilibrium in one-dimensional stochastic Ising models", Annals of Prob., 13 (1985) 72-89.

[6] Holley, R. and Stroock, D., "L^2 theory of the stochastic Ising Model", Z. Wahr. verw. Geb., 35 (1976) 87-101.

[7] Holley, R. and Stroock, D., "Applications of the stochastic Ising model to the Gibbs states", Commun. Math. Phys., 48 (1976) 249-265.

[8] Liggett, T. M., "The stochastic evolution of infinite systems of interacting particles", Lecture Notes in Mathematics 598 (1977), 187-248. Berlin-Heidelberg-New York: Springer.

[9] Liggett, T. M., "Interacting Particle Systems", Grundlehren der mathematischen Wissenschaften, 276 (1985) Berlin-Heidelberg-New York: Springer.

[10] Preston, C., "Random Fields", Lecture Notes in Mathematics, 534 (1976) Berlin-Heidelberg-New York: Springer.

DEPARTMENT OF MATHEMATICS
UNIVERSITY OF COLORADO
BOULDER, COLORADO 80309

Contemporary Mathematics
Volume 41, 1985

FIRST-PASSAGE PERCOLATION AND A HIGHER DIMENSIONAL GENERALIZATION

Harry Kesten[1]

ABSTRACT. After a short survey of classical results in first-passage percolation we list some recent results from [9] and [10]. The new results are: (i) An extension of the Cox-Durrett result on the asymptotic shape of the set of points which can be reached from the origin in a large time t. (ii) A n.a.s.c. for the time constant to be zero. (iii) An estimate on the rate of convergence of various passage times to the time constant. (iv) Asymptotic behavior of the time constant in large dimensions. We also describe a generalization of first-passage percolation in which one studies "surfaces of minimal value" instead of paths of minimal passage time. These minimal surfaces are useful in the study of maximal flows and some results on such flows are stated.

1. THE CLASSICAL SETUP. First-passage percolation was introduced by Hammersley and Welsh [7]. One assigns to each edge e of the lattice $\mathbb{Z}^d$ a nonnegative random variable, $t(e)$, which is interpreted as the time it takes to pass through this edge. All random variables $t(e)$, $e \in \mathbb{Z}^d$, are taken as independent and with the same distribution function F. A <u>path</u> on $\mathbb{Z}^d$ (from v to w) is an alternating sequence $v_0, e_1, v_1, \dots, e_n, v_n$ of vertices v_i and edges e_i such that v_i and v_{i+1} are adjacent on $\mathbb{Z}^d$ and e_{i+1} the edge between them (and $v_0 = v$ and $v_n = w$). The <u>passage time</u> of the path $r = (v_0, e_1, \dots, e_n, v_n)$ is

$$T(r) = \sum_1^n t(e_i) .$$

For any sets of vertices F_0 and F_1 the passage time from F_0 to F_1 is defined by

$$T(F_0, F_1) = \inf \{T(r) : r \text{ a path from some } v \in F_0 \text{ to some } w \in F_1\} .$$

1980 Mathematics Subject Classification. Primary 60K35; Secondary 60K10, 60C05.

[1]Supported by the NSF through a grant to Cornell University.

When F_0 consists of the single vertex v we write $T(v,F_1)$ instead of $T(\{v\},F_1)$ and similarly for F_1. $\underline{0}$ denotes the origin. The passage times most studied are

$$a_{0,n} := T(\underline{0}, (n,0,\dots,0)) \quad \text{and} \quad b_{0,n} := T(\underline{0},H_n),$$

where H_n is the hyperplane $\{\underline{x} = (x_1,\dots,x_d): x_1 = n\}$. These are called the _point to point and point to line passage times_ ($b_{0,n}$ should be called the point to hyperplane passage time when $d > 2$, but we adhere to the traditional terminology). It was proved in a series of papers that if

$$(1.1) \qquad E\, t(e) = \int x dF(x) < \infty$$

then

$$(1.2) \qquad \lim_{n\to\infty} \frac{a_{0,n}}{n} = \lim_{n\to\infty} \frac{b_{0,n}}{n} = \mu < \infty \quad \text{w.p.1.},$$

for some $\mu = \mu(F,d)$; μ is called the _time-constant_. An excellent survey of these results and their proofs can be found in [12]. (1.2) deals with the distance one can travel in the direction of the positive first coordinate axis in a large time. There is little reason to single out this direction, and one therefore considers the set of all points which can be reached in time t:

$$\tilde{B}(t) := \{v: T(0,v) \le t\} .$$

To obtain a theorem which is simple to state we fatten up $\tilde{B}$ a little by filling up a unit cube around each point. That is, we define

$$\overline{U} = \{x = (x(1),\dots,x(d)): -\tfrac{1}{2} \le x(i) \le \tfrac{1}{2},\ 1 \le i \le d\}$$

and

$$B(t) = \{v + \overline{U}: v \in \tilde{B}(t)\} .$$

The principal theorem, proven by Richardson [11] and Cox and Durrett [2] is as follows.

THEOREM 1.1. _Assume that_

$$(1.3) \qquad E \min\{t_1^d,\dots,t_{2d}^d\} < \infty ,$$

where $t_1,\dots,t_{2d}$ _are i.i.d. random variables with distribution_ F. _Then there exists a nonrandom convex set_ $B_0 \subset \mathbb{R}^d$, _which is invariant under_

permutations of the coordinates and under reflections in the coordinate hyperplanes, has nonempty interior and which is either compact or equals all of $\mathbb{R}^d$, and has the following property:
If B_0 is compact, then for all $\varepsilon > 0$

$$(1.4) \qquad (1-\varepsilon)B_0 \subset \frac{1}{t}B(t) \subset (1+\varepsilon)B_0 \quad \text{eventually w.p.1.}$$

If $B_0 = \mathbb{R}^d$, then for all $\varepsilon > 0$

$$(1.5) \qquad \{x: |x| \leq \varepsilon^{-1}\} \subset \frac{1}{t}B(t) \quad \text{eventually w.p.1.}$$

If (1.3) fails then

$$(1.6) \qquad \limsup_{v \to \infty} \frac{1}{|v|} T(\underline{0},v) = \infty \quad \text{w.p.1.}$$

REMARK. M. Eden [5] proposed the following growth model. At time 0 there is one cell at the origin. At time 1 the cell splits and fills one of the 2d sites adjacent on $\mathbb{Z}^d$ to the origin, each site being picked with probability 1/4. At time n, n cells will occupy a connected set, A_n, of n sites. At time (n+1) another empty site becomes occupied, chosen from the unoccupied sites adjacent to A_n; the probability that a given site will be filled is proportional to the number of edges which connect it to a site of A_n. Eden simulated this process in the plane and the shape of A_n looked very circular for large n. Richardson [11] observed that A_n has the same distribution as $\tilde{B}(t_n)$ for a certain sequence of stopping times t_n when F is the exponential distribution. Thus Theorem 1.1 implies that A_n has an asymptotic shape. However, not many people believe anymore that it is a Euclidean disc when d = 2. (See also Sect. 5 for large d.)

2. RESULTS WITHOUT MOMENT ASSUMPTIONS. Wierman [13] showed that in dimension 2 the moment assumption (1.1) can be reduced at the expense of changing (1.2) to

$$(2.1) \qquad \frac{a_{0,n}}{n} \to \mu \quad \text{in probability,} \quad \frac{b_{0,n}}{n} \to \mu \quad \text{w.p.1.}$$

Cox and Durrett [2] showed that in dimension 2 one can actually drop all moment assumptions by replacing $a_{0,n}$ and $b_{0,n}$ by slightly differing passage times $\hat{a}_{0,n}$ and $\hat{b}_{0,n}$. It turns out that this can be done in all dimensions and we give here the essence of the result. For details see [9], Sect. 2.

LEMMA 2.1. *There exists a number* $M = M(F,d) < \infty$ *such that* w.p.1. *for each vertex* $v \in \mathbb{Z}^d$ *there exists a finite "shell* $S(v)$ *of vertices around* v" *with the following properties*:

(i) *All edges* e *incident to a vertex in* $S(v)$ *have* $t(e) \le M$.

(ii) $S(v)$ *is connected on* $\mathbb{Z}^d$.

(iii) $S(v)$ *separates* v *from* ∞ *in the sense that any path on* $\mathbb{Z}^d$ *from* v *to* ∞ *intersects* $S(v)$.

(iv) *For any path* $r = (v_0, e_1, \ldots, e_n, v_n)$ *from* v *to* w (i.e., *with* $v_0 = v$, $v_n = w$) *one can connect* $S(v)$ *to* $S(w)$ *by a path which contains only vertices in* $\bigcup_0^n S(v_i)$.

(v) *The random sets* $\{S(v) - v \colon v \in \mathbb{Z}^d\}$ *are identically distributed*.

One defines

$$\hat{T}(v,w) = \inf \{T(r)\colon r \text{ a path from } S(v) \text{ to } S(w)\} .$$

It is easy to obtain from (v) above that

$$\text{(2.2)} \qquad \textit{the family } \{|\hat{T}(v,w) - T(v,w)| \colon v,w \in \mathbb{Z}^d\} \textit{ is tight.}$$

Thus the $\hat{T}$ are good replacements for T. Accordingly we introduce

$$\hat{a}_{0,n} = \hat{T}(\underline{0},(n,0,\ldots,0)), \ \hat{b}_{0,n} = \inf_{w \in H_n} \hat{T}(\underline{0},w),$$

$$\hat{B}(t) = \{v + \bar{U} \colon \hat{T}(0,v) \le t\} .$$

It is easy to see that

$$b_{0,n} - \hat{b}_{0,n} \textit{ is positive, and } \text{w.p.1. } \textit{bounded in } n.$$

As a replacement of (1.2) and Theorem 1.1 one has

THEOREM 2.1. *There exists a nonrandom convex set* $\hat{B}_0 \subset \mathbb{R}^d$, *which is invariant under permutations of the coordinates and under reflections in the coordinate hyperplanes, has nonempty interior and is either compact or equals all of* $\mathbb{R}^d$, *and has the following property*:

If $\hat{B}_0$ *is compact, then for all* $\varepsilon > 0$

$$\text{(2.3)} \qquad (1-\varepsilon)\hat{B}_0 \subset \frac{1}{t}\hat{B}(t) \subset (1+\varepsilon)\hat{B}_0 \quad \text{eventually w.p.1.}$$

If $\hat{B}_0$ *is* $\mathbb{R}^d$, *then for all* $\varepsilon > 0$

(2.4) $$\{x:|x| \le \varepsilon^{-1}\} \subset \frac{1}{t}\hat{B}(t) \text{ eventually w.p.1.}$$

Furthermore, there exists a constant $\hat{\mu} = \hat{\mu}(F,d) < \infty$ *such that*

(2.5) $$\lim_{n\to\infty} \frac{1}{n} a_{0,n} = \lim_{n\to\infty} \frac{1}{n} b_{0,n} \to \hat{\mu} \text{ w.p.1.}$$

If (1.3) holds then $\hat{B}_0 = B_0$, *and if (1.1) holds, then* $\hat{\mu} = \mu$.

In view of the last result we shall not distinguish between μ and $\hat{\mu}$.

3. PROPERTIES OF μ, B AND $\hat{B}$. The principal open question is to determine μ and B or $\hat{B}$ as functions of F and d. So far very little progress has been made on this question. Farily simple are the following general estimates.

THEOREM 3.1. *Let*

(3.1) $$\lambda = \lambda(F) = \inf \{x:F(x) > 0\}$$

be the left endpoint of the support of F. *If* F *is not concentrated on one point, then*

(3.2) $$\mu < \int x dF(x) = Et(e).$$

If in addition $\lambda(F) > 0$, *then*

(3.3) $$\lambda(F) < \mu .$$

One can also prove that μ is a continuous function of F.

THEOREM 3.2. *If* F_n *converges to* F *weakly, then* $\mu(F_n,d) \to \mu(F,d)$.

Last, but not least, one can distinguish between the cases where $\mu = 0$ (which is equivalent to $\hat{B}_0 = \mathbb{R}^d$) and $\mu > 0$ (which is equivalent to $\hat{B}_0$ compact). This dichotomy depends only on the atom of F at the origin. Some concepts from Bernoulli percolation are needed to formulate the result. In Bernoulli percolation edges are allowed to be in one of two states, open or closed. We associate with each open edge a 1 and with each closed edge a 0. If $\mathcal{E}$ denotes the edge set of $\mathbb{Z}^d$ then $\Omega := \{0,1\}^{\mathcal{E}}$ can be taken to be the space of configurations of open and closed edges. We denote by P_p the product measure on Ω according to which all edges are independent and

$$P_p\{e \text{ is open}\} = p = 1-P_p\{e \text{ is closed}\}.$$

We denote by W the open cluster of the origin, i.e., the set of all edges which belong to a path starting at $\underline{0}$ and having all its edges open. $\#W$ denotes the number of edges in W.

$$\theta(p) = P_p\{\#W = \infty\},$$

$$p_H = p_H(d) = \sup\{p:\theta(p) = 0\},$$

$$p_T = p_T(d) = \sup\{p:E_p\{\#W < \infty\}\}.$$

The critical probabilities p_H and p_T are unknown in general; only for $d = 2$ do we know

$$p_H(2) = p_T(2) = \frac{1}{2} .$$

THEOREM 3.3. $\mu(F,d) = 0$ *if and only if* $F(0) \geq p_T(d)$.

Even less is known about $\hat{B}_0$ than about μ. From (2.5) it follows that $\hat{B}_0$ contains the points

$$(0,\ldots,0, \pm\frac{1}{\mu},0,\ldots,0), \tag{3.4}$$

and in fact these points are the furthest points on the coordinate axes contained in $\hat{B}_0$. Thus

$$\hat{B}_0 \subset [-\frac{1}{\mu}, \frac{1}{\mu}]$$

and $\hat{B}_0$ contains the convex hull of the points (3.4) on the coordinate axes (the diamond in Fig. 1).

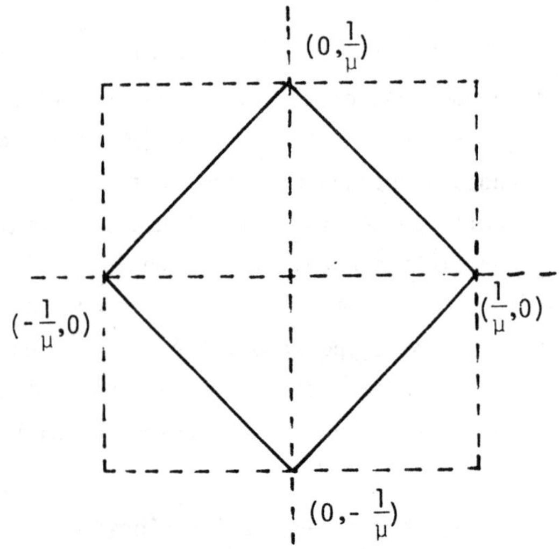

Fig. 1. $\hat{B}_0$ lies between the solidly drawn diamond and the dashed square.

It is also obvious that

$$\hat{B}_0 \subset \{x \in \mathbb{R}^d : |x| \le 1/\lambda\}, \tag{3.5}$$

where λ is as in (3.1). Durrett and Liggett [4] show that part of the boundary of $\hat{B}_0$ can actually lie on $\{x : |x| = 1/\lambda\}$, which is as far out as allowed by (3.5). For this the atom of F at λ has to be large enough. The precise bound for λ is given by an analogue of p_H for directed percolation; we denote it here by $\hat{p}_H(d)$ but do not give its definition. Similarly we shall use in Theorem 3.5 an analogue $\hat{p}_T$ of p_T for directed percolation without definition. The only important fact here is that $\hat{p}_H$ and $\hat{p}_T$ lie strictly between 0 and 1; the reader can find a formal definition in [9].

<u>THEOREM 3.4</u>. <u>If</u> $d = 2$ <u>and</u> $\lambda(F) > 0$ and $F(\lambda(F)) > \hat{p}_H(2)$ <u>then there exists an</u> $\alpha = \alpha(F) > 0$ <u>such that</u>

$$\{x : |x| = 1/\lambda,\ |x(1) - x(2)| \le \alpha(F)/\lambda(F)\} \subset \hat{B}_0 \tag{3.6}$$

<u>(see Fig. 2)</u>.

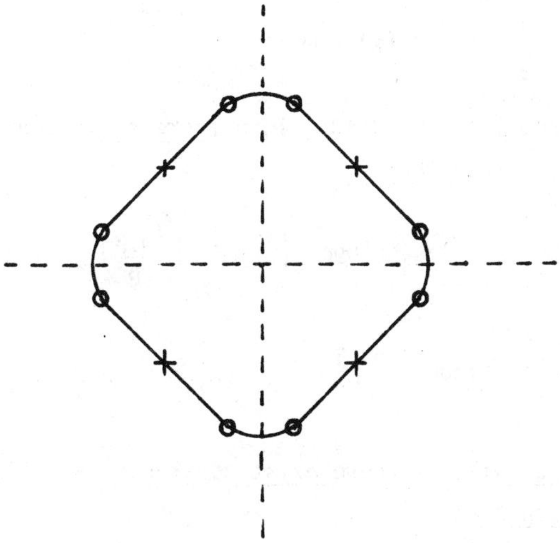

Fig. 2. Illustration of a $\hat{B}_0$ with straight edges. The points marked by a + are $(\pm(2\lambda)^{-1}, \pm(2\lambda)^{-1})$ and the points marked by o are $(\pm(1+\alpha)(2\lambda)^{-1}, \pm(1-\alpha)(2\lambda)^{-1})$ and $(\pm(1-\alpha)(2\lambda)^{-1}, \pm(1+\alpha)(2\lambda)^{-1})$.

In the opposite direction of Theorem 3.4 one has the following result (in any dimension).

THEOREM 3.5. *If* $\lambda(F) > 0$ *and* F *is not concentrated on* $\{\lambda\}$, *then* $\hat{B}_0$ *is strictly contained in the "diamond".*

$$\{x \in \mathbb{R}^d : |x| = 1/\lambda\} \quad . \tag{3.7}$$

If in addition $F(\lambda(F)) < \hat{p}_T(d)$, *then* $\hat{B}_0$ *lies entirely in the interior of the diamond (3.7).*

4. RATE OF CONVERGENCE TO THE TIME CONSTANT. It is natural to ask for second order results for the convergence in (1.2) and (2.5). Ideally one would like a distributional limit theorem for $(\theta_{0,n} - \delta(n))/\gamma(n)$ for some constants $\gamma(n) \to \infty$, $\delta(n) = o(n)$ and $\theta = a$ or b. We have not had any success with these. The next theorem gives a rate of convergence for (1.2) but it is much weaker than we expect to be true. Probably the factor $(\log n)^p$ can be replaced by a factor n^q for some $q > 0$. The proof of Theorem 4.1 is based on some large deviation estimates given in Theorem 4.2. Other large deviation estimates are in Theorem 4.3.

THEOREM 4.1. *Assume that*

$$\int_0^\infty e^{\gamma x}\, dF(x) < \infty \tag{4.1}$$

for some $\gamma > 0$ *and let* $p = 9d+3$. *Then there exist constants* $0 < C_i = C_i(F,d) < \infty$ *such that*

$$\mu \le \frac{1}{n} Ea_{0,n} \le \mu + C_1 (\log n)^{-1/p}, \sigma^2 \left(\frac{a_{0,n}}{n}\right) \le C_2 (\log n)^{-2/p}$$

and

$$\limsup\ (\log n)^{1/p} \left|\frac{1}{n}\theta_{0,n} - \mu\right| < \infty \quad \text{w.p.1.}$$

for $\theta = a$ or b.

THEOREM 4.2. (a) *For all* ε *there exist constants* $A_1 = A_1(\varepsilon,F,d)$ *and* $B_1 = B_1(\varepsilon,F,d) > 0$ *such that*

$$P\{a_{0,n} < n(\mu-\varepsilon)\} \le e^{-B_1 n}, \ n > 0, \tag{4.2}$$

and

$$\lim_{n\to\infty} -\frac{1}{n} \log P\{a_{0,n} < n(\mu-\varepsilon)\} = B_1, \tag{4.3}$$

and

$$P\{b_{0,n} < n(\mu-\varepsilon)\} \le A_1 e^{-B_1 n}, \ n \ge 0. \tag{4.4}$$

(b) If (4.1) holds for some $\gamma > 0$ then

$$B_1(\varepsilon,F,d) \geq \exp(-C_3\varepsilon^{-p}),\ \varepsilon > 0,$$

for some $C_3 > 0$ and $p = 9d+3$.

REMARK. Some further properties of B_1 as a function of ε are given in [9], Theorem 5.2c.

THEOREM 4.3. (a) Assume that (4.1) holds for some $\gamma > 0$. Then for each $\varepsilon > 0$ there exist constants $A_2 = A_2(\varepsilon,F,d)$ and $B_2 = B_2(\varepsilon,F,d) > 0$ such that for $\theta = a$ or b

$$P\{\theta_{0,n} > n(\mu+\varepsilon)\} \leq A_2 e^{-B_2 n},\quad n \geq 0. \tag{4.5}$$

(b) If (4.1) holds for all $\gamma > 0$, then for all $\varepsilon > 0$ and $\theta = a$ or b

$$\lim -\frac{1}{n}\log P\{\theta_{0,n} > n(\mu+\varepsilon)\} = \infty$$

(so that the left hand side of (4.5) decreases faster than exponentially).

(c) If $t(e)$ is bounded w.p.1, then for all $\varepsilon > 0$ there exist constants $A_3 = A_3(\varepsilon,F,d)$ and $B_3 = B_3(\varepsilon,F,d) > 0$ such that for $\theta = a$ or b

$$P\{\theta_{0,n} > n(\mu+\varepsilon)\} \leq A_3\exp(-B_3 n^d),\ n \geq 0.$$

5. ASYMPTOTICS FOR LARGE d. We discuss the behavior for large d of μ and

$$\mu^* = \mu^*(F,d) := \lim_{n\to\infty} \frac{1}{n} \inf\{T(r): r \text{ a path from } \underline{0} \text{ to the hyperplane } \{x: x(1)+x(2)+\ldots+x(d) = n\}\} \text{ w.p.1}.$$

By comparing these two we find that B_0 is not a Euclidean ball for large d for a large class of distribution functions F.

The class of distribution functions which we consider is defined for $a > 0$ and $b > 0$ as follows.

DEFINITION. $\mathcal{C} = \mathcal{C}(a,b)$ is the class of distribution functions with $F(0) = 0$ which have a density f on the interval $[0,b]$ which satisfies

$$f(x) = a\left(1+o\left(\frac{1}{|\log x|}\right)\right),\ x \downarrow 0.$$

THEOREM 5.1. There exists a universal constant $0 < C < \infty$ such that for all $F \in \mathcal{C}(a,b)$ with $\int x dF(x) < \infty$ one has for large d

$$C \frac{\log d}{ad} \le \mu(F,d) \le 11 \frac{\log d}{ad} .$$

Furthermore for such F

$$\frac{1}{6ea} \le \liminf_{d\to\infty} d\mu^*(F,d) \le \limsup_{d\to\infty} d\mu^*(F,d) \le \frac{1}{2a} .$$

Simple convexity arguments show that the furthest point in the positive direction on the line $x(1) = x(2) = \dots = x(d)$ of B_0 is the point

$$(\frac{1}{d\mu^*},\dots,\frac{1}{d\mu^*}).$$

For large d the Euclidean distance of this point to the origin behaves like $\sqrt{d}$. On the other hand, as we saw in Sect. 3, the furthest point on the positive first coordinate axis of B_0 is

$$(\frac{1}{\mu},0,\dots,0)$$

which has a distance of the order $d(\log d)^{-1}$ to the origin. We therefore have the following

COROLLARY. If F satisfies the hypotheses of Theorem 5.1, then B_0 is not a Euclidean ball for large d. When F is the exponential distribution, then this conclusion holds for $d \ge 10^6$.

6. MAXIMAL FLOWS AND MINIMAL SURFACES. Thus far we have interpreted $t(e)$ as a passage time of the edge e. We now want to interpret $t(e)$ as a capacity, i.e., as the maximal amount of fluid which can pass through e per unit time. Let $B = B(n_1,n_2,\dots,n_{d-1},m)$ be the box $[0,n_1]\times[0,n_2]\times\dots\times[0,n_{d-1}]\times[0,m]$ and denote its "bottom" and "top" by

$$F_0 = [0,n_1]\times\dots\times[0,n_{d-1}]\times\{0\}$$

and

$$F_m = [0,n_1]\times\dots\times[0,n_{d-1}]\times\{m\} ,$$

respectively. A permissible flow from F_0 to F_m in B is an assignment of a number $0 \le f(e) \le t(e)$ and a direction to each edge e in B, which satisfies for each vertex $v \notin F_0 \cup F_m$

(6.1) $$\textstyle\sum^{+} f(e) - \sum^{-} f(e) = 0,$$

where $\sum^{+}$ ($\sum^{-}$) is the sum over all edges e in B incident to v and directed towards v (away from v). $f(e)$ should be interpreted as the amount of fluid flowing through e per unit time in the direction assigned to e. (6.1) says that the total inflow at v equals the total outflow at v. The total flow from F_0 to F_m is

$$\textstyle\sum_0^{-} f(e) - \sum_0^{+} f(e),$$

where $\sum_0^{+}$ ($\sum_0^{-}$) is the sum over all edges with one endpoint in F_0 and one endpoint outside F_0 and directed towards (away from) the endpoint in F_0. The <u>maximal flow</u> from F_0 to F_m in B is

$$\Phi(n_1, n_2, \ldots, n_{d-1}, m) := \text{maximum of the flow from } F_0 \text{ to } F_m \text{ over all permissible flows.}$$

(See [6] for more details.)

A set E of edges <u>separates</u> F_0 <u>from</u> F_m <u>in</u> B if there is no path on $\mathbb{Z}^d$ in B from F_0 to F_m which avoids E. E is called an (F_0,F_m)-<u>cut</u> if it is a minimal separating set of this kind. The <u>capacity</u> or <u>value</u> of any set of edges E is taken to be

$$V(E) = \sum_{e \in E} t(e).$$

The max-flow min-cut theorem ([6]) states that

(6.2) $$\Phi(n_1, \ldots, n_{d-1}, m) = \min\{V(E)\colon E \text{ an } (F_0,F_m)\text{-cut}\}.$$

For $d = 2$ the cut-sets of $B(n,m) = [0,n] \times [0,m]$ are easy to describe. Let $\mathcal{L}^*$ be the dual graph of $\mathbb{Z}^2$. We may think of $\mathcal{L}^*$ as $\mathbb{Z}^2 + (\frac{1}{2},\frac{1}{2})$ which has vertices at the points $(i+\frac{1}{2}, j+\frac{1}{2})$, $i,j \in \mathbb{Z}$, and $(i+\frac{1}{2}, j+\frac{1}{2})$ and $(k+\frac{1}{2}, \ell+\frac{1}{2})$ are connected by an edge of $\mathcal{L}^*$ if and only if $|i-k| + |j-\ell| = 1$. (See Fig. 3.)

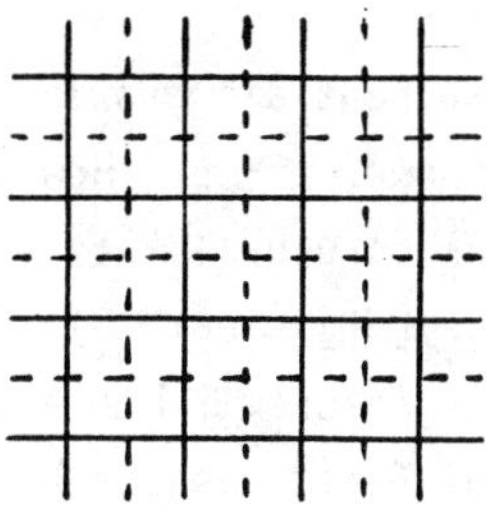

Fig. 3. $\mathbb{Z}^2$ and $\mathcal{L}^*$. The edges of $\mathbb{Z}^2$ are drawn as solid segments and those of $\mathcal{L}^*$ as dashed segments.

By the definition of a dual graph (see also Fig. 3) there is a 1-1 correspondence between the edges of $\mathbb{Z}^2$ and the edges of $\mathcal{L}^*$. In fact, each edge e of $\mathbb{Z}^2$ intersects a unique edge e^* of $\mathcal{L}^*$ and vice versa. We call such a pair of edges associated and set

$$t(e^*) = t(e)$$

if e and e^* are associated. Also we call a set of edges E^* of $\mathcal{L}^*$ an (F_0,F_m)-cut if the set of edges associated to some edge of E^* is an (F_0,F_m)-cut. The (F_0,F_m)-cuts of $B(n,m)$ are then precisely the paths without double points on $\mathcal{L}^*$ in $[-\frac{1}{2},n+\frac{1}{2}] \times [+\frac{1}{2},m+\frac{1}{2}]$ from $\{-\frac{1}{2}\} \times [\frac{1}{2},m-\frac{1}{2}]$ to $\{n+\frac{1}{2}\} \times [\frac{1}{2},m-\frac{1}{2}]$ (see Fig. 4)

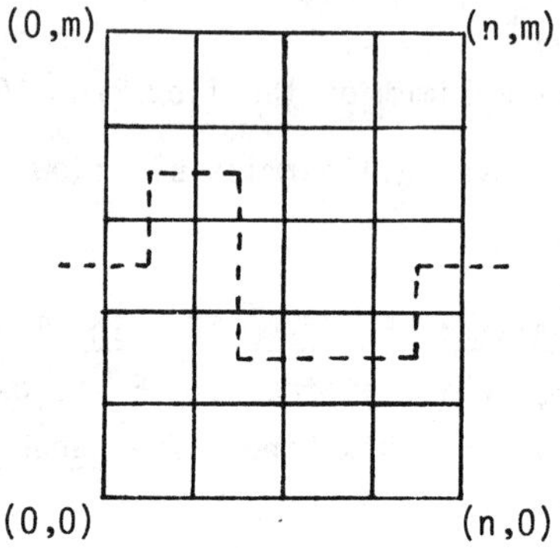

Fig. 4. The dashed path is a cut for $B(n,m)$.

Using this description of cutsets the following result was proved in [8].

THEOREM 6.1. $(d = 2)$ If $n \to \infty$, $m \to \infty$ in such a way that

$$\frac{1}{n} \log m \to 0, \tag{6.3}$$

then

$$\frac{1}{n} \Phi(n,m) \to \mu(F,2) \quad \text{w.p.1.} \tag{6.4}$$

REMARK. (6.3) allows any "subexponential" growth for m as a function of n. For example m may grow like a power of n. However, (6.3) cannot be relaxed further, since the limit in (6.4) is usually strictly less than μ if m grows faster than allowed by (6.3) (see [8], Sect. 5).

When one tries to generalize Theorem 6.1 to higher dimensions one is lead to interesting generalizations of first passage percolation. So far one has made only a small start with this theory, and only for $d = 3$. We shall therefore restrict ourselves to $d = 3$ in the remainder. Following Aizenman et al.

[1] we replace the dual edges by "plaquettes". These are the faces of the unit cubes with centers in $\mathbb{Z}^3$, i.e., plaquettes are unit squares with corners in $\mathcal{L}^* := \mathbb{Z}^3 + (\frac{1}{2},\frac{1}{2},\frac{1}{2})$. One easily sees that again each edge e of $\mathbb{Z}^3$ intersects a unique plaquette π^* and vice versa. Again we call e and π^* associated when they intersect and set

$$v(\pi^*) = t(e)$$

in this case. Analogously to the two dimensional case we call a collection E^* of plaquettes an (F_0,F_m)-cut in $B = B(n_1,n_2,m)$ if the collection E of edges which are associated to some π^* in E^* form an (F_0,F_m)-cut. Also for any set E^* of plaquettes we define

$$V(E^*) = \sum_{\pi^* \in E^*} v(\pi^*) . \tag{6.5}$$

The problem now is that we can no longer give a simple description of cut-sets of plaquettes. Intuitively, these are some kind of surfaces between the bottom and top of B. Since, by (6.2) and (6.5)

$$\Phi(n_1,n_2,m) = \inf\{V(E^*) : E^* \text{ is an } (F_0,F_m)\text{-cut}\},$$

we now are looking for "surfaces" E^* whose value $V(E^*)$ is minimal. (One should not take the term "surface" too literally here. We really mean an (F_0,F_m)-cut when we say surface. Without modification these cuts don't have to be 2-manifolds.)

It is ambiguous what the analogues of the passage times $a_{0,n}$ and $b_{0,n}$ should be. Recall that $a_{0,n}$ is defined as the inf of $T(r)$ over the collection of paths r whose endpoints are fixed at $\underline{0}$ and at $(n,0,\dots,0)$. Aizenman et al. [1], Sect. 1(ii), argue that this suggests as an analogue of $a_{0,n}$ the inf of $V(E^*)$ over a collection of surfaces E^* with fixed boundary. Here the boundary of E^* is defined as

$$\partial E^* = \text{collection of edges of } \mathcal{L}^* \text{ which belong to an odd number of plaquettes in } E^*.$$

Similarly $b_{0,n}$ was defined as an inf of $T(r)$ over paths with one endpoint completely fixed but the other endpoint only partially restricted. Note that $a_{0,n}$ as well as $b_{0,n}$ is an inf over paths running from $\underline{0}$ to H_n. In dimension 2 H_n is just the vertical line $\{x(1) = n\}$ and any path from $\underline{0}$ to H_n separates $[\frac{1}{2},n-\frac{1}{2}] \times \{-N\}$ from $[\frac{1}{2},n-\frac{1}{2}] \times \{N\}$ in $[\frac{1}{2},n-\frac{1}{2}] \times [-N,N]$ for some N (see Fig. 5).

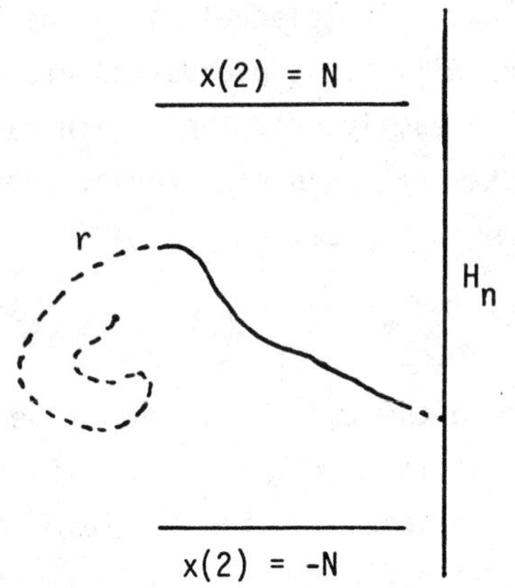

Fig. 5. The solidly drawn part of r separates the horizontal segments at $x(2) = -N$ and $x(2) = N$.

With this in mind we opted for the following definitions. Fix n_1 and n_2 for the time being and set

$$G_{-N} = [0,n_1] \times [0,n_2] \times \{-N\}, \quad G_N = [0,n_1] \times [0,n_2] \times \{N\} ,$$

$$B_N = [0,n_1] \times [0,n_2] \times [-N,N].$$

The analogues of $a_{0,n}$ and $b_{0,n}$ are then taken to be

(6.6) $\alpha(n_1,n_2) = \inf\{V(E^*) : E^*$ is a collection of plaquettes such that ∂E^* consists exactly of the edges in the perimeter of $[-\frac{1}{2},n_1+\frac{1}{2}] \times [-\frac{1}{2},n_2+\frac{1}{2}] \times \{\frac{1}{2}\}$ and such that E^* separates G_{-N} from G_N in B_N for some $N\}$,

and

(6.7) $\beta(n_1,n_2) = \inf\{V(E^*) : E^*$ is a connected set of plaquettes which contains the point $\{-\frac{1}{2},-\frac{1}{2},\frac{1}{2}\}$ and which separates G_{-N} from G_N in B_N for some $N\}$,

respectively.

REMARK. It can be shown that any set E^* whose boundary ∂E^* satisfies the condition in the right hand side of (6.6) automatically separates G_{-N} from G_N in B_N for some N, so that the separation requirement can be dropped from the definition (6.6). Note also that once E^* separates G_{-N} from G_N in B_N for some N, then it has this separation property also for all larger N.

The principal result of [10] proves an analogue of (1.2) with the above replacements for $a_{0,n}$ and $b_{0,n}$, and this may be viewed as an indication that (6.6) and (6.7) are the "right" definitions. The results are formulated in Theorems 6.2 and 6.3. Unfortunately we can prove Theorem 6.2 only when $F(0)$, the atom of F at the origin, is sufficiently small. Lemma 6.1 serves to define the interval of $F(0)$ values for which Theorem 6.2 works. The hope is that someone will prove this result for all $F(0) < 1-p_T(3))$. If that can be done then one will obtain a fairly complete analogue of the first-passage percolation results for the "minimal surfaces" and maximal flows.

LEMMA 6.1. *There exists a* $p_0 \geq 1/27$ *with the following property: For every distribution function* F *with*

$$F(0-) = 0,\ F(0) < p_0 \tag{6.8}$$

there exists constants $\Theta = \Theta(F) > 0,\ 0 < C_i = C_i(F) < \infty$ *such that*

$$\begin{aligned} P\{&\text{there exists a connected set } E^* \text{ of } n \text{ plaquettes of } \mathcal{L}^* \\ &\text{which contains the point } (-\tfrac{1}{2},-\tfrac{1}{2},\tfrac{1}{2}) \text{ and with } V(E^*) \leq \Theta n\} \\ &\leq C_1 e^{-C_2 n},\quad n \geq 0. \end{aligned} \tag{6.9}$$

THEOREM 6.2. *Assume* F *satisfies (6.8) or more generally* $F(0-) = 0$ *and (6.9). If in addition*

$$Ee^{\gamma V(\pi^*)} = \int_{[0,\infty)} e^{\gamma x} dF(x) < \infty \tag{6.10}$$

for some $\gamma > 0$, *then there exists a number* $\nu = \nu(F) \leq \int_{[0,\infty)} x dF(x) < \infty$ *which is strictly positive and such that*

$$\lim_{n_1,n_2\to\infty} \frac{1}{n_1 n_2} \alpha(n_1,n_2) = \lim_{n_1,n_2\to\infty} \frac{1}{n_1 n_2} \beta(n_1,n_2) = \nu \quad \text{w.p.1.} \tag{6.11}$$

Moreover, if $n_1, n_2, m \to \infty$ *in such a way that*

$$(\max\{n_1,n_2\})^{-1+\delta} \log m \to 0 \quad \text{for some} \quad \delta > 0$$

then also

$$\lim_{n_1,n_2\to\infty} \frac{1}{n_1 n_2} \Phi(n_1,n_2,m) = \nu \quad \text{w.p.1.}$$

THEOREM 6.2. *Assume that*

$$F(0-) = 0, \quad F(0) > 1-p_T(3) \quad \text{and} \quad \int_{[0,\infty)} x^6 dF(x) < \infty.$$

Then

$$\limsup_{n_1,n_2\to\infty} \frac{1}{n_1+n_2} \theta(n_1,n_2) < \infty \quad \text{w.p.1}$$

for $\theta = \alpha$ or β. A fortiori (6.11) holds with $\nu = 0$.

Furthermore, there exists a constant $C_3 = C_3(F) < \infty$ such that w.p.1.

$$\Phi(n_1,n_2,m) = 0 \quad \text{for all sufficiently large } n_1,n_2,$$

whenever $m(n_1,n_2) \to \infty$ as $n_1,n_2 \to \infty$ in such a way that

$$\liminf_{n_1,n_2\to\infty} \frac{m(n_1,n_2)}{\log(n_1 n_2)} > C_3.$$

BIBLIOGRAPHY

[1] M. Aizenman, J. T. Chayes, L. Chayes, J. Fröhlich and L. Russo, "On a sharp transition from area law to perimeter law in a system of random surfaces", Comm. Math. Phys. 92(1983), 19-69.

[2] J. T. Cox and R. Durrett, "Some limit theorems for percolation processes with necessary and sufficient conditions", Ann. Probab. 9 (1981), 583-603.

[3] R. Durrett, "Oriented percolation in two dimensions", Ann. Probab. 12 (1984), 999-1040.

[4] R. Durrett and T. M. Liggett, "The shape of the limit set in Richardson's growth model", Ann. Probab. 9 (1981), 186-193.

[5] M. Eden, "A two dimensional growth process", pp.223-239 in Proc. Fourth Berkeley Symp. Math. Stat. Prob., Vol. IV, Univ. of Cal. Press (1961).

[6] D. R. Fulkerson, "Flow networks and combinatorial operations research", pp. 139-171 in Studies in Graph Theory, D. R. Fulkerson, ed., Math. Assoc. of America (1975).

[7] J. M. Hammersley and D. J. A. Welsh, "First-passage percolation, subadditive processes, stochastic networks and generalized renewal theory", pp.61-110 in Bernoulli, Bayes, Laplace Anniversary Volume, J. Neyman and L. M. LeCam, eds., Springer-Verlag (1965).

[8] G. Grimmett and H. Kesten, "First-passage percolation, network flows and electrical resistances", Z. Wahrsch. verw Geb. 66 (1984), 335-366.

[9] H. Kesten, "Aspects of first-passage percolation", Notes for Summer School at St. Flour (1984), to appear in Lecture Notes in Mathematics.

[10] H. Kesten, "Surfaces with minimal random weights and maximal flows: a higher dimensional version of first-passage percolation", preprint (1985).

[11] D. Richardson, "Random growth in a tesselation", Proc. Cambr. Phil. Soc. 74 (1973), 515-528.

[12] R. T. Smythe and J. C. Wierman, "First-passage percolation on the square lattice", Lectures Notes in Math., Vol. 671, Springer-Verlag (1978).

[13] J. C. Wierman, "Weak moment conditions for time coordinates in first-passage percolation models", J. Appl. Prob. 17 (1980), 968-978.

DEPARTMENT OF MATHEMATICS
CORNELL UNIVERSITY
ITHACA, NEW YORK 14853

Contemporary Mathematics
Volume 41, 1985

USING RANDOM MATRICES TO GIVE RECURRENCE AND TRANCIENCE CRITERIA FOR RANDOM WALK IN A RANDOM ENVIRONMENT

Eric S. Key

ABSTRACT. Oseledec's Multiplicative Ergodic Theorem is used to give recurrence and transience criteria for random walk in a random environment on the integers. These criteria generalize those given by Solomon in the nearest-neighbor case. An example is computed which demonstrates the use of the criteria.

1. INTRODUCTION TO RANDOM WALK IN A RANDOM ENVIRONMENT. When one constructs a stochastic model of the motion of a particle, one may take the point of view that the randomness of the motion is due to the medium or environment in which the particle is moving. For example, if the motion is to be modeled by a Markov chain $\{X_t\}_{t=0,1,\ldots}$ on the integers, $\mathbb{Z}$, one specifies the distributions $e(x;*)$ defined by

$$e(x;*) = P\{X_{t+1}=x+*|X_t = x\}, \tag{1}$$

for every integer x. The collection

$$e = \{e(x;*), x\varepsilon\mathbb{Z}\} \tag{2}$$

represents the environment in which the particle is to move.

For example, one could have two distributions:

$$\vec{\alpha} = (\alpha(-1);\ \alpha(0);\ \alpha(1),\ \alpha(2)) = (\tfrac{720}{784};\ 0;\ \tfrac{3}{784},\ \tfrac{61}{784}) \tag{3}$$

and

$$\vec{\beta} = (\beta(-1);\beta(0);\beta(1),\beta(2)) = (\tfrac{208}{784};\ 0;\ \tfrac{419}{784},\ \tfrac{157}{784}). \tag{4}$$

A Markov chain could then be constructed which satisfies

$$e(x;*) = \begin{cases} \vec{\alpha} & \text{if } x \text{ is even} \\ \vec{\beta} & \text{if } x \text{ is odd.} \end{cases}$$

One might call such an environment a periodic environment.

Another possibility, the one we shall be investigating, is akin to constructing the Markov chain by independently and randomly assigning the value

1980 Mathematics Subject Classifications. 60J10, 60J15

$\vec{\alpha}$ or $\vec{\beta}$ to each $x\in\mathbb{Z}$:

$$e(x;*) = \begin{cases} \vec{\alpha} \text{ with probability } \frac{1}{2} \\ \vec{\beta} \text{ with probability } \frac{1}{2} \end{cases}$$

The environment now is called a random environment, and the random process which describes the motion of the particle is called a Random Walk in a Random Walk in a Random Environment, RWIRE for short.

Two natural questions to ask are: Is this process recurrent or is it transient? If it is transient, in which direction does it drift?

2. THE GENERAL SET UP:

R and L are two fixed, positive integers which represent the maximum size of a jump to the right and a jump to the left, respectively.

G is the set of all probability measures on $\{-L, \ldots, R\}$. An element $\vec{\gamma}$ of G is an $(R + L + 1)$-tuple, and we shall locate the probability of 0 with semicolons:

$$(\gamma(-L), \ldots, \gamma(-1); \gamma(0); \gamma(1), \ldots, \gamma(R))$$

$$(\tfrac{720}{784}; 0; \tfrac{3}{784}; \tfrac{61}{784}).$$

$$e = \{e(z;*), z\in\mathbb{Z}\} \tag{5}$$

is a set of independent, identically distributed (iid) G-valued, random variables, called the random environment.

Informally, a Random Walk in a Random Environment on $\mathbb{Z}$ is a process $\{X_t\}_{t=0,1,2,\ldots}$ which, given e, is a stationary, discrete time Markov chain on $\mathbb{Z}$ with transition matrix determined by e as in (1):

$$\begin{aligned} &P\{X_{t+1}=z+a \mid X_t=z,\ X_{t-1}=z_{t-1},\ldots,X_0=z_0,e\} \\ &= P\{X_{t+1}=z+a \mid X_t=z,e\} \\ &= \begin{cases} e(z;a) & \text{if } -L \le a \le R \\ 0 & \text{otherwise} \end{cases} \end{aligned}$$

The first recurrence and transcience criteria were given by F. Solomon in 1975:

THEOREM. Suppose that $R = L = 1$, and suppose that $P\{e(0;-1) + e(0;1)=1\}= 1$. Let $d = E\log[\frac{-e|0;1)}{e(0;-1)}]$.

Then:

i) $d > 0$ implies $P\{\lim X_t = \infty\} = 1$.

ii) $d = 0$ implies $P\{-\infty = \liminf_{t\to\infty} X_t < \limsup_{t\to\infty} X_t = \infty\} = 1$.

iii) $d < 0$ implies $P\{\lim_{t\to\infty} X_t = -\infty\} = 1$.

We shall generalize these criteria.

3. A ZERO-ONE LAW:

One feature of Solomon's criteria is that they take the form of a trichotomy. This behavior is quite general.

Define $G' = \{\vec{\gamma}\varepsilon G: \exists\ a <0, b> 0 \text{ S.T. } \gamma(a)\cdot\gamma(b) > 0\}$

THEOREM. The Zero-One Law: Suppose that $P\{e(0;*)\varepsilon G'\} = 1$. Then one of the following mutually exclusive possibilities holds:

1) $P\{\lim X(t) = \infty\}$

2) $P\{-\infty = \liminf_{t\to\infty} X(t) < \limsup_{t\to\infty} X(t) = \infty\} = 1$

3) $P\{\lim_{t\to\infty} X(t) = -\infty\} = 1$

4) SOME DIFFERENCE EQUATIONS WITH RANDOM COEFFICIENTS:

Now that we know what sort of behavior to expect, the question remains, when does each type of behavior occur?

For $j\varepsilon\{1,2,\dots,R\}$, define

$$f_j(y) = \begin{cases} P\{X(t) \text{ first enters } (0,R] \text{ at } j \mid X(0) = y,\ e\} & y \le 0 \\ \delta(j,y) & y > 0 \end{cases}$$

Note that for $y \le 0$,

$$\sum_{j=1}^{R} f_j(y) = P\{X(t) \text{ enters } (0,\infty) \mid X(0) = y,\ e\}. \tag{6}$$

Since $X(t)$ is a Markov chain given e, the Markov property yields for $y \le 0$,

$$f_j(y) = \sum_{a=-L}^{R} P\{X_{t+1} = y+a \mid X_t = y,\ e\} f_j(y+a) = \sum_{a=-L}^{R} e(y;a) f_j(y+a). \tag{7}$$

If $e(y;-L) > 0$ we may solve (7) for $j_j(y-L)$:

$$f_j(y-L) = \sum_{a=-L+1}^{R} \left(\frac{\delta(0,a)-e(y;a)}{e(y;-L)}\right) f\ (y+a),\ y \le 0. \tag{8}$$

A standard manipulation in the theory of difference equations is to rewrite an equation such as (8) in terms of matrices. Before giving the general formulae, let us consider an example.

Example: Let $\vec{\alpha}$, $\vec{\beta}$ be as in (3) and (4), and suppose that $P\{e(0;*) = \vec{\alpha}\} = P\{e(0;*) = \vec{\beta}\} = \frac{1}{2}$. Then $R = 2$, $L = 1$, $j = 1$ or 2 and (8) becomes

$$f_j(y-1) = \frac{-e(y;2)}{e(y;-1)} f_j(y+2) + \frac{-e(y;1)}{e(y;-1)} f_j(y+1) + \frac{1-e(y;0)}{e(y;-1)} F_j(y). \tag{9}$$

Let
$$A_y = \begin{bmatrix} 0 & 1 & 0 \\ 0 & 0 & 1 \\ \frac{-e(y;2)}{e(y;-1)} & \frac{-e(y;1)}{e(y;-1)} & \frac{1-e(y;0)}{e(y;-1)} \end{bmatrix} \tag{10}$$

so that

$$P\{A_y = \begin{bmatrix} 0 & 1 & 0 \\ 0 & 0 & 1 \\ \frac{-61}{720} & \frac{-3}{720} & \frac{784}{720} \end{bmatrix}\} = P\{A_y = \begin{bmatrix} 0 & 1 & 0 \\ 0 & 0 & 1 \\ \frac{-157}{208} & \frac{-419}{208} & \frac{784}{208} \end{bmatrix}\} = \frac{1}{2} .$$

Let $\vec{f}_{y,j} = \begin{bmatrix} f_j(y+2) \\ f_j(y+1) \\ f_j(y) \end{bmatrix}$.

It then follows from (9) and (10) that

$$\vec{f}_{y-1,j} = A_y \vec{f}_{y,j} = A_y A_{y+1} \cdots A_0 \vec{f}_{0,j} \tag{11}$$

where

$$\vec{f}_{0,2} = \begin{bmatrix} 1 \\ 0 \\ f_2(0) \end{bmatrix} \quad \text{and} \quad \vec{f}_{0,1} = \begin{bmatrix} 0 \\ 1 \\ f_1(0) \end{bmatrix}$$

The general case is as follows.

A_y is an $(R+L)\times(R+L)$ matrix defined by

$$A_y(i,r) = \begin{cases} \delta(i+1,r) & \text{if } 1 \leq i < R + L \\ \dfrac{\delta(R+1,r)-e(y;R+L-r)}{e(y;-L)} & \text{if } i = R + L. \end{cases} \tag{12}$$

For each $j\varepsilon\{1,\ldots,R\}$, $\vec{f}_{y,j}$ is the $(R+L)\times 1$ vector defined by

$$\vec{f}_{y,j}(a) = f_j(y+R+1-a).$$

Then (11) still holds:

$$\vec{f}_{y-1,j} = A_y \vec{f}_{y,j} = A_y A_{y+1} \cdots A_0 \vec{f}_{0,j} .$$

The important thing to notice about the A_y is that since the $e(y;*)$ are iid, (12) implies that the $\{A_y : y\varepsilon\mathbb{Z}\}$ are iid as well. The important thing about $\{\vec{f}_{0,j}\}_{j=1,\ldots,R}$ is that it is a set of R linearly independent vectors whose coordinates all lie between 0 and 1.

5. OSELEDECS MULTIPLICATIVE ERGODIC THEOREM. In order to use (11) to analyse the limit behavior of X(t), we use the following theorem.

Oseledec's Multiplicative Ergodic Theorem: Let $\{M_m\}_{m\geq 0}$ be an iid sequence of $r\times r$ matrices on a probability space (Ω,β,m). Suppose $E \log^+\|M_0\| < \infty$. Then there are r constants

$$-\infty \leq d_1 \leq d_2 \leq \cdots \leq d_r$$

and an increasing, nonrandom sequence of integers

$$1 = i_1 < i_2 < \cdots < i_p < i_{p+1} = r + 1$$

which mark the points of increase of $\{d_j\}_{j=1,\ldots,r}$, such that a.s. (m):

(i) $\forall \vec{v} \in \mathbb{R}^r \setminus \{\vec{0}\}$, $f(v,\omega) \equiv \lim_{n\to\infty} \frac{1}{n} \log \| M_{m-1} \cdots M_0 \vec{v} \| \in \{d_j\}_{j=1,\ldots,r}$.

(ii) For $q \leq p$, $V(q,\omega) \equiv \{v \in \mathbb{R}^r : f(\vec{v},\omega) \leq d_{i_q}\}$ is a random subspace of $\mathbb{R}^r$ of dimension $i_{q+1}-1$.

(iii) If $V(0,\omega) \equiv \{\vec{0}\}$ then $\vec{v} \in V(q,\omega)\ V(q-1,\omega)$ implies $f(v,\omega) = d_{i_q}$.

The values $\{d_j\}_{j=1,\ldots,r}$ are called Lyapunov numbers, and they are generalizations of logarithms of the absolute values of eigenvalues. In fact

$$\sum_{j=1}^{r} \lambda_j = E\log|\det M_0|. \tag{13}$$

6. THE RECURRENCE AND TRANSIENCE CRITERIA. The recurrence and transience criteria are stated in terms of the Lyapunov numbers of the sequence $\{A_{-y}, y \geq 0\}$

THEOREM: Suppose that $E\log(e(0;-L)) > -\infty$ and $E\log(e(0;R)) > -\infty$. Let $\{d_j\}_{j=1,\ldots,R+L}$ be the Lyapunov numbers of the sequence $\{A_{-y}\}_{y>0}$, and let $d = d_R + d_{R+1}$. Then

i) $d > 0$ implies $P\{\lim_{t\to\infty} X(t) = \infty\} = 1$

ii) $d = 0$ implies $P\{-\infty = \liminf_{t\to\infty} X(t) < \limsup_{t\to\infty} X(t) = \infty\} = 1$

iii) $d < 0$ implies $P\{\lim_{t\to\infty} X(t) = -\infty\}$

Remark: In the case Solomon considered,

$$A_y = \begin{bmatrix} 0 & 1 \\ \frac{-e(y_j 1)}{e(y_{j-1})} & \frac{1}{e(y_{j-1})} \end{bmatrix}$$

so $d = d_1 + d_2 = E\log|\det A_0| = E\log[\frac{e(0;1)}{e(0;-1)}]$.

7. A FEW WORDS ABOUT THE PROOF. The following lemma gives an idea of how this theorem may be proved.

LEMMA: Suppose $E\log(e(0;-L)) > -\infty$. For $\vec{v} \in \mathbb{R}^{R+L}$ let $f(\vec{v}) = \lim_{n\to\infty} \frac{1}{n} \log \| A_{1-n} \cdots A_0 \vec{v} \|$, and for $\vec{b} \in \mathbb{R}^{R+L}$ be given by $\vec{b}(a) = b^a$. Then

(i) $P\{A_0 \vec{1} = \vec{1}\} = 1$.

(ii) If b satisfies $P\{A_y \vec{b} = b\vec{b}\}$ then $f(\vec{b}) = \log|b|$, and so $\log|b|$ is a Lyapunov number of the sequence $\{A_{-y}, y \geq 0\}$

(iii) $V \equiv \{\vec{v} \in \mathbb{R}^{R+L} : f(\vec{v}) \leq 0\}$ is a random subspace of $\mathbb{R}^{R+L}$ of non-random dimension $\geq R$.

(iv) $\dim(V) = R \Rightarrow P\{X(t) \text{ enters } (0,\infty) | X(0) = y, e\} = 1$ a.s. for every $y \in \mathbb{Z}$.

The key part of this lemma is (iii), which follows from Oseledec's Theorem and the fact that for every environment, $\{\vec{f}_{0,j}\}_{j=1,\ldots,R} \subset V$, and the $\{f_{0,j}\}_{j=1,\ldots,R}$ are linearly independent. (iv) then follows from (i)-(iii) and (6) since $\vec{1} \in V$ and $\dim V = R$ implies $\vec{1} = \sum_{j=1}^{R} \vec{f}_{0,j}$.

8. BACK TO THE EXAMPLE. Recall our earlier example:

$$P\{e(0;*) = \vec{\alpha}\} = P\{e(0;*) = \vec{\beta}\} = \frac{1}{2}.$$

In this case, if $b = -\frac{1}{4}$ then $P\{A_y\vec{b} = b\vec{b}\} = 1$, so the lemma says that one of the Lyapunov numbers of $\{A_{-y}\}\ y \geq 0$ is $\log|\frac{1}{4}|$. (i) and (ii) of the lemma guarantee that 0 is another Lyapunov number. If the Lyapunov numbers are d_1, d_2 and d_3 then $d_1 + d_2 + d_3 = E\log|\det A_0| = \frac{1}{2}\log(\frac{157\cdot 61}{208\cdot 720})$ and the remaining Lyapunov number is obtained by subtraction, giving

$$\{d_1, d_2, d_3\} = \{-\log 4,\ 0,\ \frac{1}{2}\log[\frac{9577}{9360}]\}.$$

Therefore $d \equiv d_2 + d_3 = \frac{1}{2}\log[\frac{9577}{9360}] > 0$ and $P\{\lim X(t) = \infty\} = 1$.

Warning: $E(e(y;*)) = \frac{1}{2}(\vec{\alpha}+\vec{\beta}) = (\frac{464}{784};\ 0;\ \frac{211}{784},\ \frac{109}{784})$ is a probability distribution on $\{-1,0,1,2\}$ with <u>negative mean</u>. So if $S_t \equiv Y_1 + \cdots + Y_t$ is a random walk with $P\{Y_t = *\} = E(e(y;*))$, then $P\{\lim_{t\to\infty} S_t = -\infty\}$. This illustrates that replacing the random environment with its average can give behavior exactly opposite that of the original model.

BIBLIOGRAPHY

1. E. S. Key, Recurrence and Transience Criteria for Random Walk in a Random Environment, Ann. of Prob., 12 (1984), 529-560.

DEPARTMENT OF MATHEMATICAL SCIENCES
UNIVERSITY OF WISCONSIN-MILWAUKEE
MILWAUKEE, WISCONSIN 53201

Contemporary Mathematics
Volume 41, 1985

RECENT RESULTS ON THE MOVEMENT OF A TAGGED PARTICLE IN SIMPLE EXCLUSION

C. Kipnis[1]

ABSTRACT. We review here some recent results on the strong law of large numbers and the Central Limit Theorem for a particular semi-markovian process i.e. the position of one tagged particle interacting with an infinite number of indistinguishable interacting particles.

1. INTRODUCTION. The paper follows essentially a talk that was delivered at the AMS summer-school on the "Mathematics of phase transition" at Brunswick Maine in 1984. It does not contain any new result but rather aims at being a survey as well as an introduction to some open problems in this field.

Partially stimulated by the interest of mathematical physicists [6], the problem of the asymptotic behaviour of one tagged particle, interacting with an infinite number of indistinguishable particle in "global equilibrium" has been studied in the past four years. It can be considered as a special case of semi-markov process, as will be seen below, but unfortunately it lacks some of the properties that are usually assumed in this theory so that specific tools have to be developped to deal with our problem.

Our object of interest is the so called simple exclusion process, where (infinitely) many particles move on $\mathbb{Z}^d$ according to a continuous-time rate one random walk but with the interdiction of double occupancy. Therefore if a particle attempts a transition to a site already occuped the jump is supprimed and the particle remains where it was before the attempt . In other words the Markov process we are studying lives on $\{0,1\}^{\mathbb{Z}^d}$ and its pregenerator is defined on functions depending only on a finite number of coordinates by

$$Lf(\eta) = \sum_x \sum_y \eta(x)(1-\eta)(y))\ p(y-x)[f(\eta^{x,y})-f(\eta)]$$

where $\eta \in \{0,1\}^{\mathbb{Z}^d}$, ($\eta(u)$ is the occupation number of site u) and $\eta^{x,y}$ is defined by

---

[1]Centre de Mathématiques Appliquées - Ecole Polytechnique
91128 Palaiseau Cedex (France).

1980 Mathematics subject classification : 60K35.

$$\eta^{x,y}(u) = \begin{cases} \eta(u) & \text{if } u \neq x \text{ or } y \\ \eta(x) & \text{if } u = y \\ \eta(y) & \text{if } u = x \end{cases}$$

The existence of a unique Markov process which generator extends L has been proved by Liggett [7]. We now tag one of the particles located initially at x_0 and denote by ξ the relative positions of the other particles. In this description we see that if we ignore the tagging we obtain our configuration of indistinguishable particles η by the formula

$$\eta(u) \leqslant \begin{cases} 1 & \text{if } u = x_0 \\ \xi(u-x_0) & \text{if } u \neq x_0 \end{cases}$$

Now the joint generator for (x,ξ) is given by

$$Lf(x,\xi) = \sum_z p(z)(1-\xi(z))[f(x+z,\tau_z\xi)-f(x,\xi)]$$

$$+ \sum_{u\neq x} \sum_{v\neq x} \xi(u)(1-\xi(v))p(v-u)[f(x,\xi^{x,v})-f(x,\xi)]$$

where

$$\tau_z\xi(u) = \begin{cases} \xi(u+z) & \text{if } u \neq -z \text{ or } 0 \\ 0 & \text{if } u = -z \end{cases}$$

Of course one notices that by translation invariance of the random walk kernel, the process of the environments seen from the tagged particle is also a Markov process on $\{0,1\}^{\mathbf{Z}^d\setminus\{0\}}$ with generator $\bar{L}$:

$$\bar{L}f(\xi) = \sum_z p(z)(1-\xi(z))[f(\tau_z\xi)-f(\xi)]$$

$$+ \sum_{u\neq 0} \sum_{v\neq 0} \xi(u)(1-\xi(v))p(v-u)[f(\xi^{u,v})-f(\xi)]$$

On the other hand some stationary distributions are known for these processes. Define ν_ρ to be the Bernoulli distribution on $\mathbf{Z}^d$ i.e. the probability that makes the occupation variables at differents sites independent with common law, for every site $x \in \mathbf{Z}^d$,

$$\nu_\rho\{\eta(x) = 1\} = \rho .$$

Then ν_ρ is stationary for the process associated to L and every translation invariant stationary probability for this process is a mixture of the ν_ρ's [7]. One can also easily check that $\nu_\rho(\;/\eta(0) = 1) = \mu_\rho$ is stationary for $\bar{L}$ [7].

We will now study the long-time effect of the medium on a tagged particle. Unless stated differently we will start our system by taking $x_0 = 0$ and

ξ distributed according to μ_ρ. The object of interest is now the position x_t and we focus on the law of large numbers and the central limit theorem for the tagged particle.

2. A SURVEY OF THE KNOWN RESULTS. We first start with the law of large numbers

THEOREM [8] : *Almost surely* x_t/t *tends to* $(1-\rho)\sum zp(z)$ *as* t *tends to infinity.*

Sketch of proof. The proof stems from the observation that we can write

$$x_t = M_t + \int_0^t \varphi(\xi_s)ds$$

where

$$\varphi(\xi) = \sum_z (1-\xi(z))z.p(z)$$

and M_t is a martingale whose increments are stationary and in L^2. Therefore M_t/t tends to zero almost surely and we are reduced to proving the ergodicity of μ_ρ for the environment process. For this one first proves that it suffices to show that ν_ρ is ergodic for the original process generated by L.

Of course in case p is symmetric this fact was already known since it was proven that the extremal invariant probability for the symmetric simple exclusion process are the ν_ρ's. But for non-symmetric p's the same result is not known and false if $\sum zp(z) \neq 0$. Nevertheless one can prove directly that the ν_ρ's are extremal by the following argument.

Let h be a harmonic function for the semigroup P_t generated by L. Then since $P_th = h$ we also have that

$$P_t^*P_th = h \text{ where } P_t^*$$

is the semi-group adjoint to P_t in $L^2(\nu_\rho)$. It is easy to see that P_t^* is the semi-group generated by L^* obtained from L by giving to a jump of size z the rate $p(-z)$ instead of $p(z)$. From the equality $h = (P_{t/n}^*P_{t/n})^nh$ one concludes that h is harmonic for $L+L^*$. But $L+L^*$ is the generator of a symmetric simple exclusion process, hence h is constant a.s. (since the ν_ρ's are extremal for $L+L^*$), and this implies the ergodicity. ∎

We are now in position to study the central limit theorem.

In order to use the martingale central limit theorem we start from the formula

$$x_t - t.(1-\rho)\sum zp(z) = M_t + \int_0^t \sum zp(z)[(1-\xi_s(z))-(1-\rho)]ds$$

$$:= M_t + \int_0^t \psi(\xi_s)ds$$

and our concern is to replace the second term by a martingale in an appropriate sense. As we will see below one has to distinguish whether we are dealing with a transition matrix p which is symmetric or not.

Loosely speaking what one has to prove in order to have a CLT is a perhaps somewhat strengthened version of the following conditions

(1) $$\frac{1}{t}\int \mu_\rho(d\xi)E_\xi\left[\left(\int_0^t \psi(\xi_s)ds\right)^2\right] \to \sigma^2$$

(2) $$\frac{1}{t}\int \mu_\rho(d\xi)E\left[\left(\int_0^t \psi(\xi_s)ds\right)\right]^2 \to 0$$

However when p is symmetric the measure μ_ρ is not only invariant for ξ_t but also reversible so that (1) implies (2).

Moreover in order to check (1) we need to reword it in such a form that it can be checked directly in terms of the generator.

Using the self adjointness of $\bar{L}$ with respect to μ_ρ we see that (1) is implied by

(3) $$\int_0^\infty \int \mu_\rho(d\xi)E_\xi(\psi(\xi_0)\psi(\xi_t)dt < \infty$$

which can be expressed as

$$\int_0^\infty \frac{1}{x} E(dx) < \infty$$

if $E(dx)$ is the spectral measure associated to ψ and which in turn means that $\psi \in \mathcal{D}(\bar{L}^{-\frac{1}{2}})$ or :

(4) $\forall f \in \mathcal{F}$ $$\left|\int \mu_\rho(d\xi)\ \psi(\xi)f(\xi)\right| \leq C\sqrt{\int\mu_\rho(d\xi)f(\xi)\bar{L}f(\xi)}$$

where C is a finite constant independent of f.

It is then easy to check (4) and one obtains

<u>THEOREM [5]</u> : *Let* p *satisfy* $p(z) = p(-z)$ *then* $\dfrac{x_t - t\sum zp(z)(1-\rho)}{\sqrt{t}}$ *converges in distribution to a normal random variable with finite variance.*

However this theorem states only that the variance is finite but does not exclude the possibility for it to be zero. At this point one must remember that for 1-dimensional nearest neighbour symmetric simple exclusion Arratia 1 proved that the variance is zero, and even more that $\frac{x_t}{t^{\frac{1}{4}}}$ converges in distribution to a non degenerate gaussian variable. On the other hand for all the other cases i.e. when $d \geq 2$ or when the jumps are not restricted to two sites, the variance can be proved to be strictly positive [5]. These results settle the problem for symmetric p's. ∎

For non-symmetric p's the process is no longer reversible and we are lacking at the moment a reformulation of (1) and (2) easy enough to be checked directly on the generator. Nonetheless for one-dimensional nearest-neighbour non-symmetric p's one can use a correspondence with the zero-range process to prove the

THEOREM [4] : *For* $0 \leq p \leq 1$ *but* $p \neq \frac{1}{2}$, $\frac{x_t - t(2p-1)(1-\rho)}{\sqrt{t}}$ *converges in distribution to a non degenerate gaussian random variable.*

Sketch of proof. Because of the one dimensional nearest-neighbour assumption on the random walk kernel, the order of the particles remains unaltered.

Defining $\zeta_t(k)$ = number of empty sites between the k-th and the (k+1)-th particles at time t we obtain a zero-range process and x_t is then equal to the total (algebraic) number of particles that jumped from the left of the origin to its right between 0 and t. This correspondence permits to prove that x_t has stationary increments that are negatively associated i.e. for all f and g increasing function on $\mathbb{R}$ we have

$$\int f(x_t)g(x_{t+s} - x_t)\mu_\rho(d\xi) \leq \int f(x_t)\mu_\rho(d\xi) . \int g(x_s)\mu_\rho(d\xi) \tag{5}$$

this yields a central limit theorem. ∎

One needs a separate argument to verify that the limiting variance is not zero, but in this case the limiting variance was exactly computed and is equal to $|2p-1|(1-\rho)$ [2].

3. OPEN PROBLEMS.

1. From the results presented above it is clear that the first and most challenging open problem is to prove the central limit theorem for the case of a non-symmetric p. There is no reason to doubt that the theorem holds but to prove it one will need to understand how to express in terms of the generator conditions (1) and (2) when L is no longer self-adjoint.

2. The second problem is to understand what is the relationship between the equilibrium and the non-equilibrium behaviour. To understand what we mean

let us introduce an ε , that has to be thought as small and consider an initial distribution of particles that is no longer an equilibrium distribution but a slowly varying profile of density. By this we mean that we take initially a function $\varphi : \mathbb{R}^d \to [0,1]$ and define for all $\varepsilon > 0$ μ^ε as a product measure on $\{0,1\}^{\mathbb{Z}^d}$ with

$$\mu^\varepsilon\{\eta(x) = 1\} = \varphi(\varepsilon x) \quad \text{for all} \quad x \in \mathbb{Z}^d .$$

We then start our test particle at the origin and denote by x_t its position at time t .

We now know two facts :

a. if $\varphi(u) \equiv \rho$ then the tagged particle starts with a medium in equilibrium and we know at least for p symmetric that $\varepsilon\, x_{t/\varepsilon^2} \overset{d}{\to} a(\rho)B_t$ where $a(\rho)$ is the square root of variance matrix and is density dependent.

b. If $\varphi(u)$ is not a constant function it is easy to prove when p is symmetric that for all $a \in \mathbb{R}^d$

$$\tau_{a/\varepsilon}\, \mu^\varepsilon P_{t/\varepsilon^2} \overset{d}{\to} \text{Bernoulli } (\varphi(a,t))$$

where the function $\varphi(a,t) : \mathbb{R}^d \times \mathbb{R}_+ \to [0,1]$ satisfie $\frac{\partial}{\partial t} \varphi = \frac{1}{2} \Delta\varphi$ with initial condition $\varphi(a,0) = \varphi(u)$.

Therefore it is conjectured when p is symmetric that because both phenomena happen on the same time-scale t/ε^2 the processes ξ_t^ε defined as $\xi^\varepsilon := \varepsilon x_{t/\varepsilon^2}$ under the probability $\mathbb{P}_{\mu_\varepsilon}$ converge in distribution to the solution at time t of the stochastic differential equation :

$$d\xi_s = a(\varphi(t,\xi_s))dB_s + (\varphi(s,\xi_s))^{-1} . \vec{\text{grad}}\, \varphi(s,\xi_s)ds$$

where dB_s is the standard brownian motion in $\mathbb{R}^d$. It should be noted that for the one-dimensional nearest-neighbour symmetric simple exclusion the result of Arratia says that $a(\rho) = 0$ for all $\rho \neq 0$ and indeed the method he uses can be extended to non-equilibrium situation (see for instance [8]) to prove that the tagged particle obeys the ordinary differential equation conjectured above.

Of course if a direct proof is not available it might be necessary to get some information on $a(\rho)$.

3. Perhaps of more technical nature is the following question. In $d = 1$ for non symmetric simple exclusion one could prove that the increments of the displacement of the tagged particle were negatively associated [4]. Does such a fact occur in general (bearing in mind that this would imply the central limit theorem) ?

BIBLIOGRAPHY

[1]. R. Arratia, "The motion of a tagged particle in the simple symmetric exclusion system on $\mathbb{Z}$ ", Ann. Prob. 11 (1983) 362-.

[2]. A. de Masi, P. Ferrari : "Self diffusion in one-dimensional lattice gases in the presence of an external field", J. Stat. Phys., 38 (1985) (to appear).

[3]. P. Ferrari, E. Presutti, M.E. Vares, Hydrodynamics of a zero-range model, preprint 1984.

[4]. C. Kipnis, "Central limit theorems for infinite series of queues and applications to simple exclusion", to appear in Ann. Prob. (1985).

[5]. C. Kipnis, S.R.S. Varadhan, "Central limit theorem for additive functionals of reversible Markov processes and application to simple exclusion", to appear in Comm. Math. Phys. (1985).

[6]. J. Lebowitz, H. Spohn, "Self diffusion", J. Stat. Phys., 28 (1982) 539-.

[7]. T. Liggett, Interacting particle systems -Chap. 8- (to be published by Springer Verlag in 1985).

[8]. E. Saada, "Position of a tagged particle in a simple exclusion process", preprint (1984).

Contemporary Mathematics
Volume 41, 1985

ON AN INVERSE PROBLEM FOR RANDOM SCHRÖDINGER OPERATORS

Shinichi Kotani

ABSTRACT. In one-dimensional Schrödinger operators, the existence of random shift invariant potentials for a given spectral data is discussed. The results can be viewed as a generalization of those obtained in periodic potentials.

1. INTRODUCTION. Schrödinger operators with random shift invariant ergodic potentials can be regarded as a natural extension of periodic Schrödinger operators. One of the merits in considering Schrödinger operators in this general framework is that one can see clearly the change of the spectral properties as the increase of the randomness of potentials. What is more important to remark is that this generalization does not make the problem too complicated at least in one-dimension. In fact, surprisingly enough, it is known that the Lyapounov exponents can tell everything about the absolutely continuous spectrum (see THEOREM 1). Moreover this general point of view simplifies the situation even in periodic case. For instance, the characterization problem of potentials with finite-zones turns to be very simple and transparent as compared with the original ones [8],[9] , which was pointed out by [1]. Therefore, although the study of random Schrödinger operators has been motivated by solid states physics, we are now on a stage that we should regard it as a mathematical object and investigate the problems more throughly. The purpose pf this paper is to formulate an inverse spectral problem in this general framework and to give a partial answer to it.

Let $\Omega = L^2_{real}(R, \frac{dx}{1+|x|^3})$ and $\mathcal{B}$ be the topological Borel field of Ω. Then it is proved in section 2 that for any $q \in \Omega$, $L(q) = -\frac{d^2}{dx^2} + q$ defined on $C_0^2(R)$ has a unique self-adjoint extension in $L^2(R,dx)$. Let $\{T_x; x \in R\}$ be shift operators defined by $T_x q(\cdot) = q(\cdot + x)$ and P be a probability measure on $(\Omega, \mathcal{B})$ satisfying

(1.1) $\{T_x\}$ -invariance; $P(T_xA) = P(A)$ for any $x \in R$ and A .

(1.2) $\{T_x\}$ -ergodicity; $P(T_xA \ominus A) = 0$ for any $x \in R$ implies $P(A) = 0$ or 1.

(1.3) The finiteness of the second moment; $E_P^{(1)}(\int_0^1 q(x)^2 dx) < \infty$.

(1) E_P denotes the expectation by P and the abbreviation E is often used.

It is needless to say that every ergodic stationary process with finite second moment induces a probability measure P on $(\Omega, \mathcal{B})$ satisfying (1.1) $\sim$ (1.3). In particular, for a given real L^2- function f on R^d/Z^d and a rationally independent vector $\alpha \in R^d$, a familiy of potentials $\{f(x\alpha + \omega);\ x \in R, \omega \in R^d/Z^d\}$ defines one typical example of the above P if we consider these potentials under the normalized Lebesgue measure on R^d/Z^d.

For $q \in \Omega$, and $C_+ = \{\mathrm{Im}\ \lambda > 0\}$, let $g_\lambda(x,y,q)$ be the Green function of $L(q) - \lambda$. Set

$$(1.4) \qquad w(\lambda) = -\frac{1}{2} E_P\ g_\lambda(0,0,q)^{-1} .$$

Denoting $\mathcal{R} = \{h\ ;\ C_+ \to C_+ \text{ holomorphic}\}$, we see in section 2

$$(1.5) \qquad w\ ,\ -iw \text{ and } w' \in \mathcal{R}.$$

If w satisfies (1.5), then it is known (see for instance [4]) that

$$n(\xi) = \lim_{\varepsilon \downarrow 0} \mathrm{Im}\ w(\xi + i\varepsilon) \quad \text{and} \quad \gamma(\xi) = -\lim_{\varepsilon \downarrow 0} \mathrm{Re}\ w(\xi + i\varepsilon)$$

exist finitely for any $\xi \in R$ and by definition of w γ is non-negative and n is non-decreasing. Moreover we see that n is locally log-Hölder continuous. n is called <u>the integrated density of states</u> for this random system and γ is called <u>the Lyapounov exponent</u> (or index) because γ is known to be identical with

$$\gamma(\xi) = \lim_{x \to \infty} \frac{1}{x} \log \| U_\xi(x,q) \| ,$$

where $U_\xi(x,q)$ is a unique solution of

$$\frac{dU}{dx} = \begin{pmatrix} 0 & , 1 \\ q(x)-\xi, & 0 \end{pmatrix} U \ , \quad U(0) = I .$$

As will be observed in section 2, all the theorems in [4] are valid for this wider space Ω too. Especially we have

THEOREM 1. Let $\Sigma(q)$ ($\Sigma_{a.c}(q)$) be the spectrum (the absolutely continuous spectrum respectively) for $L(q)$. Then for almost every $q \in \Omega$ with respect to P,

$$\Sigma(q) = \mathrm{supp}\ dn(\cdot) \quad \text{and} \quad \Sigma_{a.c}(q) = \overline{\{\xi \in R\ ;\ \gamma(\xi) = 0\}}^{\,ess.} ,$$

where the right-hand side of the second formula means the essential closure with respect to Lebesgue measure on R.

It is also possible to prove that if the system $(\Omega, \ , P, \{T_x\}, q(x))$ is non-deterministic, then $\Sigma_{a.c}(q) = \phi$ a.e.[4]. Moreover if $\mathrm{supp}\ P_1 \subset \mathrm{supp}\ P_2$

(supp. means the topological support and the topology on Ω is defined by L^2-local metric), then $\overline{\{\gamma_{P_1}(\xi) = 0\}}^{\text{ess.}} \supset \overline{\{\gamma_{P_2}(\xi) = 0\}}^{\text{ess.}}$ (see [5]).

Now for a moment we consider periodic potentials. Let $q(x)$ be a locally L^2-integrable periodic function with period m. Then the above w is

$$w(\lambda) = -\frac{1}{2m}\int_0^m g_\lambda(x,x,q)^{-1}dx \quad .$$

Suppose that $L(q)$ defines a non-negative definite self adjoint operator on $L^2(R,dx)$. Then w has a holomorphic extension to $C_- = \{\text{Im}\,\lambda > 0\}$ through $R_- = (0,\infty)$ because $w(\lambda) < 0$ on R_-. Set

$$\theta(z) = -iw(z^2) \quad .$$

Define a set of slits H in C_+ with constant interval π/m and height $h_j = \gamma(L_j^2)$ where $L_j = \inf\{\xi \geqq 0 \; ; \; n(\xi^2) = j\pi/m\}$ for $j \geqq 0$ and $L_{-j} = L_j$ by

$$H = \{ j\pi/m + tih_j \; ; \; j = 0, \pm 1, \pm 2, \ldots, 0 < t \leqq 1 \} \quad .$$

Then it is known that

THEOREM 2.([7]) θ maps conformally C_+ onto $C_+ - H$ and $C_+ - H$ becomes an image of θ generated by a periodic potential q with period m and L^2-loc. integrable if and only if

$$\sum_{j=0}^{\infty} (jh_j)^2 < \infty \text{ or equivalently } \int_0^\infty \sqrt{\xi}\gamma(\xi)d\xi < \infty \, .$$

Our next results are a generalization of THEOREM 2. Let

$$\Omega_+ = \{q \in \Omega; \; (L(q)\phi,\psi) \text{ defines a non-negative definite form on } C_0^2(R). \} \quad .$$

THEOREM 3. Suppose that a probability measure on $(\Omega,\mathcal{B})$ with the properties (1.1) $\sim$ (1.3) lies on Ω_+. Then the corresponding w-function satisfies

(i) w, $-iw$ and $w' \in \mathcal{R}$,

(ii) $w(\lambda) < 0$ on R_- ,

(iii) $\lim_{\lambda \to -\infty} \dfrac{w(\lambda)}{-\sqrt{-\lambda}} = 1$,

(iv) $\gamma(\xi)$ is bounded on $R_+ = (0,\infty)$,

(v) $\int_0^\infty \sqrt{\xi}\,\gamma(\xi)d\xi < \infty$.

For the existence of P we need much more properties of w than (i) $\sim$ (v). Namely we have

THEOREM 4. Suppose that a holomorphic function w on C_+ satisfies the (i) $\sim$ (iii) of Theorem 3 and there exists a $\xi_0 > 0$ such that $\gamma(\xi) = 0$ on (ξ_0, ∞). (In this case the conditions (iv) and (v) are satisfied automatically .) Then there exists a probability measure P on $(\Omega_+, \mathcal{B})$ satisfying (1.1), (1.3) and (1.4).

To choose an ergodic one among the above probability measures does not seem easy. What we have is

THEOREM 5. In addition to the conditions in Theorem 4 , assume

$$\int_0^\infty \gamma(\xi) n(d\xi) = 0 \quad .$$

Then there exists a pobability measure P on Ω_+ satisfying (1.1) $\sim$ (1.3) and (1.4).

There is no reason to doubt that Theorem 5 remains valid under the conditions (i) $\sim$ (v) in Theorem 3.

2. NECESSARY CONDITIONS. In this section we explain how Theorem 1 and 3 are obtained. For $q \in \Omega$, define $L(q) = -\frac{d^2}{dx^2} + q$ on $C_0^2(R)$. Since $q \in L^2_{loc}(R)$,$L(q)$ defines a symmetric operator on $L^2(R,dx)$.

LEMMA 2.1. $L(q)$ has a unique self-adjoint extension in $L^2(R,dx)$.

PROOF. We prove that the two boundaries $\pm\infty$ are of limit point type due to Weyl's classification of boundaries. To do this , first we show that for any real-valued solution u of $L(q)u = 0$ on R_+ satisfying $u \in L^2(R_+,dx)$,

$$\int_1^\infty \frac{u'(x)^2}{x^2} dx < \infty \quad . \tag{2.1}$$

Setting $H(x) = \int_1^x \frac{u'(y)^2}{y^2} dy$, we have

$$\begin{aligned} -\frac{u(x)u'(x)}{x^2} &= -u(1)u'(1) - \int_1^x \frac{u'(y)^2}{y^2}dy - \int_1^x \frac{u(y)u''(y)}{y^2} dy + 2\int_1^x \frac{u(y)u'(y)}{y^3}dy \\ &= c_1 - H(x) - \int_1^x \frac{u(y)^2 q(y)}{y^2}dy + 2\int_1^x \frac{u(y)u'(y)}{y^3}dy \\ &\leqq c_1 - H(x) - \int_1^x \frac{u(y)^2 q(y)}{y^2}dy + 2\left(\int_1^x \frac{u(y)^2}{y^4}dy\right)^{\frac{1}{2}}\left(\int_1^x \frac{u'(y)^2}{y^2}dy\right)^{\frac{1}{2}} \\ &\leqq c_1 - H(x) - \int_1^x \frac{u(y)^2 q(y)}{y^2}dy + c_2(H(x))^{\frac{1}{2}} \quad . \end{aligned}$$

Hereafter c_j denotes positive constant independent of x.

However, using $Q(x) = \int_1^x \frac{q(y)^2}{y^3} dy$, we see

$$-\int_1^x \frac{u(y)^2 q(y)}{y^2} dy \leqq \left(\int_1^\infty u(y)^2 dy\right)^{\frac{1}{2}} \left(\int_1^x \frac{u(y)^2}{y} Q'(y) dy\right)^{\frac{1}{2}}$$

and moreover

$$\int_1^x \frac{u(y)^2}{y} Q'(y) dy = \frac{u(x)^2}{x} Q(x) - 2\int_1^x \frac{u(y)u'(y)}{y} Q(y) dy + \int_1^x \frac{u(y)^2}{y^2} Q(y) dy$$

$$\leqq c_3 \frac{u(x)^2}{x} + c_4 \left(\int_1^x \frac{u'(y)^2}{y^2} dy\right)^{\frac{1}{2}} + c_5$$

$$= c_3 \frac{u(x)^2}{x} + c_4 (H(x))^{\frac{1}{2}} + c_5 .$$

Finally $\frac{u(x)^2}{x}$ can be estimated as

$$\frac{u(x)^2}{x} = u(1)^2 + 2\int_1^x \frac{u(y)u'(y)}{y} dy - \int_1^x \frac{u(y)^2}{y^2} dy$$

$$\leqq c_6 + c_7 (H(x))^{\frac{1}{2}} .$$

Combining the above estimates, we have

$$-\frac{u(x)u'(x)}{x^2} \leqq c_1 - H(x) + c_2 (H(x))^{\frac{1}{2}} + (c_8 + c_9 (H(x))^{\frac{1}{2}})^{\frac{1}{2}} .$$

Suppose $H(+\infty) = +\infty$. Then from the above estimate we see that $u(x)u'(x) > 0$ for every sufficiently large x. However this contradicts $u \in L^2(R_+, dx)$. Hence we have (2.1).

Now assume the boundary $+\infty$ is of limit circle type. Then the equation $L(q)u = 0$ on R_+ has two linearly independent (real-valued) solutions u_1, u_2 belonging to $L^2(R_+, dx)$. Since the Wronskian of u_1 and u_2 is constant,

$$u_1(x) \frac{u_2'(x)}{x} - u_2(x) \frac{u_1'(x)}{x} = \frac{\text{const.}}{x}$$

is valid. However (2.1) and $u_1, u_2 \in L^2(R_+, dx)$ imply the const. = 0, which contradicts the linear independence of u_1, u_2. This completes the proof.

For $\lambda \in C$ let $U_\lambda(x,q)$ be a unique solution of

$$\frac{d}{dx} U = \begin{pmatrix} 0 & , 1 \\ q(x)-\lambda, & 0 \end{pmatrix} U \quad \text{with} \quad U(0) = I$$

and set $U_\lambda(x,q) = \begin{pmatrix} \phi_\lambda(x,q), \psi_\lambda(x,q) \\ \phi_\lambda'(x,q), \psi_\lambda'(x,q) \end{pmatrix}$. Then for any $b > 0$ and $\lambda \in C_+$

$$\Delta_b(\lambda,q) = \{z \in C_+ ; \int_0^b |\phi_\lambda(x,q) + z\psi_\lambda(x,q)|^2 dx \leqq \frac{\text{Im } z}{\text{Im } \lambda}\}$$

defines a closed disc in C_+ and its boundary is described by

$$\Delta_b(\lambda,q) = \{-\frac{\phi_\lambda(b,q)\cos\theta - \phi'_\lambda(b,q)\sin\theta}{\psi_\lambda(b,q)\cos\theta - \psi'_\lambda(b,q)\sin\theta}\ ;\ 0 \leqq \theta < \pi\} \quad .$$

Obviously $\Delta_b(\lambda,q)$ decreases as b increases. Weyl's limit point and circle theorem says that if the boundary $+\infty$ is of limit point type , then $\Delta_b(\lambda,q)$ decreases to one point. Therefore from Lemma 2.1 we can define $h_+(\lambda,q)$ as a unique common element of all $\Delta_b(\lambda,q)$s. $h_-(\lambda,q)$ can be defined similarly by considering the boundary $-\infty$. It is easy to see that $h_\pm$ are holomorphic functions taking values in C_+ , hence $h_\pm \in \mathcal{R}$. We can show the continuity of $h_\pm$ with respect to q.

LEMMA 2.2. Suppose a sequence $\{q_n\}$ of Ω converges to q of Ω in each $L^2((-k,k),dx)$ weakly. Then $h_\pm(\lambda,q_n)$ converge to $h_\pm(\lambda,q)$ compact uniformly on C_+ .

PROOF. Since $\Delta_b(\lambda,q)$ decreases to $h_+(\lambda,q)$, for the point-wise convergence of $h_+(\lambda,q_n)$ it is sufficient to show that for any fixed $\lambda \in C$

$$U_\lambda(x,q_n) \quad \to \quad U_\lambda(x,q) \quad \text{compact uniformly in } x \in R_+.$$

Then the compact uniform convergence of $h_+(\lambda,q_n)$ follows immediately because a family $\{h_+(\lambda,q_n)\}$ is always normal. To see the above convergence, set $A_\lambda(x,q) = \begin{pmatrix} 0 & , & 1 \\ q(x)-\lambda, & & 0 \end{pmatrix}$. Then $U_\lambda(x,q)$ satisfies

$$U_\lambda(x,q) = I + \int_0^x A_\lambda(y,q)U_\lambda(y,q)dy \quad . \tag{2.2}$$

Then it is not difficult to see by induction that

$$\|U_\lambda(x,q)\| \leqq \exp(\int_0^x \|A_\lambda(y,q)\| dy) \quad . \tag{2.3}$$

Since the weak convergence of $\{q_n\}$ in $L^2((0,m),dx)$ implies the $L^2((0,m),dx)$ -boundedness of $\{q_n\}$, from (2.3) it follows that

$$\|U_\lambda(x,q_n)\| \leqq c_1 \quad \text{for all } x \in [0,m] \quad , \tag{2.4}$$

with a constant c_1. Therefore

$$\int_0^m \left\|\frac{d}{dx}U_\lambda(x,q_n)\right\|^2 dx = \int_0^m \|A_\lambda(x,q_n)U_\lambda(x,q_n)\|^2 dx$$
$$\leqq c_1^2 \int_0^m \|A_\lambda(x,q_n)\|^2 dx$$
$$\leqq c_2 \quad ,$$

with another constant c_2 . Ascoli-Arzelà's theorem says that there exists a subsequence $\{n_i\}$ such that $U_\lambda(x,q_{n_i}) \to U(x)$ uniformly on $[0,m]$. Since it is easy to see that $U(x)$ also satisfies (2.2) , from the uniqueness of solutions of (2.2) we can conclude $U(x) = U_\lambda(x,q)$ on $[0,m]$. This implies $U_\lambda(x,q_n) \to U_\lambda(x,q)$ compact uniformly on R.

Set $f_\lambda^{\pm}(x,q) = \phi_\lambda(x,q) \pm h_\pm(\lambda,q)\psi_\lambda(x,q)$. Then by definition $f_\lambda^{\pm}(\cdot\,,q) \in L^2(R_\pm,dx)$ and satisfy

(2.5) $$\operatorname{Im} h_\pm(\lambda,q) \geqq (\operatorname{Im}\lambda)\int_{R_\pm} |f_\lambda^{\pm}(y,q)|^2 dy \quad .$$

(Actually we have an identity here.) Moreover, since $h_\pm(\lambda,q) = \mp \lim_{x\to\pm\infty} \frac{\phi_\lambda(x,q)}{\psi_\lambda(x,q)}$, it is easily seen that

(2.6) $$h_\pm(\lambda,T_xq) = \pm \frac{(f_\lambda^{\pm}(x,q))'}{f_\lambda^{\pm}(x,q)} \quad ,$$

(2.7) $$f_\lambda^{\pm}(x,q) = \exp\left(\pm\int_0^x h_\pm(\lambda,T_yq)dy\right) \quad .$$

Especially (2.6) implies

(2.8) $$\pm \frac{d}{dx} h_\pm(\lambda,T_xq) = q(x) - \lambda - h_\pm(\lambda,q)^2 \; .$$

Denoting by $g_\lambda(x,y,q)$ the Green function of $L(q) - \lambda$, which is equal to $-(h_+(\lambda,q)+h_-(\lambda,q))^{-1} f_\lambda^{+}(x,q)f_\lambda^{-}(y,q)$ for $x \geqq y$, we see from (2.8)

(2.9) $$h_+(\lambda,T_xq) - h_-(\lambda,T_xq) = -\frac{d}{dx}\log g_\lambda(0,0,T_xq) \; ,$$

(2.10) $$\frac{-\operatorname{Im}\lambda}{\operatorname{Im} h_\pm(\lambda,T_xq)} - 2\operatorname{Re} h_\pm(\lambda,T_xq) = \pm\frac{d}{dx}\log(\operatorname{Im} h_\pm(\lambda,T_xq)) \; ,$$

(2.11) $$g_\lambda(0,0,T_xq) + \frac{1}{2}\frac{d}{d\lambda}\, g_\lambda(0,0,T_xq)^{-1} = -\frac{1}{2}\frac{d}{dx}\left(g_\lambda(0,0,T_xq)\frac{d}{d\lambda}(h_+(\lambda,T_xq) - h_-(\lambda,T_xq))\right) \quad .$$

The details can be found in [2] , [4].

Since we have to take the expectation of $h_{\pm}(\lambda,q)$ and $(\mathrm{Im}\, h_{\pm}(\lambda,q))^{-1}$ to define the w-function , we check the integrability of these functions.

LEMMA 2.3.([3]) For any $q \in \Omega$, $\lambda \in C_+$,

$$(\mathrm{Im}\,\lambda)M(\lambda,q)^{-1} \leqq \mathrm{Im}\, h_+(\lambda,q) \leqq |h_+(\lambda,q)| \leqq M(\lambda,q) \quad ,$$

where $M(\lambda,q) = c_1 + c_2(1+|\lambda|)(1+\frac{1}{\mathrm{Im}\,\lambda})(1+\int_0^\infty e^{-2x}q(x)^2 dx)$ with positive constants c_1 , c_2 independent of q and λ.

Since (1.3) is equivalent to $E_p(\int_0^\infty e^{-2x}q(x)^2 dx) < \infty$, from the above lemma we can define

$$w(\lambda) = w_P(\lambda) = E_P(-\tfrac{1}{2}\, g_\lambda(0,0,q)^{-1}) = \tfrac{1}{2}\, E_P(h_+(\lambda,q) + h_-(\lambda,q)) \quad .$$

From the identities (2.9) $\sim$ (2.11) it follows immediately that

LEMMA 2.4.([2],[4])

(i) $$w(\lambda) = E_P h_{\pm}(\lambda,q) \quad ,$$

(ii) $$\frac{-2\mathrm{Re}\, w(\lambda)}{\mathrm{Im}\,\lambda} = E_P \frac{1}{\mathrm{Im}\, h_{\pm}(\lambda,q)} \quad ,$$

(iii) $$w'(\lambda) = E_P\, g_\lambda(0,0,q) \quad .$$

By Lemma 2.4 we see that all the theorems in [4] are valid in this wider space Ω too. Hence we have Theorem 1 . The property (i) of Theorem 3 is a direct consequence of Lemma 2.4.

Hereafter we restrict ourselves to potentials of Ω_+ .

LEMMA 2.5. For $q \in \Omega_+$ and $\lambda < 0$, we have

$$h_+(\lambda,q) = -\sqrt{-\lambda} - \int_0^\infty q(x)g_\lambda(x)\, dx + (G_\lambda^+(q)qg_\lambda, qg_\lambda)_{L^2(R_+,dx)} \quad ,$$

where $g_\lambda(x) = e^{-\sqrt{-\lambda}x}$ and $G_\lambda^+(q)$ is the Green operator of $L(q)-\lambda$ on $L^2(R_+,dx)$ with Dirichlet boundary condition at $x = 0$.

PROOF. The proof goes similarly to that of Lemma 2.4 . We have only to note identities :

$$h_+(\lambda,q) = \lim_{\substack{0<x<y \\ y\to 0}} \frac{\partial^2 g_\lambda^+(x,y,q)}{\partial x \partial y} \quad ,$$

$$G_\lambda^+(q) = G_\lambda^+(0) - G_\lambda^+(0)qG_\lambda^+(0) + G_\lambda^+(0)qG_\lambda^+(q)qG_\lambda^+(0) \quad ,$$

where $g_\lambda^+(x,y,q)$ is the kernel of $G_\lambda^+(q)$.

Set

$$m = E_P\int_0^1 q(x)dx \quad , \quad \sigma^2 = E_P\left(\int_0^1 q(x)^2 dx\right) \quad .$$

Then from Lemma 2.5 it follows easily that

LEMMA 2.6. For any $\lambda < 0$,

$$\frac{-m}{2\sqrt{-\lambda}} \leqq w(\lambda) + \sqrt{-\lambda} \leqq \frac{-m}{2\sqrt{-\lambda}} + \frac{\sigma^2}{-2\lambda\sqrt{-\lambda}} \quad .$$

As a corollary of this lemma we have the property (iii) of Theorem 3 .(ii) is trivial from the definition of w and $q \in \Omega_+$. For a given holomorphic function w on C_+ satisfying (i) $\sim$ (iii) of Theorem 3 , set

$$\theta(z) = -iw(z^2) \quad .$$

$w' \in \mathcal{R}$ implies that w is conformal on C_+ , and hence θ maps C_+ conformaly into C_+ . Since $\mathrm{Im}\theta(\xi+i0) = \gamma(\xi^2)$ on R and $-\overline{\theta(-\bar{z})} = \theta(z)$, we see

$$\theta(z) = \beta z + \frac{1}{\pi}\int_R\left\{\frac{1}{\xi - z} - \frac{\xi}{1+\xi^2}\right\}\gamma(\xi^2)d\xi \quad ,$$

$$= \beta z + \frac{z}{\pi}\int_{R_+}\frac{\gamma(\xi)}{\sqrt{\xi}(\xi - z^2)}d\xi \quad ,$$

with a constant $\beta \geqq 0$. In view of (iii) , $\beta = 1$, therefore

$$w(\lambda) = -\sqrt{-\lambda} - \frac{\sqrt{-\lambda}}{\pi}\int_{R_+}\frac{\gamma(\xi)}{\sqrt{\xi}(\xi-\lambda)}d\xi \quad . \tag{2.12}$$

Combining Lemma 2.6 and (2.12) , we have

LEMMA 2.7.

(i) $$\frac{2}{\pi}\int_{R_+}\frac{\gamma(\xi)}{\sqrt{\xi}}d\xi = m$$

(ii) $$\frac{2}{\pi}\int_{R_+}\sqrt{\xi}\gamma(\xi)d\xi \leqq \sigma^2 \quad .$$

The above (ii) implies (v) of Theorem 3. In (ii) we may have an identity in general. In fact if the potentials are smooth , by expanding $g_\lambda(0,0,q)$

in a neighbourhood $\lambda = -\infty$, we have the equality in (ii).

The property (iv) of Theorem 3 can be checked by

LEMMA 2.8. For any probability measure P on Ω satisfying (1.1) (1.3) (P is not necessarily lying on Ω_+),

$$\gamma(\xi) \leqq 1 + E_P\left(\int_0^1 |q(x)|dx\right)$$

is valid for any $\xi \geqq 0$.

PROOF. Fix $\xi \geqq 0$, $q \in \Omega$ and let

$$A(x) = \begin{pmatrix} 0 & 1 \\ q(x)-\xi, & 0 \end{pmatrix} \quad \text{and} \quad U(x) = U_\xi(x,q) \quad .$$

Then U satisfies

$$U(x) = I + \int_0^x A(y)U(y)dy \quad .$$

Therefore clearly we have $\|U(x)\| \leqq \exp\left(\int_0^x \|A(y)\| dy\right)$, which implies

$$\gamma(\xi) = \lim_{x \to +\infty} \frac{1}{x} \log\|U(x)\| \leqq E_P\left(\int_0^1 \{1 \vee |\xi - q(x)|\}dx\right)$$

$$\leqq 1 + E_P\left(\int_0^1 |q(x)|dx\right)$$

for any $0 \leqq \xi \leqq 1$. To get an estimate for $\xi \geqq 1$, we set

$$A_0 = \begin{pmatrix} 0 & ,1 \\ -\xi & ,0 \end{pmatrix} \quad , \quad A_1 = \begin{pmatrix} 0 & , & 0 \\ 1 & , & 0 \end{pmatrix} \quad \text{and} \quad \mathfrak{U}(x) = e^{-xA_0}U(x) \; .$$

Then $\mathfrak{U}$ satisfies

$$\mathfrak{U}(x) = I + \int_0^x \mathfrak{A}(y)\mathfrak{U}(y)dy$$

with $\mathfrak{A}(x) = q(x)e^{-xA_0}A_1e^{xA_0}$. Since it is not difficult to see $\|\mathfrak{A}(x)\| \leqq |q(x)|$ for any $\xi \geqq 1$, we have $\|\mathfrak{U}(x)\| \leqq \exp\left(\int_0^x |q(y)|dy\right)$. Since e^{-xA_0} is bounded on R, we can conclude the lemma.

3. SUFFICIENT CONDITIONS. In this section we prove Theorem 4 and 5 by approximating w by w-functions corresponding to periodic potentials.

LEMMA 3.1. Suppose a holomorphic function w on C_+ satisfies all the conditions in Theorem 4. Then there exists a sequence of smooth periodic potentials $\{q_k(x)\}$ whose w-functions w_k also satisfy all the conditions and

$$w_k(\lambda) \to w(\lambda) \quad \text{compact uniformly on } C_+ ,$$

$$\int_{R_+} \gamma_k(\xi)\sqrt{\xi}d\xi \leqq c \quad \text{independently of } k .$$

PROOF. We can assume that w has smooth γ , n and $n'(\xi) > 0$ on R_+ , because we have only to take a convolution $w_\varepsilon(\lambda) = \int_R w(\lambda-x)\rho^\varepsilon(x)dx$ with $\rho^\varepsilon(x) = \frac{1}{\varepsilon}\rho(\frac{x}{\varepsilon})$ satisfying $C^\infty(R) \ni \rho(x) \geqq 0$, $\rho(x) = 0$ for $x \in R-(0,1)$ and $\int_0^1 \rho(x)dx = 1$ and to make a convex combination $\alpha w_\varepsilon(\lambda) + (1-\alpha)(-\sqrt{-\lambda})$. Now fix $m > 0$. Let ξ_j $(j \geqq 0)$ be a unique ξ such that $n(\xi^2) = \pi j/m$ and $h_j = \gamma(\xi_j{}^2) = h_{-j}$. Set

$$H_m = \{\frac{\pi j}{m} + ith_j ; j = 0 , \pm 1 , \pm 2 , \ldots, 0 < t \leqq 1\}$$

and construct a unique conformal mapping θ_m from C_+ onto C_+-H_m satisfying $\theta_m(iy)/iy \to 1$ as $y \to +\infty$. Since

$$\theta_m(z) = z + \frac{z}{\pi}\int_{R_+} \frac{\gamma_m(\xi)}{\sqrt{\xi}(\xi-z^2)} d\xi \qquad , \tag{3.1}$$

it is easy to see that θ_m has a unique holomorphic extension to $C-[-L_m^2,L_m^2]$, where $L_m = \sup \operatorname{supp}\gamma_m(\xi) < +\infty$. We want to show the convergence of θ_m to $\theta(z) = -iw(z^2)$ preserving the boundedness of the moments. To see this we consider the inverse functions $z_m(\theta)$ of $\theta_m(z)$ instead of θ_m themselves. θ_m maps $C-\{H_m \cup \overline{H}_m\} = D_m$ conformaly into C. It is not difficult to see that z_m forms a normal family on $C-\{ \pm n(\xi^2) + i\gamma ; -\gamma(\xi^2) \leqq \gamma \leqq \gamma(\xi^2)\} = D$, because on each four half-plane ($\operatorname{Im}\theta \gtrless 0$, $\operatorname{Re}\theta \gtrless 0$) $\operatorname{Im} z_m(\theta)$ or $\operatorname{Re} z_m(\theta)$ have a definite sign. They can not converge to ∞ because $\gamma_m(\xi) \leqq \sup_{\xi>0} \gamma(\xi)$ for any $\xi \geqq 0$ and m . Therefore there exists a subsequence $\{m_k\}$ such that $z_{m_k}(\theta) \to z(\theta)$ compact uniformly on D. Moreover since z_m has only a pole at $\theta = \infty$ with the coefficient 1 , z also should have the same property. This implies, in particular , that z is not a constant and hence z should be conformal. On the other hand , since the property $w' \in \mathcal{R}$ assures the injectivity of w on C_+ , θ also defines a conformal mapping on C_+ . Therefore , in view of $\theta(z) \sim z$ as $|z| \to +\infty$, $z(\theta)$ should coincide with the inverse of $\theta(z)$. Now all the coefficients of the Tayler expansion at ∞ of $z_{m_k}(\theta)$ converge to those of $z(\theta)$, which implies that all the coefficients of the Tayler expansion at ∞ of $\theta_{m_k}(z)$ converge to those of $\theta(z)$. This concludes the lemma , because $\int_{R_+} \gamma_{m_k}(\xi)\sqrt{\xi}d\xi$ appears as the third coefficient of the Tayler expansion of $\theta_{m_k}(z)$ at $z = \infty$.

LEMMA 3.2. Let H be a separable Hilbert space and $\mathcal{B}$ be the topological Borel field. Suppose that we are given a sequence of probability measures P_n on $(H, \mathcal{B})$ satisfying

$$\int_H \|x\|^2 P_n(dx) \leqq c \quad \text{independently of } n .$$

Then there exists a subsequence $\{n_k\}$ and a probability measure P on $(H, \mathcal{B})$ such that for any function $f(x)$ on H satisfying

(i) $\quad f(x_n) \to f(x)$ if $x_n \to x$ weakly ,

(ii) $\quad |f(x)| \leqq c_1 + c_2\|x\|$ with constants c_1 , c_2 ,

$$\int_H f(x)P_{n_k}(dx) \to \int_H f(x)P(dx)$$

is valid as $n_k \to \infty$.

PROOF. Since for any fixed $\delta > 0$, $\{x \in H ; \|x\| \leqq \delta\} \subset H$ is weakly compact , the lemma follows by an ordinary argument.

PROOF OF THEOREM 4. Let $H = L^2_{real}(R, \frac{dx}{1+|x|^3})$ and P_k be the probability measure on H induced by a periodic potential corresponding to w_k of Lemma 3.1 . Then P_k satisfies the assumption of Lemma 3.2 , because if the potential is smooth then we have the equality in (ii) of Lemma 2.7. Take $f(q)=-h_+(\lambda,q)$ for a fixed $\lambda<0$. Then Lemma 2.5 says

$$0 \leqq f(q) \leqq \sqrt{-\lambda} + \int_{R_+} q(x)e^{-2\sqrt{-\lambda}x}dx ,$$

which guarantees the conditions (i) and (ii) of Lemma 3.2. In this way we can apply Lemma 3.2 and we obtain Theorem 4.

To show the last theorem , we need

LEMMA 3.3. Let w_1 , w_2 be two holomorphic functions on C_+ satisfying the conditions (i) , (ii) and (iii) of Theorem 4. Set $n_j(\xi) = \operatorname{Im} w_j(\xi+i0)$, $\gamma_j(\xi) = -\operatorname{Re} w_j(\xi+i0)$ for $j = 1,2$. Asssume that γ_j vanish for sufficiently large ξ. Then

$$\int_{C_+} |w_1'(\lambda) - w_2'(\lambda)|^2 dxdy = -\int_{R_+} (\gamma_1(\xi)-\gamma_2(\xi))(n_1(d\xi)-n_2(d\xi)) .$$

PROOF. Set $w(\lambda) = w_1(\lambda) - w_2(\lambda)$. Then

$$w(\lambda) = -\frac{\sqrt{-\lambda}}{\pi}\int_{R_+} \frac{\gamma(\xi)}{\sqrt{\xi}(\xi-\lambda)} d\xi$$

with $\gamma(\xi) = \gamma_1(\xi) - \gamma_2(\xi)$. Since γ is bounded and has a compact support in

$[0,\infty)$ it is easily seen that there exist c, $\delta > 0$ such that

$$|w(x+iy)w'(x+iy)| \leqq \frac{c}{x^2+y^2} \tag{3.2}$$

holds for any $y > 0$ and $|x| \geqq \delta$. Therefore for any $\alpha > 0$ and $a < b$

$$\int_\alpha^\infty dy \int_a^b dx |w'(x+iy)|^2 = \int_a^b u(x,\alpha)v_x(x,\alpha)dx + \int_\alpha^\infty \{u_x(b,y)u(b,y)-u_x(a,y)u(a,y)\}dy.$$

Letting $\alpha \downarrow 0$, we see by (3.2) that the first term converges to $\int_a^b \gamma(\xi)n(d\xi)$ $(n = n_1 - n_2)$, and the second term tends to zero as $-a$, $b \to +\infty$ uniformly with respect to $\alpha > 0$. Hence we have the lemma.

PROOF OF THEOREM 5. From Theorem 4 we already know that there exists at least one probability measure P on Ω_+ satisfying (1.1) $\sim$ (1.3) and (1.4) for a given w. Now let $(\Omega, \overline{\mathcal{B}}, \overline{P})$ be the completion of the probability space $(\Omega, \mathcal{B}, P)$. Since Ω is a complete separable metric space, $(\Omega, \overline{\mathcal{B}}, \overline{P})$ turns to be an abstract Lebesgue space. Therefore the flow $\{T_x ; x \in R\}$ on $(\Omega, \overline{\mathcal{B}}, \overline{P})$ has an ergodic decomposition in the following sense: there exists a probability space $(S, \mathcal{J}, \mu)$ and for any $\alpha \in S$ there corresponds a probability measure P_α on $(\Omega, \overline{\mathcal{B}}, \overline{P})$ which is $\{T_x\}$- invariant and ergodic and

$$\overline{P} = \int_S P_\alpha \mu(d\alpha) .$$

P_α is a probability measure on Ω_+ satisfying (1.1) $\sim$ (1.3) for almost all $\alpha \in S$ with respect to μ. Hence we see by Theorem 3 that the corresponding w-function w_α satisfies (i) $\sim$ (iii) of Theorem 3. Obviously

$$w = \int_S w_\alpha \mu(d\alpha) . \tag{3.3}$$

Since we are assuming $\gamma(\xi) = 0$ on (ξ_0, ∞) for a $\xi_0 > 0$, γ_α also has the same property for almost all $\alpha \in S$. Fix such an α and apply Lemma 3.3 to $w_1 = w$ and $w_2 = w_\alpha$. Then we have

$$\begin{aligned}\int_{C_+} |w'(\lambda) - w'_\alpha(\lambda)|^2 dxdy &= - \int_{R_+} (\gamma(\xi) - \gamma_\alpha(\xi))(n(d\xi) - n_\alpha(d\xi)) \\ &\leqq \int_{R_+} \gamma(\xi) n_\alpha(d\xi) + \int_{R_+} \gamma_\alpha(\xi) n(d\xi)\end{aligned}$$

Now integrate the both sides by μ. From (3.3) we obtain

$$\int_S \mu(d\alpha) \int_{C_+} |w'(\lambda) - w'_\alpha(\lambda)|^2 dxdy \leqq 2 \int_{R_+} \gamma(\xi) n(d\xi) = 0 .$$

To get the last identity we used the assumption of Theorem 5. Hence this implies $w = w_\alpha$ for a.e. $\alpha \in S$. Then we have only to choose one of such αs to conclude the theorem.

REMARK. In Lemma 2.1 , one can approximate any w-function satisfying (i) $\sim$ (iv) of Theorem 3 by w-functions corresponding to smooth periodic potentials with uniform bounded Lyapounov exponents on R_+ . However it is not clear if we can choose w_k keeping $\int_{R_+} \gamma_k(\xi)\sqrt{\xi}d\xi$ uniformly bounded.

REMARK. If we assume that the spectrum consists only of absolutely continuous part , then $\int_{R_+} \gamma(\xi)n(d\xi) = 0$ is valid. In this connection, it should be remarked that [6] gave a sufficient condition for the existence of almost periodic potentials in terms of a sequence of spectral gaps.

REMARK. We can conjecture that the smoothness of potentials may be described by the finiteness of the moments of $\gamma(\xi)/\sqrt{\xi}$ on R_+ . In fact , in periodic case , [7] showed that q belongs to $H^n(R/Z)$ if and only if $\gamma(\xi)/\sqrt{\xi}$ has the n+1 moment on R_+ .

BIBLIOGRAPHY

[1] R. Johnson, private communication.

[2] R. Johnson and J. Moser, The rotation number for almost periodic potentials, Comm. Math. Phys., 84(1982) , 403-438.

[3] W. Kirsch , S. Kotani and B. Simon, Absence of absolutely continuous spectrum for some one dimensional random but deterministic Schrödinger operators , to be submitted to Ann. Inst. H. Poincaré.

[4] S. Kotani, Ljapunov indices determine absolutely continuous spectra of stationary random one-dimensional Schrödinger operators, Proc. Taniguchi Symp . SA Katata 1982 , 225-247.

[5] ________, Support theorems for random Schrödinger operators, to appear in Comm. Math. Phys..

[6] B.M. Levitan, An inverse problem for the Sturm-Liouville operators in the case of finite-zone and infinite-zone potentials, Trans. Moscow Math. Soc., 1(1984) , 1-34.

[7] V.A. Marčenko and I.V. Ostrovskii, A characterization of the spectrum of Hill's operator, Math. USSR Sbornik , 26(1975), 493-554.

[8] H.P. McKean and P. van Moerbeke, The spectrum of Hill's equation, Invent. Math., 30(1975), 217-274.

[9] S.P. Novikov, The periodic problem for the Korteweg-de Vries equation , Funct. Anal. Appl., 8(1974), 236-246.

DEPARTMENT OF MATHEMATICS
KYOTO UNIVERSITY
KYOTO , JAPAN .

Contemporary Mathematics
Volume 41, 1985

INTERSECTIONS OF SIMPLE RANDOM WALKS

Gregory F. Lawler[1]

ABSTRACT. Let $f(n)$ be the probability that the paths of two simple random walks of length n starting at the origin in $\mathbb{Z}^d$ do not intersect. Methods to estimate $f(n)$ are considered, and in particular it is shown that $f(n)$ decays like $(\log n)^{-1/2}$ for $d = 4$.

In order to motivate the problem I will discuss, let us consider a more difficult problem. Let C_n denote the number of self-avoiding walks on $\mathbb{Z}^d$, i.e. the number of nearest neighbor random walks starting at the origin in $\mathbb{Z}^d$ with the constraint that no lattice point may be visited more than once. It is well known (see [4]) that

$$\lim_{n\to\infty} C_n^{1/n} = \beta_d ,$$

but the rate of convergence is not well understood. If we define $g(n)$ by

$$C_n = \beta_d^n \, g(n), \tag{1}$$

then it is conjectured that

$$g(n) \approx \begin{cases} \text{constant} & d \geq 5 \\ (\log n)^{\alpha} & d = 4 \\ n^{\alpha_d} & d = 2,3 \end{cases} \tag{2}$$

However, not even the $d \geq 5$ case has been proved rigorously.

Although we cannot prove (2) for self-avoiding walks, we would at least like to prove the corresponding result for simple random walks, i.e. walks without the self-avoiding constraint. At first, no non-trivial analogue appears since the number of simple random walks of length n is exactly $(2d)^n$. We can look at $g(n)$ differently, however. By (1),

$$C_{2n} = C_n \cdot C_n \cdot \tilde{g}(n) , \tag{3}$$

1980 Mathematics Subject Classification 60J15.
[1]Research supported by the National Science Foundation.

where $\tilde{g}(n) = \frac{g(2n)}{[g(n)]^2}$. If one could determine the behavior of decay of $\tilde{g}$, one could analyze g. By (3) we can interpret $\tilde{g}(n)$ as the probability that two self-avoiding walks of length n have no points of intersection. We will consider the corresponding problem for simple walks, i.e. we consider

$f(n)$ = the probability that the paths of two simple random walks of length n starting at the origin in $\mathbb{Z}^d$ do not intersect.

Let S_1, S_2 denote independent simple random walks in $\mathbb{Z}^d$ starting at the origin and let Π_i denote the paths

$$\Pi_i[a,b] = \{S_i(j): a \le j \le b\}$$

$$\Pi_i(a,b) = \{S_i(j): a < j < b\}$$

and similarly for $\Pi_i[a,b)$ and $\Pi_i(a,b]$. We then set

$$f(n) = P\{\Pi_1[0,n] \cap \Pi_2(0,n] \neq \emptyset\} .$$

In order to analyze $f(n)$ we start by making a straightforward computation of the <u>expected</u> number of intersections, i.e.,

$$e_n = E[\#\{(i,j): 0 \le i \le n, 0 < j \le n, S_1(i) = S_2(j)\}]$$

$$\begin{aligned} e_n &= E[\sum_{i=0}^{n} \sum_{j=1}^{n} \delta(S_1(i) - S_2(j))] \\ &= \sum_{i=0}^{n} \sum_{j=1}^{n} P\{S_1(i) = S_2(j)\} \\ &= \sum_{i=0}^{n} \sum_{j=1}^{n} P\{S_1(i + j) = 0\} \\ &\sim \sum_{i=0}^{n} \sum_{j=1}^{n} \tilde{c}(i + j)^{-d/2} \\ &\sim \begin{cases} c & d \ge 5 \\ c(\log n) & d = 4 \\ c\, n^{(4-d)/2} & d \le 3 \end{cases} \end{aligned} \tag{4}$$

Here we have used the well known estimate $P\{S_1(j) = 0\} \sim c\, j^{-d/2}$. From (4) it is not difficult to show

$$\lim_{n\to\infty} f(n) > 0, \qquad d \ge 5.$$

A modified Borel-Cantelli Lemma can be used, see e.g. [7], to show

$$\lim_{n\to\infty} f(n) = 0, \qquad d \le 4 .$$

However, this method gives no information about the rate of decay of $f(n)$.

Estimating $f(n)$ is considerably more difficult than estimating e_n, since we are actually looking at a large deviations problem--what is the probability of obtaining the smallest possible value for the "intersections" random variable? One estimate for $f(n)$ can be given in terms of e_n: there exists a $c_1 > 0$ such that

$$f(n) \geq c_1[e_n]^{-1}$$

To prove this, we first note that there exists a constant $c_1 > 0$ such that (for any d)

$$P\{\Pi_1[0,2n] \cap \Pi_2(n,2n] = \emptyset \text{ and } \Pi_1(n,2n] \cap \Pi_2(0,2n] = \emptyset\} \geq c_1 \tag{5}$$

(One can show this for $d = 1$, which clearly suffices, by noting that the above set contains

$$\{\Pi_1[0,2n] \subset [-\sqrt{n},\infty), \Pi_1[n,2n] \subset [2\sqrt{n},\infty),$$
$$\Pi_2(0,2n] \subset (-\infty,\sqrt{n}], \Pi_2(n,2n] \subset (-\infty,-2\sqrt{n}]\} ,$$

whose probability can be estimated using the reflection principle). Now, call (σ_1,σ_2) a *-last intersection for (S_1,S_2) if $S_1(\sigma_1) = S_2(\sigma_2)$ and

$$S_1(j) \neq S_2(k) \quad \sigma_1 \leq j \leq 2n, \quad \sigma_2 < k \leq 2n .$$

Clearly, every pair of paths has at least one *-last intersection, although some paths may have many. By (5) we can conclude that with probability at least c_1, there is a *-last intersection with $0 \leq \sigma_1 \leq n$, $0 \leq \sigma_2 \leq n$. For $0 \leq \sigma_1 \leq n$, $0 \leq \sigma_2 \leq n$, the probability that (σ_1,σ_2) is a *-last intersection is bounded by $P\{S_1(\sigma_1) = S_2(\sigma_2)\}f(n)$. Hence

$$c_1 \leq \sum_{\sigma_1=0}^{n} \sum_{\sigma_2=0}^{n} P\{(\sigma_1,\sigma_2) \text{ is a *-last intersection}\}$$
$$\leq f(n)e_n .$$

An argument similar to this was first given by Erdös and Taylor [5], who asserted incorrectly that $[e_n]^{-1}$ gave the correct decay rate of $f(n)$. The problem is that the first inequality above is not sharp; a particular pair of paths can have many *-last intersections.

There is a similar problem for which $[e_n]^{-1}$ does give the correct rate. Let S_1, S_2, S_3 be three independent simple random walks starting at the origin and let $F(n)$ denote the probability that the first walk intersects neither of the other two walks, i.e.

$$F(n) = P\{\Pi_1[0,n] \cap (\Pi_2(0,n] \cup \Pi_3(0,n]) = \emptyset\}$$

then there exist $c_2,c_3 > 0$,

$$c_2[e_n]^{-1} \leq F(n) \leq c_3[e_n]^{-1} \tag{6}$$

For $d = 4$ this was proved in [8]. For lower dimensions, a related result (for exponentially killed random variables) follows from work of Aizenman [2] and Felder and Fröhlich [6]. Schwarz inequality on the indicator functions of

$\{\Pi_1[0,n] \cap \Pi_3(0,n]\}$ and $\{\Pi_1[0,n] \cap \Pi_2(0,n]\}$ can be used to show

$$F(n) \leq f(n) \leq [F(n)]^{1/2}, \tag{7}$$

which then gives us an upper bound on f, as well as the lower bound proved above.

For $d = 2$ and 3, the bounds in (7) are all that is known. For $d = 1$, $f(n)$ and $F(n)$ can be estimated easily using the well known result that the probability that a walk is always positive (or always negative) decays like $n^{-1/2}$; we get $F(n) \sim cn^{-3/2}$ and $f(n) \sim c\,n^{-1}$. Note in the case $d = 1$, neither inequality in (7) is sharp.

We now consider the case $d = 4$. By (4), (6), and (7) we have

$$c_4(\log n)^{-1} \leq f(n) \leq c_5(\log n)^{-1/2}$$

for some $c_4, c_5 > 0$. In fact, in this case we can prove that the right hand inequality is almost sharp. More precisely, for any $r > 1/2$,

$$\lim_{n\to\infty} (\log n)^r f(n) = \infty. \tag{8}$$

Below I will sketch some of the ideas to derive (8); details can be found in [8,9].

Let S_i and Π_i be defined as above and define events

$$A_n = \{\Pi_1[0,n] \cap \Pi_2(0,n] = \emptyset\}$$

$$B_n = \{\Pi_1[0,n] \cap \Pi_2(n,2n] = \emptyset \text{ and }$$
$$\Pi_1(n,2n] \cap \Pi_2(0,2n] = \emptyset\}.$$

Then

$$\begin{aligned} A_{2n} &= A_n \cap B_n \\ P(A_{2n}) &= P(A_n)P(B_n|A_n) \\ f(2n) &= f(n)P(B_n|A_n) . \end{aligned} \tag{9}$$

Hence the decay rate of f is closely related to the probability of "long-range" intersections of the paths.

Such long-range probabilities were analyzed in [8]. Consider the following set (which is a little easier than B_n to analyze)

$$D_n = \{\Pi_1[an,bn] \cap \Pi_2(0,\infty) \neq \emptyset\}$$

where $0 < a < b < \infty$. Let I_n be the random variable which measures the number of intersections,

$$I_n(\omega) = \sum_{i=an}^{bn} \sum_{j=1}^{\infty} \delta(S_1(i,\omega) - S_2(j,\omega)) .$$

Then it is clear that

$$P(D_n)\ E(I_n | I_n \geq 1) = E(I_n) \tag{10}$$

A straightforward computation (similar to that done to derive (4)) gives

$$E(I_n) \sim c_6 \log(b/a)\ . \tag{11}$$

The constant $c_6 > 0$, which comes from the Green's function of the random walk, is independent of a,b,n. A similar computation for two walks starting at the same point, i.e. for

$$J_n(\omega) = \sum_{i=0}^{bn} \sum_{j=1}^{\infty} \delta(S_1(i,\omega) - S_2(j,\omega))$$

gives

$$E(J_n) \sim c_6(\log n). \tag{12}$$

The tricky part is to calculate $E(I_n | I_n \geq 1)$. Suppose $\Pi_1[an,bn] \cap \Pi_2(0,\infty) \neq \emptyset$ and take the first intersection (where first is taken on the S_2 time scale), i.e. choose a time τ_2 with $S_2(\tau_2) \in \Pi_1[an,bn]$ and $\Pi_2[0,\tau_2) \cap \Pi_1[an,bn] = \emptyset$.

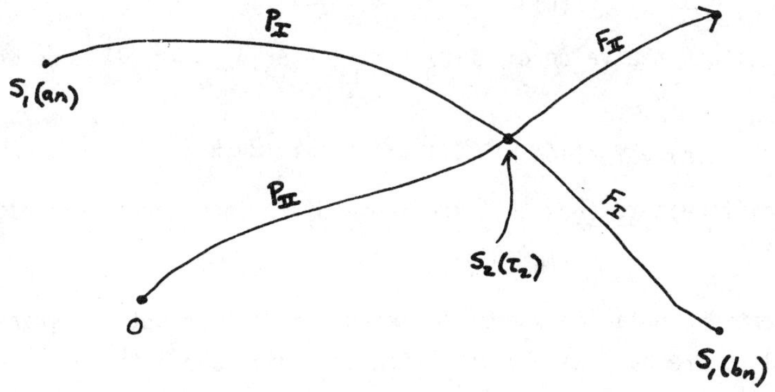

Then in the above picture, $P_{II} \cap P_I = \emptyset$ and $P_{II} \cap F_I = \emptyset$. Also, F_{II} is a simple walk independent of the choice τ_2, and the contributions to I_n come from the intersections of F_{II} with P_I and the intersections of F_{II} with F_I. Therefore, the expected number of intersections after τ_2 should look like twice the number of intersections of two walks starting at the same point, i.e., using (12),

$$E(I_n | I_n \geq 1) \sim 2\ c_6(\log n) \tag{13}$$

Although there are technical problems to making these ideas rigorous, a proof can be given. The main problem lies in the fact that there is conditioning placed on P_1 and F_1 by the choice of τ_2, that is, P_I and F_I (considered as walks starting at $S_2(\tau_2)$) are not simple walks. However, one can show that there is sufficient uniformity in the path (in four dimensions) for (13) to hold. Finally from (10)-(13) we conclude

$$P(D_n) \sim \frac{1}{2} \log(\frac{b}{a})(\log n)^{-1} \tag{14}$$

In a similar fashion we can estimate $P(B_n)$; however, in order to use (9) we will need $P(B_n|A_n)$ which is considerably more difficult to analyze.

In order to prove (8), it is technically convenient to consider times scaled by powers of the logarithm. Fix $\beta > 0$ and let

$$\tilde{A}_n = \{\Pi_1[0,n(\log n)^{-\beta}] \cap \Pi_2(0,n(\log n)^{-\beta}] = \emptyset\}$$

$$\tilde{B}_n = \{\Pi_1[0,n] \cap \Pi_2(n(\log n)^{-\beta},n] = \emptyset \text{ and}$$
$$\Pi_1(n(\log n)^{-\beta},n] \cap \Pi_2(0,n] = \emptyset\} .$$

Then, as in (9),

$$f(n) = P(\tilde{A}_n)\ P(\tilde{B}_n|\tilde{A}_n)$$
$$f(n) = f(n(\log n)^{-\beta})P(\tilde{B}_n|\tilde{A}_n) . \tag{15}$$

The techniques to derive (14) are sufficient to give the asymptotic value of the unconditional probability, $P(\tilde{B}_n)$, which is (compare with (14))

$$1 - P(\tilde{B}_n) \sim \frac{1}{2}\beta\left(\frac{\log\log n}{\log n}\right) . \tag{16}$$

If we could substitute the value of $P(\tilde{B}_n)$ for $P(\tilde{B}_n|\tilde{A}_n)$ in (15), we would get

$$f(n) \approx f(n(\log n)^{-\beta})(1 - \frac{1}{2}\beta \frac{\log\log n}{\log n}) .$$

It is then not difficult to show that the above difference equation implies

$$f(n) \approx (\log n)^{-1/2} .$$

(Here I have adopted a notation where $\sim$ means asymptotic, but $\approx$ means only that the two values are basically alike, with no claims about the exact nature of the relationship.)

The key step, therefore, in proving (8) is to show that $P(\tilde{B}_n) \approx P(\tilde{B}_n|\tilde{A}_n)$, i.e. that "long-range" intersections are independent of "short-range" intersections. Technically this is the trickiest part of the argument. It also demonstrates a peculiarity of $d = 4$ as the critical dimension--the property that long-range intersections are asymptotically independent of short-range intersections holds for $d \geq 4$; but, as mentioned before, $f(n)$ goes to zero for $d \leq 4$.

The bulk of the argument comes in proving an intermediate step of showing that $P(\tilde{B}_n|\tilde{C}_n) \approx P(\tilde{B}_n)$ where

$$\tilde{C}_n = \{\Pi_1[0,n(\log n)^{-2\beta}] \cap \Pi_2(0,n(\log n)^{-2\beta}] \neq \emptyset\} .$$

In fact, it is not very difficult to conclude $P(\tilde{B}_n|\tilde{A}_n) \approx P(\tilde{B}_n)$ if we know $P(\tilde{B}_n|\tilde{C}_n) \approx P(\tilde{B}_n)$.

How can we analyze $P(\tilde{B}_n|\tilde{C}_n)$? We formalize the notion that intersections are a local property--let α be a number satisfying $\beta < \alpha < 2\beta$ (for technical reasons, we eventually will need to assume that $2\beta - \alpha > 7$, and hence that $\beta > 7$). Let $S_1(n)$, $S_2(n)$ be the random walks and let $\eta(i)$ denote the two-sided walk

$$\eta(i) = \begin{cases} S_1(-i) & i \le 0 \\ & \\ S_2(i) & i \ge 0 \end{cases}, \quad -\infty < i < \infty .$$

We say η is "loop-free at step i" if the walk translated so that $\eta(i)$ is the origin has no intersections up through time $n(\log n)^{-2\beta}$, i.e. if

$$\eta(j) \ne \eta(k), \; i - n(\log n)^{-2\beta} \le j \le i < k \le i + n(\log n)^{-2\beta}.$$

Then, if $|i - j| \ge 2n(\log n)^{-2\beta}$, it is clear from the definition that the events "loop-free at step i" and "loop-free at step j" are independent. In fact if

$$L(\eta) = \sum_{i=-n(\log n)^{-\alpha}}^{n(\log n)^{-\alpha}} 1_{\{\text{loop-free at } i\}}(\eta),$$

then one can show that L satisfies a weak law, i.e.

$$\frac{L}{(EL)} \to 1 \tag{17}$$

in probability.

This fact suggests the following method of computing the conditional probability. Let Λ denote the set of all two-sided random walks η which have no intersections up through time $\eta(\log n)^{-2\beta}$, i.e.

$$\Lambda = \{\eta\colon \eta(j) \ne \eta(k), -n(\log n)^{-2\beta} \le j \le 0 < k \le n(\log n)^{-2\beta}\} ,$$

and let $\hat{P}$ denote the conditional probability measure on Λ derived from simple random walk. Then the goal is to compute $\hat{P}(\tilde{B}_n)$. We can consider a different measure $\tilde{P}$ on Λ defined as follows:

- Start with the usual random walk measure on the set Ω of <u>all</u> two-sided paths η (not just those in Λ)
- For any path $\eta \in \Omega$, look at all integers i with $-n(\log n)^{-\alpha} \le i \le n(\log n)^{-\alpha}$, such that η is loop-free at i
- Choose one such loop-free point at random (with equal probability to all such i) and shift the path η such that $\eta(i)$ is the origin.

This produces a measure $\tilde{P}$ on Λ. Since L satisfies the weak law (17), the procedure of picking a loop-free point at random turns out to produce an almost uniform measure on Λ, i.e. $\tilde{P}$ is close to $\hat{P}$.

Finally, we note that $\tilde{P}$ is derived from the simple random walk measure by making a translation of at most $n(\log n)^{-\alpha}$ on each path. Since $n(\log n)^{-\alpha}$ is small compared to $n(\log n)^{-\beta}$, the estimates of [8] can be used to show that $\tilde{P}(B_n)$ is close to $P(B_n)$. This essentially finishes the argument; the details can be found in [9].

Remarks

1. An analogous problem to the intersection of two random walks in four dimensions is that of the intersections of three random walks in three dimensions. The analogy can be seen if one recalls that the path of a random walk (or a Brownian motion) is an "almost 2-dimensional" set. Four is the critical dimension for two walks, because two almost 2-dimensional sets in $\mathbb{R}^4$ will "just barely" intersect or not intersect. Similarly, two almost 2-dimensional sets in $\mathbb{R}^3$ will intersect in an almost 1-dimensional set; this set will then just barely intersect or not intersect with another almost two-dimensional set. Some of the results described have been derived for three walks in $\mathbb{Z}^3$ [10]. Let S_1, S_2, S_2 denote independent walks starting at the origin. Then

$$P\{\Pi_1[0,n] \cap \Pi_2[0,n] \cap \Pi_3[n,2n] = \emptyset\} \approx c(\log n)^{-1}$$

$$c_1(\log n)^{-1} \leq P\{\Pi_1[0,n] \cap \Pi_2(0,n] \cap \Pi_3(0,n]\} \leq c_2(\log n)^{-1/4}$$

We expect that the second probability is of order $(\log n)^{-1/4}$, and that the methods to prove (8) can be adapted to prove this.

2. One of the main motivations for studying the above questions comes from the random walk representations of ϕ_4^4 field theory of Aizenman [1] and Brydges, et al. [3]. In order to prove the triviality (i.e. Gaussian behavior) of these theories, it would suffice to have a result analogous to (14) for these walks. Unfortunately, these walks are not simple walks, and it is not clear how to extend the results to this case (although heuristically one expects similar results). Recently Aizenman [2] and Felder and Fröhlich [6] independently have given proofs of results similar to (14) using renormalization group ideas. Their methods, while again giving good heuristic evidence for results for general walks, as of yet can be made rigorous for simple walks.

BIBLIOGRAPHY

1. M. Aizenman, Geometric analysis of ϕ^4 fields and Ising models, Commun. Math. Phys. 86 (1982), 1.

2. M. Aizenman, The intersection of Brownian paths as a case study of a renormalization group method for quantum field theory, to appear in Commun. Math. Phys.

3. D. Brydges, J. Fröhlich and T. Spencer, The random walk representation of classical spin systems and correlation inequalities, Commun. Math. Phys. 83 (1982), 123-150.

4. C. Domb, Self-avoiding random walk on lattices, in Stochastic Processes in Chemical Physics, ed. K. E. Shuler, John Wiley & Sons (1969), 229-260.

5. P. Erdos and S. J. Taylor, Some intersection properties of random walk paths, Acta Math. Sci. Hung. 11 (1960), 231-248.

6. G. Felder and J. Frohlich, Intersection properties of simple random walks: a renormalization group approach, to appear in Commun. Math. Phys.

7. G. Lawler, A self-avoiding random walk, Duke Math. Jour. 47 (1980), 655-694.

8. G. Lawler, The probability of intersection of independent random walks in four dimensions, Commun. Math. Phys. 86 (1982), 539-554.

9. G. Lawler, Intersections of random walks in four dimensions II, to appear in Commun. Math. Phys.

10. G. Lawler, The probability of intersection of three random walks in three dimensions, to appear.

DEPARTMENT OF MATHEMATICS
DUKE UNIVERSITY
DURHAM, NORTH CAROLINA 27706

Contemporary Mathematics
Volume 41, 1985

A Process in a Randomly Fluctuating Environment

NEAL MADRAS

ABSTRACT. For every integer x, construct a stationary continuous-time Markov process $\gamma(x;t)$, with state space $\{-1,+1\}$ (all processes independent, having the same distributions). This environment determines the following sequence of random variables: $S_o = 0$, $S_{n+1} = S_n + \gamma(S_n;n)$. S_n is not Markov, nor are its increments stationary, yet this process has much in common with the classical random walk, including zero-one laws, a strong law of large numbers, and an invariance principle. Using coupling arguments, one can verify the natural conjecture that the process S_n is recurrent if and only if $P\{\gamma(0;0) = +1\} = \frac{1}{2}$. We also show how the FKG inequality can be used to investigate this process.

1. Introduction; main properties. Let $\{\gamma(t),\ t \geq 0\}$ be a stationary two-state continuous time Markov process on the state space $\{-1,+1\}$, with transition probability matrix

$$(1.1)\qquad \begin{pmatrix} p_{-1,-1}(t) & p_{-1,+1}(t) \\ p_{+1,-1}(t) & p_{+1,+1}(t) \end{pmatrix} = \begin{pmatrix} q + p\, r^t & p - p\, r^t \\ q - q\, r^t & p + q\, r^t \end{pmatrix}$$

where

$$(1.2)\qquad p = \beta/(\beta + \delta)\ ,\quad q = \delta/(\beta + \delta)\ ,\quad \text{and } r = e^{-(\beta + \delta)}\ .$$

(β and δ are positive parameters in this model.)

By stationarity, $\{\gamma(0) = +1\}$ has probability p. On a probability space $(\Omega, \mathcal{F}, P)$, construct a family of indpendent copies of $\gamma(t)$, indexed by

1980 Subject Classification: Primary 60K99; Secondary 60J27

$x \in Z$; call these processes $\gamma(x;t)$. Now, think of a particle moving along the real line at unit speed, starting at time $t = 0$, from position $x = 0$. Its speed remains constant, but its direction is determined by the $\gamma(x;t)$ processes, as follows: if $\gamma(0;0) = +1$, then the particle begins by moving to the right; if $\gamma(0;0) = -1$, then it moves to the left. The direction doesn't change at non-integral times. In general, if the particle is at position $x' \in Z$ at time t (t an integer), then it moves in the $\gamma(x';t)$ direction.

Explicitly, let S_t be the position of the particle at time $t \geq 0$; let $X_n = S_n - S_{n-1}$. We define

$$(1.3)\qquad \begin{aligned} S_0 &= 0 \\ X_{i+1} &= \gamma(S_i;i)\ , \quad i = 0,1,2,\ldots \\ S_t &= S_i + (t-i)X_{i+1}\ , \quad i < t \leq i+1\ . \end{aligned}$$

We define a <u>walk</u> W to be a finite sequence of points in Z, $W = (w_0,w_1,\ldots,w_n)$, such that $w_0 = 0$ and $|w_i-w_{i-1}| = 1$ for $i=1,2,\ldots,n$. W will also denote the event $\{S_0=w_0,\ldots, S_n=w_n\}$. If $V = (v_0,\ldots,v_m)$ is a second walk, we define a third walk $V * W = (v_0,\ldots,v_m,v_m + w_1,\ldots v_m + w_n)$. The probability of $W = (w_0,\ldots,w_n)$ can be computed from (1.1). For example: $P((0,1,0,1,2)) = p\,q(p + q\,r^2)(p - p\,r^2)$.

Although the process $\{S_n\}$ is obviously not Markovian, we can use the following lemma to show that it shares many properties with classical random walks and Markov chains. (The proof of this lemma, as well as those of all other results discussed here, are contained in Madras (1985).)

LEMMA 1. There exist constants L and U ($0 < L < 1 < U < \infty$), depending only on β and δ, such that for any walks V and W we have

$$L\ P(V)\ P(W) \leq P(V * W) \leq U\ P(V)\ P(W)\ .$$

(Note that in the classical random walk, L=U=1.) Using this lemma, we can show that the following are equivalent:

(a) $P\{S_n = 0 \text{ for some } n > 0\} = 1$;

(b) $P\{S_n = 0 \text{ infinitely often}\} = 1$;

(c) $P\{S_n \text{ visits every point infinitely often}\} = 1$;

(d) $E(\text{cardinality } \{n:S_n=0\}) = \infty$.

(If (a) through (d) hold, we say that S_n is recurrent.)

Let $\mathcal{F}_\infty$ be the sub-sigma-algebra of $\mathcal{F}$ generated by $\{X_1, X_2, \ldots\}$, and let $\mathcal{F}'$ be the tail sigma-algebra of that sequence. We now define the shift operator θ to be the set transformation $\theta: \mathcal{F}_\infty \to \mathcal{F}_\infty$ satisfying

$$\theta A = \{\omega \in \Omega : \exists\, \omega' \in A \text{ such that, for all } k \in \mathbb{N},\ X_{k+1}(\omega) = X_k(\omega')\} .$$

In particular, $\theta^n\{X_1 = +1\} = \{X_{n+1} = +1\}$. It can be shown that $P(\theta^n \,\cdot)$ converges (as $n \to \infty$) to a probability measure $P^\infty(\cdot)$ on $(\Omega, \mathcal{F}_\infty)$, with respect to which $\{X_n\}$ is stationary and exponentially ϕ-mixing. This is proven either by applying Theorem 2.4 of Norman (1968), or by a slight refinement of the coupling argument of Harris (1955). The key is the following observation: for any positive integers k and n, and any fixed $a_{n+1}, \ldots, a_{n+k} \in \{-1,+1\}$, the difference between the sup and inf of

$$\{P\{X_{n+k} = a_{n+k} \,|\, X_{n+k-1} = a_{n+k-1}, \ldots, X_1 = a_1\}: a_1, \ldots, a_n \in \{-1,+1\}\}$$

is at most r^k.

Moreover, Lemma 1 implies that P and P^∞ are equivalent measures. From this, we can deduce the following: a zero-one law for $\mathcal{F}'$; a strong law of large numbers ($\lim_{n\to\infty} S_n/n = \hat{\mu}$ a.s., where $\hat{\mu} = \int_\Omega X_1(\omega)P^\infty(d\omega)$); and an invariance principle (for each n, let $W_n(t) = (S_{nt} - nt\hat{\mu})/\sigma\sqrt{n}$ for $t \in [0,1]$, where $\sigma^2 > 0$ is the limiting P^∞-variance of $S_n/\sqrt{n}$; then as $n \to \infty$, W_n converges in distribution to the standard Wiener process in $C[0,1]$).

The preceding may easily be generalized to processes with non-nearest-neighbor jumps X_n, or to higher dimensions. The following material, however, relies heavily on the one-dimensional nearest-neighbor characteristic. Let $\Lambda = \{(x,t) \in Z \times \{0,1,\ldots\}: (x+t)/2 \in Z\}$. For $(z,T) \in \Lambda$, we define $\{S_t^{(z,T)}, t \geq T\}$, the path of a particle starting from position z at time T:

$$S_T^{(z,T)} = z$$

$$S_{i+1}^{(z,T)} = S_i^{(z,T)} + \gamma(S_i^{(z,T)}; i) \ , \ i = T, T+1, \ldots$$

(So $S_t \equiv S_t^{(0,0)}$.) Observe that the sequences $\{S_{T+n}^{(z,T)} - z): n \geq 0\}$ and $\{S_n: n \geq 0\}$ have the same distributions. By construction, these processes <u>coalesce</u>: i.e., if $S_u^{(z,T)} = S_u^{(y,T)}$ for some $u \geq T$, then $S_t^{(z,T)} = S_t^{(y,T)}$ for all $t \geq u$. Moreover, these processes are naturally coupled, in the following sense: if z

$< y$, then $S_t^{(z,T)} \leq S_t^{(y,T)}$ for all $t \geq T$. These ideas are used to prove the following theorem, which is our main result on recurrence:

THEOREM 1. S_n is recurrent if and only if $\beta = \delta$.

The proof of the "if" direction is easy: apply the zero-one law to $\{\lim_{n\to\infty} S_n = +\infty\}$, and use symmetry. The converse, however, is harder to prove; we will give a brief outline of the steps.

Fix $\beta > \delta$. We construct a coupled family of processes $\gamma'(x;t)$ with parameters $p = 1/2$ and $r = e^{-2\delta}$ such that $\gamma'(x;t) \leq \gamma(x;t)$ for all x and t. Then we define the processes S_n' as in (1.3), except with γ replaced by γ'. It follows that $S_n' \leq S_n$ for every n. We know that S_n' is recurrent, and that $\lim_{n\to\infty} S_n'/n = 0$; hence $\hat{\mu} \equiv \lim_{n\to\infty} S_n/n \geq 0$. The theorem is a consequence of the fact that this inequality is strict, which can be proved in two steps.

Step 1: Show that $\lim_{n\to\infty} E(S_n) = +\infty$. Fix a large integer A. Then for sufficiently large M, it is unlikely that $S_n - S_n' \leq A$ for all $n \leq M$; and once these processes get far apart, it can be shown that they remain far apart (in expectation) at later time.

Step 2: Let $S_0^* = 0$, and let S_n^* "run naturally" for a long (fixed) time M (i.e., $S_{n+1}^* = S_n^* + \gamma(S_n^*;n)$ for $n=1,\ldots,M$). Then let S^* take a step to the left at each integral time, for a short (random) time T_1, which is long enough to "forget the past". Then let S^* continue to run naturally for another M time units; then it moves left for T_2 time units, and so on. Done correctly, $\{S^*_{kM+t_1+\ldots+T_k} : k \geq 1\}$ is a sequence of partial sums of i.i.d. random variables with positive mean (using Step 1). Since $S_n^* \leq S_n$ for all n, we are done.

2. The FKG inequality, with applications.

Let B be a finite subset of Λ, and let Γ_B be the sigma-algebra generated by $\{\gamma(x;t) : (x,t) \in B\}$. Define the random set $B_+ = \{(x,t) \in B : \gamma(x;t) = +1\}$. If we write $P(C) = P\{B_+ = C\}$ for $C \subset B$, then P is a probability measure on the lattice of subsets of B. (We denote this lattice by L_B.) As usual, we identify functions on L_B with Γ_B-measurable random variables. For example, if B is a subset of Λ which contains $\Lambda \cap ([-n,n] \times [0,n])$, then $S_n^{(0,0)}$ is an increasing function on L_B.

PROPOSITION (FKG Inequality.) If f and g are both increasing functions on L_B (for some finite $B \subset \Lambda$), then $E(fg) \geq E(f)\,E(g)$.

This follows from Fortuin, Kasteleyn, and Ginibre (1971), by verifying the "convexity condition": for any $C,D \in L_B$, $P(C \cap D)\ P(C \cup D) \geq P(C)\ P(D)$.

COROLLARY. Let $W = (w_0,\ldots,w_n)$ be a walk such that

$$(2.1) \qquad w_n > w_i \qquad \text{for each } i = 0,1,\ldots,n-1 .$$

Let Q be an increasing function on L_B, where B is a finite subset of $\Lambda \cap \{(x,t): t \geq n\}$. then $E(Q|W) \geq E(Q)$.

PROOF: Let $v = \min\{w_0,\ldots,w_n\}$, $\Delta W_i = w_i - w_{i-1}$, and $t_W(x) = \max\{j: w_j = x\}$. Then (3.1) implies $\Delta W_{t_W(x)+1} = +1$ for each $x = v,v+1,\ldots,w_{n-1}$. So $W \subset W'$, where $W' = \{\gamma(x;t_W(x)) = +1$ for each $x = v,v+1,\ldots,w_n-1\}$. Moreover, the Markov property of the processes $\gamma(x;\ )$ implies that $E(Q|W) = E(Q|W')$. $1_{W'}$ (the indicator function of W') and Q are both increasing functions, so, by the Proposition, $E(Q1_{W'}) \geq E(Q)E(1_{W'})$. Equivalently, $E(Q|W') \geq E(Q)$. The Corollary follows, since $E(Q|W) = E(Q|W')$. Q.E.D.

For $(x,T) \in \Lambda$ and $y \in Z$, define $\tau_y^{(x,T)} = \min\{i \geq T: S_i^{(x,T)} = y\}$; let $\tau_y = \tau_y^{(0,0)}$. We are now in a position to prove

THEOREM 2. If $\beta = \delta$, then $E(\tau_1) = \infty$.

PROOF: Suppose $\beta = \delta$. First, we will prove the following claim: for any positive integers x and y, $E(\tau_{x+y}) \leq E(\tau_x) + E(\tau_y)$. To prove this, decompose the event $\{\tau_{x+y} \geq k\}$ according the path of S_t up to its first visit to x:

$$(2.2) \qquad P\{\tau_{x+y} \geq k\} = \Sigma_n \Sigma_W\ P(\{\tau_{x+y} \geq k\} \cap W) ,$$

where the outer sum is over all integers n, and the inner sum is over all walks W of length n such that $w_n = x$ and $w_i < x$ for each $i < n$. Now

$$P(\{\tau_{x+y} \geq k\} \cap W) = P(\{\tau_{x+y}^{(x,n)}\} \geq k\} \cap W) \leq P\{\tau_{x+y}^{(x,n)} \geq k\}\ P(W)$$

by the Corollary, since W satisfies (2.1), and since the indicator function of $\{\tau_{x+y}^{(x,n)} \geq k\}$ is a decreasing function (on L_B, where $B = \Lambda \cap ([-k,k] X [n,k])$). We substitute this into (2.2):

$$P\{\tau_{x+y} \geq k\} \leq \Sigma_n P\{\tau_{x+y}^{(x,n)} \geq k\}\ \Sigma_W\ P(W)$$

$$= \Sigma_n P\{\tau_y \geq k-n\}\ P\{\tau_x = n\} .$$

Thus, if τ_x' is a random variable with the same distribution as τ_x, but independent of τ_y, then we have

$$(2.3) \qquad P(\tau_{x+y} \geq k\} \leq P\{\tau_y + \tau_x' \geq k\} .$$

Summing over all k=1,2,... in (2.3) gives $E(\tau_{x+y}) \leq E(\tau_y + \tau_x')$, which proves the claim.

Now, by the strong law of large numbers, $\lim_{n\to\infty}(S_{\tau_n}/\tau_n) = 0$; therefore $\lim_{n\to\infty}(\tau_n/n) = +\infty$ a.s. By Fatou's Lemma, $\lim_{n\to\infty} E(\tau_n/n) = +\infty$. But $E(\tau_n) \leq nE(\tau_1)$ for all $n \geq 1$, by the claim, hence $E(\tau_1) = +\infty$. Q.E.D.

The final result concerns a particular case of the "Gambler's Ruin Problem" for the process S_n. The proof uses the above Corollary.

THEOREM 3. Suppose $\beta = \delta$. Then for any non-zero integer c, and any positive integer k,

$$\frac{1}{k+1} \leq P\{\tau_{(kc)} < \tau_{-c}\} \leq \frac{1}{L(k+1)}$$

(Note that the left hand side is the exact value for the classical simple symmetric random walk.)

Bibliography

Fortuin, C.M., Kasteleyn, P.W., and Ginibre, J. (1971). Correlation inequalities on some partially ordered sets. Communications in Mathematical Physics, 22, 89-103.

Harris, T.E. (1955). On chains of infinite order. Pacific J. Math, 5, 707-724.

Madras, N. (1985). A Process in a Randomly Fluctuating Environment. To appear in Annals of Probability.

Norman, M.F. (1968). Some convergence theorems for stochastic learning models with distance diminishing operators. Journal of Mathematical Psychology, 5, 61-101.

COURANT INSTITUTE OF MATHEMATICAL SCIENCES
NEW YORK UNIVERSITY
NEW YORK, NEW YORK 10012

Contemporary Mathematics
Volume 41, 1985

INFINITE CHAINS AND CLUSTERS IN ONE DIMENSIONAL DIRECTED AND UNDIRECTED PERCOLATION

C. M. Newman[1] and L. S. Schulman

ABSTRACT. In one dimensional independent bond percolation with probability p_j of an occupied bond between site i and site $i \pm j$, positive density infinite clusters occur, with probability one, if $\liminf j^2 p_j > 1$ and (for fixed $p_2, p_3, \ldots$) p_1 is sufficiently close to 1. In one dimensional independent directed bond percolation with probability p_j of an occupied directed bond from site i to site $i + j$, where $p_j = 0$ for $j \leq 0$, positive (lower) density infinite chains occur, with probability one, if $\liminf j^s p_j > 0$ for some $s < 2$ and p_1 is sufficiently close to 1.

1. MAIN RESULTS. The purpose of this paper is to report some results of [1] concerning one dimensional percolation and to sketch their proofs. The further results of [2] on this subject will be discussed here only in passing. We begin with directed (or oriented) percolation. The main directed percolation result is Theorem 1.1, which goes somewhat beyond [1]; its proof is sketched in Section 2. The main undirected percolation result is Theorem 1.2; its proof is sketched in Section 3. For those details of the proofs which are omitted here, see [1].

The one dimensional directed percolation models we will consider may be defined in terms of 0 or 1 valued site random variables, ρ_i, and 0 or 1 valued bond random variables, μ_{ij}, with i, j in $\{1,2,\ldots\}$ and $i \neq j$. The site i is said to be alive or living (otherwise dead) if $\rho_i = 1$ and the directed bond from i to j is said to be occupied

1980 Mathematics Subject Classification. 82A25,60K35,05C80,82A43.

[1]John S. Guggenheim Memorial Fellow. Research Supported in Part by NSF Grant MCS-8019384.

(otherwise _vacant_) if $\mu_{ij} = 1$; we make the convention that $\mu_{ij} = 0$ automatically unless i and j are alive. All our directed models will have $\mu_{ij} = 0$ unless $i < j$. A _chain_ in such a model is a strictly increasing (finite or infinite) sequence of sites, $i_1 < i_2 < \ldots$ such that each i_k is alive and each bond from i_k to i_{k+1} is occupied. The _lower density_ of an infinite chain is defined as

$$\liminf_{j\to\infty} j/i_j \, . \tag{1.1}$$

If the ρ_i's are independent with $P(\rho_i = 1) = \lambda$ and if, conditional on the ρ_i's, the μ_{ij}'s (with i alive, j alive and $i < j$) are independent with $P(\mu_{ij} = 1) = p_{j-i}$, we will call the model the _independent directed percolation model with parameters_ $\lambda, p_1, p_2, \ldots$. The following theorem somewhat extends the results of [1] by showing that for appropriate choices of parameters, not only do infinite chains exist but they also have positive lower density.

THEOREM 1.1 _Suppose_ $p_2, p_3, \ldots$ _in_ $[0,1]$ _are fixed and satisfy_

$$\text{for some } s < 2, \quad \liminf_{j\to\infty} j^s p_j > 0. \tag{1.2}$$

Then given any $\varepsilon > 0$, _there exists_ $p_1 = p_1(\varepsilon)$ _in_ $[0,1)$ _and some_ $\lambda = \lambda(\varepsilon)$ _in_ $(0,1)$ _so that in the independent directed percolation model with parameters_ $\lambda, p_1, p_2, \ldots$, _there is, with probability one, an infinite chain with lower density at least_ $1 - \varepsilon$.

The one dimensional (undirected) percolation models we will consider are also defined in terms of 0 or 1 valued ρ_i's and μ_{ij}'s except that now i, j are in $\{\ldots -1, 0, 1, \ldots\}$ (with $i \neq j$) and $\mu_{ij} = \mu_{ji}$. Associated with such a model is a random (undirected) graph whose vertices are the living sites $(\rho_i = 1)$ and whose edges correspond to the occupied bonds $(\mu_{ij} = \mu_{ji} = 1)$. A (finite or infinite) _cluster_ is the set of (living) vertices in a maximal connected subgraph of this random graph. The _lower density_ of an infinite cluster C is defined as

$$\liminf_{n\to\infty} |C \cap \{-n, \ldots, n\}| / (2n+1), \tag{1.3}$$

where $|A|$ denotes the cardinality of A. The _independent (undirected) percolation model with parameters_ $\lambda, p_1, p_2, \ldots$ is (as in the directed case) the model in which the ρ_i's are independent with $P(\rho_i = 1) = \lambda$, and conditional on the ρ_i's, the μ_{ij}'s (with i alive, j alive and $i < j$) are independent with $P(\mu_{ij} = 1) = p_{j-i}$.

The next theorem, taken from [1], concerns the occurence of positive lower density infinite clusters in independent percolation models with $\liminf j^2 p_j > 1$. We remark that it follows from arguments of [3] that if, with positive probability, there is such an infinite cluster in an independent model (with each $p_j < 1$) then with probability one there is an infinite cluster with well defined positive density. A possibility, discussed in [2] (see also [3]), is that when models with $\limsup j^2 p_j < \infty$ have their parameters just at the threshhold for infinite clusters to occur, there may be infinitely many distinct infinite clusters, all of zero density.

It is a trivial fact that if

$$\lambda > 0 \quad \text{and} \quad \sum_{j=1}^{\infty} p_j = \infty ,$$

then there are infinite clusters with probability one. Other than that and prior to [1], it was known on the one hand [4] that if

$$\sum_{i=1}^{\infty} \sum_{j=1}^{\infty} p_{i+j} < \infty \text{ (with each } p_j < 1),$$

then with probability one, there are no infinite clusters and on the other hand [5] that if

$$\lambda = 1, \sum_{j=1}^{\infty} p_j = \infty \text{ and the greatest common divisor of } \{j \geq 1 : p_j > 0\} \text{ is 1,}$$

then with probability one, the associated random graph is totally connected. The only result then known that applied to the intermediate region (e.g. $p_j \sim \beta j^{-s}$ as $j \to \infty$ with $\beta > 0$ and $1 < s \leq 2$) was [4] that if

$$\sum_{j=1}^{\infty} p_j < 1/2,$$

then with probability one, there are no infinite clusters. It follows from the directed percolation result of Theorem 1.1 that if (1.2) is valid in the undirected case, then there are (for p_1 and λ sufficiently close to 1) positive lower density infinite clusters. The following theorem gives the same conclusion with a weaker hypothesis than (1.2). A result of [2], which is complementary to and demonstrates the essential optimality of the following theorem, states that if $\limsup j^2 p_j < 1$ (and each $p_j < 1$), then with probability one there are no infinite clusters. See [1] for a partial resolution of those situations where $\liminf j^2 p_j < 1$ but $\limsup j^2 p_j > 1$.

THEOREM 1.2 *Suppose* $p_2, p_3, \ldots$ *in [0,1] are fixed and satisfy*

$$\liminf_{j\to\infty} j^2 p_j > 1 . \tag{1.4}$$

Then given any $\varepsilon > 0$, *there exists some* $p_1 = p_1(\varepsilon)$ *in [0,1) and some* $\lambda = \lambda(\varepsilon)$ *in (0,1) so that in the independent (undirected) percolation model with parameters* $\lambda, p_1, p_2, \ldots$, *there is, with probability one, an infinite cluster with lower density at least* $1 - \varepsilon$.

Sections 2 and 3 of this paper present reasonably complete proofs of Theorems 1.1 and 1.2 but with some missing details which may be found in [1]. We conclude this section with some comments about the proofs. The meaning of these comments should become clearer as (or if) the reader studies the proofs themselves.

The proofs have a renormalization group structure. We choose a sequence K_n of increasing length scales and define renormalized parameters. Suppose the initial parameters satisfy $p_j \sim \beta j^{-s}$ with $\beta > 0$ and $s \leq 2$. The key renormalized parameters at scale K_n are λ_n, β_n and (for $s = 2$) p_1^n. λ_n is roughly the probability of occurence, on the scale K_n, of chains or clusters with a prescribed lower density. Because $(y-x)^{-s}dy\,dx$ transforms, under the scaling $(x,y) = (Kx', Ky')$, to $K^{2-s}(y'-x')^{-s}dy'\,dx'$, β_n is "essentially" $(K_n)^{2-s}\beta$. For $s < 2$ (Theorem 1.1 and Section 2), $\beta_n \to \infty$ which forces all the renormalized p_j's toward 1 which in turn forces λ_n toward 1 and thus implies the existence of positive lower density infinite chains. The actual proofs in Section 2 are not unreasonably long and would be rather shorter if we treated only $s < 2$ undirected percolation. On the other hand, for $s = 2$ (Theorem 1.2 and Section 3), β_n is roughly constant, and the renormalized p_j's (for $j \geq 2$) are not forced toward 1. However, p_1^n is forced to 1 because of the logarithmic divergence of

$$(1.5) \qquad \int_0^{1-\xi/K} \int_{1+\xi/K}^{2} (y'-x')^{-2}\, dy'\, dx'$$

as $K \to \infty$ for fixed ξ. Most of the lengthy Section 3 is devoted to showing (see Prop. 3.3) that p_1^n tending to 1 (together with $\beta > 1$) is sufficient to force λ_n toward 1 and thus imply the existence of positive density infinite clusters. The reason for needing $\beta > 1$ (i.e. $\liminf j^2 p_j > 1$) when $s = 2$ is somewhat subtle. See the discussion of this point near the beginning of the proof of Prop. 3.3; also see Prop. 3.4 and [2].

2. DIRECTED PERCOLATION. The proof of Theorem 1.1 (and of Theorem 1.2) is based on a block variable or coarse graining or renormalization group approach. We will choose a sequence of scaling sizes $C_1, C_2, \ldots$ (each an integer ≥ 2) and a sequence θ_n in $(0,1)$ (roughly speaking, these are prescribed lower densities) and inductively define a sequence of directed percolation models,

$$(2.1) \quad \{\rho_i^n, \mu_{ij}^n\} = \mathcal{F}_{\theta_n, C_n}(\{\rho_i^{n-1}, \mu_{ij}^{n-1}\}) = \mathcal{F}_{\theta_n, C_n}(\cdots \mathcal{F}_{\theta_1, C_1}(\{\rho_i^0, \mu_{ij}^0\})\cdots),$$

where $\{\rho_i^0, \mu_{ij}^0\}$ is the independent model of Theorem 1.1 with parameters

λ, p_1, p_2,... . For $n \neq 0$, $\{\rho_i^n, \mu_{ij}^n\}$ will not be an independent percolation model, but we will compare it to an independent model.

Given θ, C and some model, $\{\rho_i, \mu_{ij}\}$, $\{\rho_i', \mu_{ij}'\} = \tilde{F}_{\theta,C}(\{\rho_i, \mu_{ij}\})$ is defined as follows. $\rho_i' = 1$ (otherwise $= 0$) if in the block of sites,

$$H_i = \{(i-1)C+1,\ldots,iC\}$$

in the $\{\rho_i,\mu_{ij}\}$ model, there are at least $\theta^{1/2}C$ living sites and all the living sites in H_i form a chain. If $i < j$ and $\rho_i' = \rho_j' = 1$, we set $\mu_{ij}' = 1$ (otherwise $= 0$) if $\mu_{k\ell} = 1$ for some k in H_i and some ℓ in H_j (k and ℓ are necessarily alive if $\mu_{k\ell} = 1$ according to our convention) with the fraction of living sites in H_i to the right of k and the fraction of living sites in H_j to the left of ℓ each bounded above by $(1-\theta^{1/2})/2$. Note that if $i_1<\ldots<i_k$ is a chain of the primed model, then there is a chain of the unprimed model contained in $H_{i_1} \cup\cdots\cup H_{i_k}$ and containing at least $(k-2)\ \theta C + 2\theta^{1/2}C(1+\theta^{1/2})/2 \geq k\theta C$ sites.

The definition of $\tilde{F}_{\theta,C}$ guarantees that if $\rho_i^n = 1$, then in the ith block of the initial $\{\rho_i^0, \mu_{ij}^0\}$ model of length $K_n = C_1\cdots C_n$, denoted by

$$(2.2)\qquad H_{n,i} = \bigcup_{j=(i-1)C_n+1}^{iC_n} H_{n-1,j} = \{(i-1)K_n+1,\ldots,iK_n\},$$

there is a chain containing at least

$$(2.3)\qquad (\theta_n)^{1/2}C_n \prod_{k=1}^{n-1} \theta_k C_k \geq (\prod_{k=1}^{n} \theta_k)K_n \equiv \psi_n K_n$$

sites. If $\rho_1^n = 1$ for every $n = 0,1,2,\ldots$ and

$$(2.4)\qquad \psi \equiv \lim_{n\to\infty} \psi_n = \prod_{k=1}^{\infty} \theta_k > 0,$$

then there is an infinite chain $i_1 < i_2 <\cdots$ of the initial $\{\rho_i^0, \mu_{ij}^0\}$ model with $i_1 = 1$ and

$$(2.5)\qquad i_{[\psi_n K_n]} \leq K_n \qquad \text{for every } n = 1,2,\ldots,$$

where $[u]$ denotes the greatest integer $\leq u$. It follows from (2.5) that $\limsup j/i_j \geq \psi > 0$, i.e. the infinite chain has positive upper density. If (2.4) is valid, then to show positive lower density, it would suffice to have, as a replacement for $\rho_1^n = 1$ for every n, that

$$(2.6)\qquad \frac{1}{k}\sum_{i=1}^{k} \rho_i^n \geq (1-\delta) \quad \text{for all } k \leq C_{n+1} \text{ and all } n = 0,1,2,\ldots$$

since this would imply

(2.7) $$\liminf_{j\to\infty} j/i_j \geq (1-\delta)\psi \ .$$

The following proposition is a straightforward consequence of the above arguments, a standard zero-one law and the evident fact that if $\{\rho_i^0, \mu_{ij}^0\}$ is the independent directed model with parameters λ, p_1, $p_2,\cdots$, then for fixed n, i, $\{C_k\}$, $\{\theta_k\}$, and $\{p_j:j\geq 2\}$,

$$P(\rho_i^n = 1) \to 1 \quad \text{as} \quad (\lambda,\ p_1) \to (1,\ 1).$$

PROPOSITION 2.1 <u>In order to prove Theorem 1.1, it suffices to show that for the given</u> $p_2,p_3,\ldots$ <u>and any</u> $\varepsilon > 0$, <u>sequences</u> C_k <u>and</u> θ_k <u>may be chosen with</u> $\psi \geq 1 - \varepsilon/2$ <u>so that for sufficiently large</u> p_1 <u>and</u> λ <u>in</u> $(0,1)$,

(2.8) $$\sum_{n=0}^{\infty} P\Big(\frac{1}{k}\sum_{i=1}^{k} \rho_i^n < (1-\varepsilon/2) \ \underline{\text{for some}}\ k \leq C_{n+1}\Big) < \infty.$$

The next step in our renormalization group approach is to compare the nonindependent $(\rho_i^n,\ \mu_{ij}^n)$ models to a sequence of independent models with parameters defined inductively by

(2.9) $$\{\overline{\lambda}_n,\overline{p}_1^n,\overline{p}_2^n,\ldots\} = F_{\theta_n,C_n}(\{\overline{\lambda}_{n-1},\overline{p}_1^{n-1},\overline{p}_2^{n-1},\ldots\})$$

where $\{\overline{\lambda}_0,\overline{p}_1^0,\ldots\}$ is identical to the initial parameter set of Theorem 1.1, $\{\lambda,p_1,p_2,\ldots\}$. Given θ, C and some $\{\overline{\lambda},\overline{p}_1,\overline{p}_2,\ldots\}$ and its corresponding $\{\rho_i,\mu_{ij}\}$, $\{\overline{\lambda}',\overline{p}_1',\overline{p}_2',\ldots\} = F_{\theta,C}(\{\overline{\lambda},\overline{p}_1,\overline{p}_2,\ldots\})$ is defined in terms of $\{\rho_i,\mu_{ij}\}$ and $\{\rho_1',\mu_{ij}'\} = \tilde{F}_{\theta,C}(\{\rho_i,\mu_{ij}\})$ as follows.

(2.10) $$\overline{\lambda}' = P(\rho_i' = 1)$$

(2.11) $$\overline{p}_{j-i}' = \min_{\gamma\varepsilon\Gamma_{ij}} P(\mu_{ij}' = 1|\gamma),$$

where Γ_{ij} denotes the set of configurations γ of $\{\rho_k,\ \mu_{kk'}:$ either k and k' are both in H_i or k and k' are both in $H_j\}$ for which $\rho_i' = 1 = \rho_j'$; i.e. p_{j-i}' is the worst case conditional probability for μ_{ij}' to be 1 given a configuration inside H_i and H_j for which $\rho_i' = 1 = \rho_j'$. Tracing through the definitions one finds that

(2.12) $$1 - \overline{p}_{j-i}' = \max_{S,S'} \prod_{k\varepsilon S} \prod_{\ell\varepsilon S'} (1 - \overline{p}_{\ell-k})$$

where S is a subset of H_i and S' is a subset of H_j containing (approximately) $(\theta^{1/2}(1-\theta^{1/2})/2)C$ sites with all the sites of S located in the rightmost portion of H_i occupying a fraction (approximately) $(1-\theta^{1/2}) + \theta^{1/2}(1-\theta^{1/2})/2$ and all the sites of S' located in the corresponding leftmost portion of H_j.

It is immediate from (2.9) and (2.10) that $\overline{\lambda}_1 = P(\rho_i^1 = 1)$. However, the key point of the definitions (2.9) - (2.11) is an inequality, namely that for every $n \geq 1$, $\overline{\lambda}_n \leq P(\rho_i^n = 1)$. Similarly, the probabilities appearing in (2.8) are bounded above by the corresponding probabilities in which $\{\rho_i^n : i = 1,\ldots\}$ is replaced by the set of (independent) site variables of the $(\overline{\lambda}_n, \overline{p}_1^n, \ldots)$-model. These facts are straightforward consequences of the worst case nature of definition (2.11).

The next step in our approach is to replace the general independent models with infinitely many parameters, which appear in Theorem 1.1, by a special class of models with very few parameters. This will enable us to replace the $F_{\theta,C}$'s by mappings on finite dimensional spaces. For a given $s < 2$, we define the canonical model with parameters λ and β to be the independent directed percolation model with parameters $\lambda, p_1, p_2, \ldots$ where

$$p_j = p_j(\beta) = 1-\exp(-\beta \int_0^1 \int_j^{j+1} |y-x|^{-s}\, dy\, dx) \quad \text{for } j = 1,2,\ldots . \tag{2.13}$$

Note that $\lim j^s p_j = \beta$. There is a direct argument (see [1]) which shows that in order to prove Theorem 1.1, it suffices to take the initial model as such a canonical model. More specifically, it suffices to show that for any $s < 2$, (2.8) will be valid for λ sufficiently close to 1 and β sufficiently large.

In order to replace the $(\overline{\lambda}_n, \overline{p}_1, \ldots)$-models by canonical models, we do a further worst case modification of (2.11) to define

$$(\lambda_n, \beta_n) = f_{\theta_n, C_n}(\lambda_{n-1}, \beta_{n-1}) \tag{2.14}$$

with λ_0, β_0 the initial parameters of the canonical model. Given $s < 2$, θ, C and some (λ, β) and its corresponding $\{\lambda, p_1(\beta), p_2(\beta), \ldots\}$ (as defined in (2.13)), $(\lambda', \beta') = f_{\theta,C}(\lambda,\beta)$ is defined in terms of $\{\overline{\lambda}', \overline{p}_1', \overline{p}_2', \ldots\} = F_{\theta,C}(\{\lambda, p_1(\beta), p_2(\beta), \ldots\})$ by

$$\lambda' = \overline{\lambda}' \tag{2.15}$$

$$\beta' = \sup\{\overline{\beta} : \overline{p}_j' \geq p_j(\overline{\beta}) \text{ for } j = 1,2,\ldots\}. \tag{2.16}$$

The following proposition is a consequence of Proposition 2.1 and the preceding discussion.

PROPOSITION 2.2. *In oder to prove Theorem 1.1, it suffices to show that for any* $s < 2$, *and any* $\varepsilon > 0$, *sequences* C_k *and* θ_k *may be chosen with* ψ *(as defined in (2.4))* $\geq 1-\varepsilon/2$, *so that for* λ_0 *sufficiently close to* 1 *and* β_0 *sufficiently large*

$$\sum_{n=0}^{\infty} P\left(\frac{1}{k} \sum_{i=1}^{k} X_i^n > \varepsilon/2 \quad \text{for some } k \geq 1\right) < \infty, \tag{2.17}$$

where for each n, $\{X_1^n, X_2^n, \ldots\}$ are i.i.d. Bernoulli random variables with $P(X_i^n = 1) = 1 - \lambda_n$, where λ_n is defined inductively by (2.14) - (2.16).

To complete the proof of Theorem 1.1, we need to obtain suitable estimates on how fast $1-\lambda_n$ tends to zero and then use these to show (2.17). The former was done in [1] and will be repeated below (see Prop. 2.3). Before doing that, we handle the latter question. Using standard large deviation estimates, we have

$$P\left(\frac{1}{k}\sum_{i=1}^{k} X_i^n > \varepsilon/2 \text{ for some } k\right) \leq P(X_i^n = 1 \text{ for some } i < \ell) + \sum_{k=\ell}^{\infty} P\left(\sum_{i=1}^{k} X_i^n > k\varepsilon/2\right)$$

$$\tag{2.18} \leq \ell\,(1-\lambda_n) + \sum_{k=\ell}^{\infty} \exp(-kW(\varepsilon/2, 1-\lambda_n))$$

$$= \ell(1-\lambda_n) + \exp(-\ell W(\varepsilon/2,1-\lambda_n))\cdot\{1-\exp(-W(\varepsilon/2,1-\lambda_n))\}^{-1}$$

where

$$W(u,p) = -\inf_{t\geq 0}\{\ell n(1-p+pe^t) - u\,t\}$$

$$\tag{2.19} = \begin{cases} 0 & , \text{ if } u \leq p \\ u\,\ell n\,(u/p) - (1-u)\ell n(\frac{1-p}{1-u}) & , \text{ if } p < u \leq 1\,. \\ +\infty & , \text{ if } u > 1 \end{cases}$$

Since $W(u,p) \geq$ const. $+ u\,\ell n(1/p)$ as $p \to 0$ for fixed $u > 0$, we see from (2.18) that to obtain (2.17), it suffices if $1 - \lambda_n \to 0$ sufficiently rapidly so that, for some ℓ

$$\tag{2.20} \sum_{n=0}^{\infty}(1-\lambda_n) + \sum_{n=0}^{\infty}\exp(+\ell(\varepsilon/2)\ell n(1-\lambda_n)) < \infty\,.$$

By choosing $\ell > 2/\varepsilon$, we see that it thus sufficies if

$$\tag{2.21} \sum_{n=0}^{\infty}(1-\lambda_n) < \infty\,.$$

That suitable C_k, θ_k, λ_0 and β_0 exist to yield, via (2.21), the requirements of Proposition 2.2, is an immediate consequence of the next proposition (by choosing θ sufficiently small).

PROPOSITION 2.3. Given $s < 2$, choose $q > 1$, r satisfying

$$\tag{2.22} r > 2q/(2-s),$$

$\theta_k = 1 - \theta k^{-q}$ (where θ is in $(0,1)$) and $C_k = [2\,k^r]$. Define (λ_n, β_n) by (2.14) - (2.16). Then for λ_0 sufficiently close to 1 and β_0 sufficiently large, and all $n \geq 1$,

$$\tag{2.23} \beta_n \geq (2H')^n(n!)^{r(2-s)-2q}\,,$$

$$1 - \lambda_n \leq (\theta/4)(n+1)^{-q}, \tag{2.24}$$

where H' is a constant depending only on s, q and r.

Sketch of Proof of Prop. 2.3. Since for fixed θ and C, $f_{\theta,C}(\lambda,\beta)\to(1,\infty)$ as $(\lambda,\beta)\to(1,\infty)$, it follows that (2.23) - (2.24) can be made valid for finitely many n by choosing appropriate λ_0, β_0. In order to complete the proof inductively, it suffices to show that the validity of (2.23) - (2.24) for n = some large N implies their validity for $n = N + 1$. This induction step is a consequence of the following two inequalities, in which $(\lambda',\beta') = f_{\theta,C}(\lambda,\beta)$, $\theta \geq 1/2$ and $1 - \theta^{1/2} \geq (3/2)(1-\lambda)$:

$$\beta' \geq \beta\, HC^{2-s}(\theta^{1/2}(1-\theta^{1/2})/2)^2, \tag{2.25}$$

$$1 - \lambda' \leq \exp(-C(1-\lambda)/16) + \binom{C}{2}\exp(-\beta(C+1)^{-s}), \tag{2.26}$$

where $H > 0$ depends only on s.

The inequality (2.25) is a consequence of (2.12) and the scaling properties of the integral appearing in (2.13), namely that under the transformation $(x,y) = (Cx',Cy')$, $\beta|y-x|^{-s}dy\,dx$ is transformed to $C^{2-s}\beta|y'-x'|^{-s}dy'dx'$.

The first term on the right hand side of (2.26) is a large deviation estimate (closely related to (2.18) - (2.19)) for the probability that a block H_i in the (λ,β)-model has less than $\theta^{1/2}$ living sites. The second term on the right hand side of (2.26) is a crude upper bound for the probability that the directed bond is vacant between some pair of living sites, j and k (with $j < k$), in a block H_i. Note that if all directed bonds between living j and k in H_i (with $j < k$) are occupied, then all living sites in H_i form a chain. Thus these two terms give an upper bound for $1 - \lambda'$ since $\lambda' = P(H_i$ has at least $\theta^{1/2}$ living sites and all the living sites form a chain).

3. UNDIRECTED PERCOLATION. As in the case of Theorem 1.1, it can be readily shown (see [1]) that it suffices to prove Theorem 1.2 for a special model we call the canonical model with parameters λ, β and p_1, which is the independent (undirected) percolation model with parameters $\lambda, p_1, p_2, \ldots$ where

$$p_j = p_j(\beta) = 1-\exp(-\beta \int_0^1 \int_j^{j+1} |y-x|^{-2} dy\, dx) \quad \text{for } j = 2,3,\ldots \tag{3.1}$$

Note that the integral in (3.1) has a logarithmic divergence for $j = 1$. Since $\lim_{j\to\infty} j^2 p_j = \beta$, (1.4) of Theorem 1.2 will be valid if $\beta > 1$.

Proceeding by arguments analogous to those of Section 2 but actually simpler (for the time being) in the undirected case, the validity of Theorem

1.2 can be reduced to the asymptotic behavior of certain iterated mappings, $g_{\theta,C}$, acting on the three dimensional (λ,β,p_1)-parameter space.

Given θ,C and $\{\rho_i,\mu_{ij}\}$ corresponding to the canonical model with parameters λ, β and p_1,

$$(\lambda',\beta',p_1') = g_{\theta,C}(\lambda,\beta,p_1) \tag{3.2}$$

is defined as follows. For each integer i, consider the original model restricted to the block H_i, $\{\rho_\ell,\mu_{\ell k}:\ell,\ k \in H_i\}$, and define, with respect to this restricted model, B_i to be the cluster in H_i with the most sites (leftmost in case of tie); i.e. B_i is the largest set of living sites in H_i which together with the occupied bonds between them form a connected graph. Define $\rho_i' = 1$(otherwise $= 0$) if $|B_i| \geq \theta C$, and define

$$\lambda' = P(\rho_i' = 1). \tag{3.3}$$

Define for i < j

$$p_{j-i}' = \min_{\gamma\in\Gamma_{ij}} P(\mu_{k\ell} = 1 \text{ for some } k \text{ in } B_i \text{ and some } \ell \text{ in } B_j|\gamma), \tag{3.4}$$

where Γ_{ij} denotes the set of configurations, of the sites and bonds either totally within H_i or totally within H_j, for which $\rho_i' = 1 = \rho_j'$. p_1' is now defined by (3.4). Finally, with $p_j(\beta)$ as given by (3.1), define

$$\beta' = \sup\{\overline{\beta}:p_j' \geq p_j(\overline{\beta}) \text{ for } j = 2,3,\ldots\}\ . \tag{3.5}$$

The mapping $g_{\theta,C}$ has now been defined by (3.2) - (3.5). The following proposition is the undirected analogue of Proposition 2.2 of the last section, and is easier to prove (see [1]).

PROPOSITION 3.1 <u>In order to prove Theorem 1.2, it suffices to show that for any</u> $\beta_0 > 1$, <u>and any</u> $\varepsilon' > 0$, <u>sequences</u> C_k <u>and</u> θ_k <u>may be chosen with</u> ψ <u>(as defined in (2.4))</u> $\geq 1 - \varepsilon'$, <u>so that for</u> λ_0 <u>and</u> p_1^0 <u>sufficiently close to</u> 1,

$$\lim_{n\to\infty} \lambda_n = 1\ , \tag{3.6}$$

<u>where</u> λ_n <u>is defined inductively by</u>

$$(\lambda_n,\beta_n,p_1^n) = g_{\theta_n,C_n}(\lambda_{n-1},\beta_{n-1},p_1^{n-1})\ . \tag{3.7}$$

The undirected analogue of Proposition 2.3 is the following which, via Proposition 3.1, implies Theorem 1.2. Note that the lower bound for β_n of (3.9), unlike that of (2.23), does not tend to ∞ ; this will make the proof of (3.11) much more difficult than the similar inequality in Prop. 2.3.

PROPOSITION 3.2. *Choose* q *and* r *satisfying*

$$1 < q < r \leq 4/3, \tag{3.8}$$

$\theta_k = 1 - \theta k^{-q}$ *(with* θ *in* $(0,1)$*)* *and* $C_k = [2k^r]$. *Define* $(\lambda_n, \beta_n, p_1^n)$ *by* (3.7). *If* $\beta_0 > 1$, *then for* θ *sufficiently small (depending on* β_0*),* λ_0 *sufficiently close to* 1 *(depending on* β_0 *and* θ*) and* p_1^0 *sufficiently close to* 1 *(depending on* β_0 *and* θ*), the following are valid for all* $n \geq 1$:

$$\beta_n \geq \theta_n^6 \beta_{n-1} \geq \psi^6 \beta_0 \text{ and } \beta_n \leq \beta_{n-1} \leq \beta_0, \tag{3.9}$$

$$1 - p_1^n \leq (4(1-\theta_n))^{\beta_n}, \tag{3.10}$$

$$1 - \lambda_n \leq (\theta/4)(n+1)^{-q}. \tag{3.11}$$

Sketch of Proof of Prop. 3.2. The order of proof is as follows. First (3.9) will be proved for all n. Since, as in the proof of Prop. 2.3, (3.10) - (3.11) can be made valid for finitely many n by choosing λ_0 and p_1^0 sufficiently close to 1, it suffices to prove their validity for all large n. Second (3.10) will be shown to follow from (3.9) providing $(1-\theta_n)C_n \geq 1$, which will be the case at least for large n. This second proof requires that $p_1^{n-1} \geq p_2(\beta_{n-1})$ which itself can be assured inductively because $\beta_n \leq \beta_0$. Finally by choosing θ sufficiently small (so that $\psi^6\beta_0 > 1$), (3.11) for large $n = N + 1$ will be shown to follow from (3.9) - (3.11) for $n = N$; this will be a consequence of Proposition 3.3 below.

To obtain (3.9) and (3.10), we consider (λ',β',p_1') and (λ,β,p_1) related as in (3.2). The analogue of (2.12) is

$$1 - p_j' = \max_{S,S'} \prod_{k\in S} \prod_{\ell\in S'} (1-p_{\ell-k}(\beta)) \quad \text{for } j = 1,2,\ldots, \tag{3.12}$$

where now S and S' are subsets of H_0 and H_j containing (approximately) θC sites. By (3.1) this leads to the bounds

$$\begin{aligned} 1 - p_j' &\geq \prod_{k\in H_0} \prod_{\ell\in H_j} (1-p_{\ell-k}(\beta)) = \exp(-\beta I(C,jC,C)) \\ &= 1 - p_j(\beta) \quad \text{for } j \geq 2, \end{aligned} \tag{3.13}$$

where

$$I(x_1,y_1,\Delta) = \int_{x_1-\Delta}^{x_1} \int_{y_1}^{y_1+\Delta} (y-x)^{-2} dy\, dx, \tag{3.14}$$

and the last equality of (3.13) uses the invariance of $(y-x)^{-2}dy\,dx$ under the scaling $(x,y) = (Cx',Cy')$. (3.13) implies $\beta' \leq \beta$ which gives the second part of (3.9).

Using (3.12) to obtain an upper bound, we find

$$1 - p_j' \leq \exp(-\beta I(\theta C,(j+1-\theta)C,\theta C)), \tag{3.15}$$

which is valid in general for $j \geq 2$ and is additionally valid for $j = 1$ if $(1-\theta)C \geq 1$ and $p_1 \geq p_2(\beta)$. The scaling invariance of $(y-x)^{-2}dy\,dx$ and further elementary estimates imply

$$I(\theta C,(j+1-\theta)C,\theta C) = I(1,j+\frac{(j+1)(1-\theta)}{\theta}, 1) \geq \theta^6 I(1,j,1) \text{ for } j \geq 2, \tag{3.16}$$

which, with (3.15), implies $p_j' \geq p_j(\theta^6\beta)$ for $j \geq 2$ and thus that $\beta' \geq \theta^6\beta$ which gives the first part of (3.9).

The $j = 1$ case of (3.15) and the identity

$$I(x_1,y_1,\Delta) = \ell n \left\{\frac{(y_1-x_1+\Delta)^2}{(y_1-x_1)(y_1-x_1+2\Delta)}\right\}, \tag{3.17}$$

together with the already derived $\beta' \leq \beta$, yield

$$1 - p_1' \leq \exp\{-\beta\, \ell n\, (\frac{(2-\theta)^2}{2(2-2\theta)})\} \leq \left\{\frac{4(1-\theta)}{(1+(1-\theta))^2}\right\}^{\beta'} \leq (4(1-\theta))^{\beta'}, \tag{3.18}$$

which gives (3.10).

We now suppose Θ has been picked so small that $\beta \equiv \psi^6\beta_0 > 1$ and hence by (3.9) $\beta_n \geq \beta > 1$ for all n, and by (3.10) $p_1^n \geq 1-(4(1-\theta_n))^\beta$ for all n. An inductive proof of (3.11) for large n (and hence for all n if λ_0 and p_1^0 are sufficiently close to 1) is an immediate consequence of the next proposition.

PROPOSITION 3.3 <u>With fixed</u> β, Θ, q <u>and</u> r <u>satisfying</u> $\beta > 1$, $\Theta > 0$, <u>and</u> $1 < q < r \leq 4/3$, <u>let</u> $(\lambda_N', \beta_N', p_1'^N) = g_{\theta,C}(\lambda,\beta,p_1)$, <u>where</u> $\theta = 1 - \Theta N^{-q}$, $C = [2N^r]$, $\lambda = 1 - (\Theta/4)(N+1)^{-q}$, <u>and</u> $p_1 = 1 - (4\Theta N^{-q})^\beta$. <u>Then</u>

$$\lim_{N\to\infty} N^q(1-\lambda_N') = 0, \tag{3.19}$$

<u>and hence</u> $\lambda_N' \geq 1 - (\Theta/4)(N+2)^{-q}$ <u>for all large</u> N.

<u>Sketch of Proof of Prop. 3.3</u>. Throughout this sketch we suppress the subscript or superscript N. Recalling the definition of λ' given by (3.3), we must obtain an appropriate uper bound for the probability, $1 - \lambda'$, that $\rho_1' = 0$, i.e. that in the block $H_1 = \{1,\ldots,C\}$, there is (with respect to the canonical (λ,β,p_1)-model) no set of θC or more living sites which, together with the occupied bonds between them, form a connected graph. We reiterate the fact that the derivation of this upper bound and hence of (3.19) will be much more difficult than that of the upper bound

(2.26) which led to (2.24), the analog of (3.19) in Prop. 2.3. That is because in obtaining (2.24), full use was made of the fact that by (2.23), β_n tended to ∞ rapidly; consequently a crude upper bound such as led to the second term on the right hand side of (2.26) was tolerable. In the present case, β in Prop. 3.3 is fixed, reflecting the inequalities of (3.9) which are in turn due to the scaling invariance of $(y-x)^{-s}dy\, dx$ when $s = 2$. Our upper bound for $1 - \lambda'$ must be based only on the small values of $1 - \lambda$ and $1 - p_1$, together with the assumption that $\beta > 1$.

The rough idea of the bound for $1 - \lambda'$ is that the smallness of $1 - \lambda$ and $1 - p_1$ insure the likely existence of many clusters within H_1, based on only nearest neighbor bonds, whose size is on an intermediate length scale $(O(N^q))$ while the assumption that $\beta > 1$ will insure the likely occupation of bonds which will connect together "essentially" all of these intermediate scale clusters to form a cluster on the long length scale of H_1 $(O(N^r))$. One of the (necessary?) complications of our analysis is that we must allow for the possibility that some of the clusters based on only nearest neighbor bonds will have anomalously short size ($O(N^{q_1})$ for some $q_1 < q$ to be chosen below) and will not participate in the long length scale cluster. In order to make all this more precise, we introduce a number of definitions.

We denote by $\overline{V}$ the <u>number</u> of dead sites in H_1 and by $\overline{R}$ the <u>set</u> of living sites in H_1. A sequence, $\overline{\Lambda} = \{k,k+1,...,k+\ell\}$, of successive living sites (in H_1) for which all nearest neighbor bonds are occupied and which is maximal (i.e. cannot be extended in H_1 to the left or right) will be called a <u>maximal run</u>. We partition $\overline{R}$ into maximal runs, $\overline{R} = \overline{\Lambda}_1 \cup ... \cup \overline{\Lambda}_M$, where each $\overline{\Lambda}_i$ is to the left of $\overline{\Lambda}_{i+1}$. Note that M is random. $\overline{\Lambda}_i$ and $\overline{\Lambda}_{i+1}$ will be said to be <u>involved</u> in an <u>obstruction</u> if every bond between a site in $\overline{\Lambda}_i$ and one in $\overline{\Lambda}_{i+1}$ is vacant. Such an obstruction will be said to be <u>bridged</u> if there is an occupied bond between some site in $\overline{\Lambda}_1 \cup ... \cup \overline{\Lambda}_i$ and some site in $\overline{\Lambda}_{i+1} \cup ... \cup \overline{\Lambda}_M$. A $\overline{\Lambda}_i$ will be called <u>long</u> if it contains more than N^{q_1} sites, and otherwise will be called <u>short</u>, where q_1 is a fixed constant satisfying

$$(3.20) \qquad 0 < q_1 < r - q ,$$

so that $N^{q_1} \leq ((1-\theta)C/2)$ for large N. We shall also require that

$$(3.21) \qquad q_1 > (r-q)/(2\beta-1);$$

as will be seen below (see (3.34), (3.43) and (3.48)) this requirement will help insure that no long $\overline{\Lambda}_i$ is involved in obstructions both to its left and to its right. Since $\beta > 1$ and hence $2\beta - 1 > 1$, q_1 can be chosen to satisfy both (3.20) and (3.21). An obstruction involving $\overline{\Lambda}_i$ and $\overline{\Lambda}_{i+1}$ will be called an <u>interior</u> obstruction if either the rightmost site in $\overline{\Lambda}_i$ or the leftmost site in $\overline{\Lambda}_{i+1}$ (or both) are in $\{[N^{q_1}]+1,\ldots,C-[N^{q_1}]\}$.

Our bound on $1 - \lambda'$ is based on estimating the probability that one or more of the following five events occur, since (as explained below) ρ_1' must be 1 if none of the events occur: $\overline{V} > (1-\theta)C/2$; more than one obstruction occurs between a long $\overline{\Lambda}_i$ and a long $\overline{\Lambda}_{i+1}$; more than one of the short $\overline{\Lambda}_i$'s is involved in an obstruction (but a single short $\overline{\Lambda}_i$ may be involved in two obstructions, one to its left and one to its right); there is simultaneously some obstruction involving two long maximal runs and some other one involving a short maximal run; and finally some interior obstruction is not bridged. We claim that if none of these five events occur, then H_1 will contain a cluster at least as big as $C-(1-\theta)C/2-N^{q_1}$ which exceeds θC (for large N) and hence $\rho_1' = 1$. This claim can be verified by assuming that none of the five events occur and working through all the possible cases. For example, if an obstruction involving two long maximal runs occurs, then it is necessarily an interior obstruction, hence bridged, and no other obstruction occurs; thus all $C - \overline{V}$ living sites belong to a single cluster and $C-\overline{V} \geq \theta C$ since $\overline{V} \leq (1-\theta)C/2$, hence $\rho_1' = 1$.

We define three random variables related to obstructions. L^1 is the number of obstructions involving two long maximal runs, L^2 is the number of short maximal runs involved in at least one obstruction, and L^3 is the number of interior obstructions which are not bridged. Noting that $\{L^1 + L^2 \geq 2\}$ is equivalent to $\{L^1 \geq 2$ or $L^2 \geq 2$ or $(L^1 \geq 1$ and $L^2 \geq 1)\}$, our basic bound which will eventually lead to (3.19) is

$$1 - \lambda' \leq P(\overline{V} > (1-\theta)C/2) + P(L^1 + L^2 \geq 2) + P(L^3 \geq 1). \tag{3.22}$$

In order to obtain (3.19) we will show that each term on the right hand side of (3.22) is $o(N^{-q})$ as $N \to \infty$. Since $\overline{V}$ has a binomial distribution with parameters $C = [2N^r]$ and $1 - \lambda = (\theta/4)(N+1)^{-q}$, we may use a large deviation estimate (see (2.18) - (2.19) and the first term on the right hand side of (2.26)) to obtain

$$\begin{aligned} P(\overline{V} > (1-\theta)C/2) &= P(\overline{V}/C > \theta N^{-q}/2) \leq P(\overline{V}/C \geq 2(1-\lambda)) \\ &\leq \exp(-C(1-\lambda)/16) = o(N^{-q}) \text{ as } N \to \infty . \end{aligned} \tag{3.23}$$

It remains to prove

$$N^q P(L^1 + L^2 \geq 2) \to 0, \tag{3.24}$$

$$N^q P(L^3 \geq 1) \to 0. \tag{3.25}$$

We first give a heuristic sketch of the proof of (3.24) (a complete proof is given in [1]) and then return below to (3.25). The $\overline{\Lambda}_i$'s $(i=1,\ldots,M)$, which are the maximal runs in H_1, are closely related to the successive maximal runs in the half space, $\{1,2,3,\ldots\}$, which we denote by $\Lambda_i(i=1,2,3,\ldots)$. Indeed M is the largest i such that Λ_i is not disjoint from H_1 and for $i \leq M$, $\overline{\Lambda}_i = \Lambda_i \cap H_1(=\Lambda_i$ for $i < M)$. The sizes, $|\Lambda_i|$, of the Λ_i's, are independent geometrically distributed random variables with

(3.26) $$E(|\Lambda_i|) = (1 - p_1\lambda)^{-1} \sim \text{const. } N^q \text{ as } N \to \infty,$$

where we have used that

(3.27) $$1 - p_1\lambda = 1 - (1 - (4\Theta N^{-q})^{\beta})\cdot(1 - (\Theta/4)(N+1)^{-q}) \sim \text{const.}' \ N^{-q}.$$

This leads readily (by another large deviation argument) to the estimate that (with probability at least $1 - o(N^{-q})$)

(3.28) $$M = O(C/E(|\Lambda_i|)) = O(N^{r-q}).$$

We define W_i^1 to be the event that $|\Lambda_i| > N^{q_1}$, $|\Lambda_{i+1}| > N^{q_1}$ and an obstruction occurs between Λ_i and Λ_{i+1}. We define W_i^2 to be the event that $|\Lambda_i| \leq N^{q_1}$ and Λ_i is involved in an obstruction. Our estimate for (3.24) is

$$P(L^1+L^2 \geq 2) = O(\sum_{j=1}^{2} \sum_{j'=1}^{2} P(W_i^j \text{ and } W_k^{j'} \text{occur for some } i < k \text{ in } \{1,\ldots,O(N^{r-q})\}))$$

(3.29)

$$= O(\sum_{j=1}^{2} \sum_{j'=1}^{2} \sum_{i=1}^{O(N^{r-q})} \sum_{k=i+1}^{O(N^{r-q})} P(W_i^j , W_k^{j'})).$$

We use (3.29) together with the facts that W_i^j and $W_k^{j'}$ are independent if $k - i > 2$ (or if $k - i > 1$ and $j = j' = 1$), that

(3.30) $P(W_i^2,W_k^2) \leq P(|\Lambda_i| \leq N^{q_1},|\Lambda_k| \leq N^{q_1})=(P(|\Lambda_1| \leq N^{q_1}))^2$ if $k-i=1$ or 2,

and that

(3.31) $P(W_i^j,W_k^{j'}) \leq P(|\Lambda_1| \leq N^{q_1})P(W_1^1)$ if $k-i=1$ or 2 and exactly one of (j,j') is 2,

to see that (3.24) would be a consequence of the following limits:

(3.32) $$N^q(N^{r-q})^2(P(W_1^1))^2 \to 0,$$

(3.33) $$N^q(N^{r-q})^2(P(W_2^2))^2 \to 0,$$

(3.34) $$N^q N^{r-q} P(W_1^1, W_2^1) \to 0,$$

(3.35) $$N^q N^{r-q} (P(|\Lambda_1| \le N^{q_1}))^2 \to 0,$$

(3.36) $$N^q N^{r-q} P(|\Lambda_1| \le N^{q_1}) P(W_1^1) \to 0.$$

We note that W_2^2 appears in (3.33) because $P(W_1^2) < P(W_2^2) = P(W_i^2)$ for $i > 2$, where the inequality is due to our working in a half-space.

To prove (3.35), we use the fact that $|\Lambda_1|$ is geometrically distributed with mean given by (3.26) to write

(3.37) $$\begin{aligned} P(|\Lambda_1| \le N^{q_1}) &= 1 - P(|\Lambda_1| > N^{q_1}) \le 1 - (p_1\lambda)^{N^{q_1}} \\ &= 1 - (1 - O(N^{-q}))^{N^{q_1}} = O(N^{q_1 - q}). \end{aligned}$$

Thus the ℓ.h.s. of (3.35) is $O(N^{-2(q-r/2-q_1)})$ and (3.35) will be valid if

(3.38) $$q_1 < q - r/2 .$$

Since it follows from (3.8) that $r - q < q - r/2$, (3.38) is a consequence of (3.20), which proves (3.35). In order to obtain (3.24), it remains to prove (3.32) - (3.34) and (3.36).

We denote by G_i the distance between the rightmost site of Λ_i and the leftmost of Λ_{i+1}. Since $G_i - 1$ is the number of dead sites between Λ_i and Λ_{i+1} and since dead sites occur with probability $1 - \lambda = O(N^{-q})$, it follows that with probability at least $1 - o(N^{-q})$, $G_1 \le 3$ and $G_2 \le 3$ (so each is $O(1)$). In bounding $P(W_1^1)$ or $P(W_2^2)$, we combine this with the estimate (compare (3.15) and (3.17))

(3.39) $$\begin{aligned} &P(\text{every bond between } \Lambda_i \text{ and } \Lambda_{i+1} \text{ is vacant} \mid |\Lambda_i| = \ell,\ G_i = d,\ |\Lambda_{i+1}| = \ell') \\ &\le \exp\left(-\beta \int_0^{\ell-1} \int_{\ell+d}^{\ell+d+\ell'-1} (y-x)^{-2}\, dy\, dx\right) \\ &= \left\{\frac{(d+1)(\ell+\ell'+d-1)}{(\ell+d)(\ell'+d)}\right\}^{\beta} \le \left\{\frac{2(d+1)}{\min(\ell,\ell')}\right\}^{\beta} . \end{aligned}$$

In bounding $P(W_1^1, W_2^1)$, we use the similar estimate,

(3.40) $$\begin{aligned} &P(\text{every bond between } \Lambda_1 \text{ and } \Lambda_2 \text{ and between } \Lambda_2 \text{ and } \Lambda_3 \text{ is vacant} \mid |\Lambda_1| = \ell_1,\ G_1 = d,\ |\Lambda_2| = \ell_2,\ |G_2| = d',\ |\Lambda_3| = \ell_3) \\ &\le \left\{\frac{2(d+1)}{\min(\ell_1,\ell_2)}\right\}^{\beta} \cdot \left\{\frac{2(d'+1)}{\min(\ell_2,\ell_3)}\right\}^{\beta} . \end{aligned}$$

It follows from these arguments and (3.37) that if we define

$Q_i = \min(|\Lambda_i|,|\Lambda_{i+1}|)$, $\tilde{Q}_i = \min(|\Lambda_i|,|\Lambda_{i+1}|,|\Lambda_{i+2}|)$, and write $I(A)$ for the indicator of the event A, then

$$N^{2r-q}\{E(Q_1^{-\beta}I(Q_1>N^{q_1}))\}^2 \to 0 \text{ implies (3.32)}, \tag{3.41}$$

$$N^{2r-q}(E(Q_1^{-\beta}))^2 \to 0 \text{ implies (3.33)}, \tag{3.42}$$

$$N^r E(\tilde{Q}_1^{-2\beta}I(\tilde{Q}_1 > N^{q_1})) \to 0 \text{ implies (3.34)}, \tag{3.43}$$

$$N^r N^{q_1-q} E(Q_1^{-\beta}I(Q_1 > N^{q_1})) \to 0 \text{ implies (3.36)} . \tag{3.44}$$

Note that the limit in (3.42) implies that in (3.41).

Since $|\Lambda_1|$, $|\Lambda_2|$ and $|\Lambda_3|$ are independent geometric random variables with mean $(1-p_1\lambda)^{-1}$, it follows that Q_1 and $\tilde{Q}_1$ are geometric with means $(1-(p_1\lambda)^2)^{-1}$ and $(1-(p_1\lambda)^3)^{-1}$. It can be shown by elementary estimates that if Q is geometric with mean γ^{-1}, then for $\bar{\beta} > 1$ and $z > 1$,

$$E(Q^{-\bar{\beta}}) \leq \gamma\bar{\beta}/(\bar{\beta}-1), \tag{3.45}$$

$$E(Q^{-\bar{\beta}}I(Q>z)) \leq \gamma/((\bar{\beta}-1)(z-1)^{\bar{\beta}-1}). \tag{3.46}$$

We use these bounds and the fact that both $1-(p_1\lambda)^2$ and $1-(p_1\lambda)^3$ are $O(N^{-q})$ to replace (3.41) - (3.44) by

$$2r - q - 2q < 0 \text{ implies (3.32) and (3.33)}, \tag{3.47}$$

$$r - q - q_1(2\beta-1) < 0 \text{ implies (3.34)}, \tag{3.48}$$

$$r + q_1 - q - q - q_1(\beta-1) = r - 2q - (\beta-2)q_1 < 0 \text{ implies (3.36)}. \tag{3.49}$$

The inequality in (3.47) follows from (3.8). The inequality in (3.48) is equivalent to (3.21) which is valid by the original choice of q_1. Since $2q - r > 0$ (by (3.8)), the inequality in (3.49) is automatically valid for $\beta \geq 2$; for $1 < \beta < 2$, it is equivalent to $q_1 < 2(q-r/2)/(2-\beta)$ which follows from (3.38) (itself a consequence of (3.8) and (3.20)) and the fact that $2/(2-\beta) \geq 1$. We have now proved (3.32) - (3.34) and (3.36) which completes the proof of (3.24). In order to complete the proof of (3.19) and hence obtain Prop. 3.3 (and thus Theorem 1.2), it remains to prove (3.25).

We begin our derivation of (3.25) by making several definitions. For k in $\{1,2,\ldots,C = [2N^r]\}$, we define U_k^3 to be the event that for some $i < M$, k is either the rightmost site in $\bar{\Lambda}_i$ or the leftmost in $\bar{\Lambda}_{i+1}$, and there is an unbridged obstruction between $\bar{\Lambda}_i$ and $\bar{\Lambda}_{i+1}$. Thus

$$P(L^3 \geq 1) \leq E(L^3) \leq \sum_{k=[N^{q_1}]+1}^{C-[N^{q_1}]} P(U_k^3) . \tag{3.50}$$

We denote by U_k the event that k is a rightmost or leftmost point of some $\bar{\Lambda}_i$. It is easy to see, since $1 - p_1$ and $1 - \lambda$ are $O(N^{-q})$, that for $k \neq 1$ or C, $P(U_k) = O(N^{-q})$. For $\ell = 1,2,\ldots$, we denote by $U_{k,\ell}$ the event that $\mu_{ij} = 0$ for every i and j with $1 \leq i \leq k-\ell$ and $k + \ell \leq j \leq C$. U_k is independent of $U_{k,2}$ since U_k depends at most on ρ_{k-1}, ρ_k, ρ_{k+1}, $\mu_{(k-1)k}$, and $\mu_{k(k+1)}$.

Since $U_k^3 \subseteq (U_k \cap U_{k,2})$, we have, from (3.50), that

$$P(L^3 \geq 1) = O(N^{-q} \sum_{k=[N^{q_1}]+1}^{C-[N^{q_1}]} P(U_{k,2})). \tag{3.51}$$

For $1 \leq i \leq j \leq C$ and $\ell = 1,2,\ldots$, we define (where again $I(A)$ denotes the indicator of the event A)

(3.52)
$$M_\ell^C(i,j) = \sum_{k=i}^{j} I(U_{k,\ell}) = \text{no. of sites } k \text{ in } \{i,i+1,\ldots,j\} \text{ such that every bond between a living site in } \{1,\ldots,k-\ell\} \text{ and a living site in } \{k+\ell,\ldots,C\} \text{ is vacant.}$$

Thus, by (3.51), in order to prove (3.25) it suffices to show that

$$\lim_{N\to\infty} E(M_2^C([N^{q_1}]+1,\ C - [N^{q_1}])) = 0\ . \tag{3.53}$$

Note that the expectation in (3.53) depends on N, through $C = [2N^r]$, $[N^{q_1}]$ and $\lambda = 1 - (\theta/4)(N+1)^{-q}$ but not through any of the p_j's since $p_j = p_j(\beta)$ for $j \geq 2$ with β fixed and p_1 does not enter into $M_\ell^C(i,j)$ for $\ell \geq 1$. Note further that for fixed $C,\ell,i,j,p_2,p_3,\ldots$, $E(M_\ell^C(i,j))$ is decreasing in λ; as a consequence of this and the fact that $\lambda^2 \lim_j j^2 p_j(\beta) = \lambda^2\beta > 1$ for sufficiently large N, we see that (3.53) will follow from the next proposition, which we state in a general form due to its independent interest.

PROPOSITION 3.4. *In an independent (undirected) percolation model with fixed parameters* $\lambda,p_1,p_2,\ldots$, *define* $M_\ell^n(i,j)$ *as in* (3.52). *If*

$$\lambda^2(\liminf_{j\to\infty} j^2 p_j) > 1, \tag{3.54}$$

then for any fixed $\ell \geq 1$,

$$\sup_{n\geq 1} E(M_\ell^n(1,n)) < \infty \tag{3.55}$$

and

(3.56) $$\lim_{n\to\infty} E(M_\ell^n(i_n,j_n)) = 0 \text{ if } i_n \to \infty \text{ and } n - j_n \to \infty .$$

Proof of Prop. 3.4. Since $E(M_\ell^n(i,j))$ is decreasing in each p_j and increasing in ℓ and since it is independent of p_j for $j < 2\ell$, we may, by standard arguments, assume without loss of generality that

(3.57) $$p_j = 1 - e^{-\beta J_j} \text{ with } J_j = j^{-2} \text{ and } \lambda^2\beta > 1 .$$

With ρ_i denoting the independent site variables of our percolation model, we have, using the notation of (3.52), that

(3.58) $$\begin{aligned} E(M_\ell^n(i',j')) &= \sum_{k=i'}^{j'} P(U_{k,\ell}) = \sum_{k=i'}^{j'} E\Big(\prod_{i=1}^{k-\ell}\prod_{j=k+\ell}^{n} (1-p_{j-i})^{\rho_i\rho_j}\Big) \\ &= \sum_{k=i'}^{j'} E(\exp\{-\beta \sum_{i=1}^{k-\ell}\sum_{j=k+\ell}^{n} J_{j-i}\rho_i\rho_j\}) . \end{aligned}$$

We define

(3.59) $$Y_{m,m'} = \sum_{i=-m}^{-\ell}\sum_{j=\ell}^{m'} J_{j-i}\rho_i\rho_j \quad\text{and}\quad Y_m = Y_{m,m} ,$$

and note that, by translation invariance and the inequality $Y_{m,m'} \geq Y_{\min(m,m')}$,

(3.60) $$\begin{aligned} E(M_\ell^n(i',j')) &= \sum_{k=i'}^{j'} E(\exp(-\beta Y_{k-1,n-k})) \\ &\leq 2 \sum_{k'=\min(i'-1,n-j')}^{[n/2]} E(\exp(-\beta Y_{k'})). \end{aligned}$$

Consequently, to obtain both (3.55) and (3.56), it suffices to show that

(3.61) $$\sum_{m=\ell}^{\infty} E(\exp(-\beta Y_m)) < \infty \quad \text{if} \quad \lambda^2\beta > 1 .$$

To help obtain (3.61), we use the identity

(3.62) $$\sum_{i=-m}^{-\ell}\sum_{j=\ell}^{m} J_{j-i}\rho_i\rho_j = \sum_{k=\ell}^{\infty}\sum_{k'=\ell}^{\infty} \mathcal{J}_{k'+k}\Big(\sum_{i'=\ell}^{\min(k',m)} \rho_{-i'}\Big)\Big(\sum_{j=\ell}^{\min(k,m)} \rho_j\Big),$$

where

(3.63) $$\mathcal{J}_k = (J_k - J_{k+1}) - (J_{k+1} - J_{k+2}) = \frac{6k^2+12k+4}{k^2(k+1)^2(k+2)^2} > 0 .$$

The identity (3.62) can itself be derived from simpler identities such as,

$$(3.64)\qquad \sum_{j=\ell}^{m} J_{j-i}\rho_j = \sum_{k=\ell}^{\infty} (J_{k-i}-J_{k-i+1})\Big(\sum_{j=\ell}^{\min(k,m)} \rho_j\Big) .$$

If it were the case that for some ℓ_m in $\{\ell,\dots,m\}$ and $\varepsilon < 1$,

$$(3.65)\qquad \sum_{j=\ell}^{k} \rho_j \geq (1-\varepsilon)\lambda(k-\ell+1) \text{ and } \sum_{i'=\ell}^{k} \rho_{-i'} \geq (1-\varepsilon)\lambda(k-\ell+1) \text{ for all } k\geq\ell_m,$$

it would then follow that

$$(3.66)\qquad \begin{aligned} Y_m &\geq \sum_{k=\ell_m}^{\infty}\sum_{k'=\ell_m}^{\infty} \mathcal{J}_{k'+k}(1-\varepsilon)^2\lambda^2(\min(k',m)-\ell+1)(\min(k,m)-\ell+1) \\ &\geq (1-\varepsilon)^2\lambda^2 \sum_{k=\ell_m}^{\infty}\sum_{k'=\ell_m}^{\infty} \mathcal{J}_{k'+k}\Big(\sum_{i'=\ell}^{\min(k',m)} 1\Big)\Big(\sum_{j=\ell}^{\min(k,m)} 1\Big) \\ &= (1-\varepsilon)^2\lambda^2 \sum_{i=-m}^{-\ell_m}\sum_{j=\ell_m}^{m} J_{j-i} , \end{aligned}$$

where both the first inequality and the final equality use (3.62).

Now, by elementary estimates (related to the equalities of (3.17) and (3.39)), it can be seen that for $J_j = j^{-2}$,

$$(3.67)\qquad \ell_m = O(m^\delta) \text{ for every } \delta>0 \text{ implies } \lim_{m\to\infty} (\log m)^{-1} \sum_{i=-m}^{-\ell_m}\sum_{j=\ell_m}^{m} J_{j-i} = 1 .$$

Let us define $A_m(\varepsilon,\ell_m)$ as the complement of the event described in (3.65). Then, by (3.66), one has

$$(3.68)\qquad E(\exp(-\beta Y_m)) \leq P(A_m(\varepsilon,\ell_m)) + \exp\Big(-\beta(1-\varepsilon)^2\lambda^2 \sum_{i=-m}^{-\ell_m}\sum_{j=\ell_m}^{m} J_{j-i}\Big).$$

Since $\beta\lambda^2 > 1$, we may choose ε so small that $\beta(1-\varepsilon)^2\lambda^2 > 1$. It then follows from (3.67) that the second term on the r.h.s. of (3.68) would be summable over m, providing $\ell_m = O(m^\delta)$ for every $\delta > 0$. Thus, in order to obtain (3.61) and complete the proof of Prop. 3.4, it suffices to show that

(3.69) there exists a sequence ℓ_m, with $\ell_m = O(m^\delta)$ for every $\delta > 0$, so that for any $\varepsilon > 0$, $\sum_{m=1}^{\infty} P(A_m(\varepsilon,\ell_m)) < \infty$.

Since $A_m(\varepsilon,\ell_m)$ is the complement of (3.65), we have, as in (2.18) above,

$$P(A_m(\varepsilon,\ell_m)) \le 2 \sum_{k=\ell_m}^{\infty} P\Big(\sum_{j=\ell}^{k} \rho_j < (1-\varepsilon)\lambda(k-\ell+1)\Big)$$

(3.70) $$= 2 \sum_{k=\ell_m}^{\infty} P\Big(\sum_{j=\ell}^{k} (1-\rho_j)/(k-\ell+1) > 1 - \lambda + \varepsilon\lambda\Big)$$

$$\le 2 \sum_{k=\ell_m}^{\infty} \exp(-(k-\ell+1)W)$$

where $W = W(1-\lambda+\varepsilon\lambda, 1-\lambda)$, as defined in (2.19), is strictly positive. Performing the last summation of (3.70), we see that

(3.71) $$P(A_m(\varepsilon,\delta_m)) = O(\exp(-\ell_m W)) \text{ as } m \to \infty .$$

Thus any choice of ℓ_m satisfying

(3.72) $$\ell_m/\ell n(m) \to \infty \text{ and } \ell_m/m^\delta \to 0 \text{ for all } \delta > 0,$$

e.g. $\ell_m = [\ell n(m)]^2$, validates (3.69). This in turn yields (3.61) which finishes the proof of Prop. 3.4 and hence of (3.25) and hence of (3.19) and hence of Prop. 3.3 and hence of Theorem 1.2!

BIBLIOGRAPHY

1. Newman, C. M., Schulman, L. S., "One dimensional $1/|j-i|^s$ percolation models: the existence of a transition for $s \le 2$" , Univ. of Arizona/Technion preprint (1985).

2. Aizenman, M., Newman, C. M., "Discontinuity of the percolation density in one dimensional $1/|x-y|^2$ percolation models", in preparation.

3. Newman, C. M. Schulman, L. S., "Infinite clusters in percolation models", J. Stat. Phys., 26 (1981), 613-628.

4. Schulman, L. S., "Long range percolation in one dimension", J. Phys. A. Lett., 16 (1983), L639-L641.

5. Grimmett, G. R., Keane, M., Marstrand, J. M., "On the connectedness of a random graph", Math. Proc. Camb. Phil. Soc., 96 (1984), 151-166.

DEPARTMENT OF MATHEMATICS
UNIVERSITY OF ARIZONA
TUCSON, ARIZONA 85721

DEPARTMENT OF PHYSICS
TECHNION
HAIFA, ISRAEL

Contemporary Mathematics
Volume 41, 1985

On the Shannon-Perez-Moy Theorem

Steven Orey[1]

ABSTRACT. The theorem named in the title is extended in two respects. First the class of processes to which it applies is enlarged. Next it is shown that convergence actually holds in the probability one sense.

Shannon's convergence theorem and its improvements by Mc-Millan and Breiman are well known (see [1] for an excellent exposition and references to the original literature). A generalisation of Shannon's theorem in a different direction was undertaken in Perez [8] and a further extension was given in Moy [6]. Moy's results are studied in Section 1, where we also recall the required notation and relations.

In Moy's work the underlying process is a Markov process with stationary transition probabilities. In [7] we stated without proof an extension applying to certain non-Markovian cases. This will be proved in Section 2. In these results one obtains convergence in L^1 (or in measure). An a.e. convergence result (related to Moy's theorem as Breiman's is to Shannon's) is proved in Section 3.

There have been other attempts to extend Moy's theorem to non-Markovian situations. The result in [9] contains an oversight. Counter-examples were given by Kieffer [5]. More recently Perez [10], [11] has again considered non-Markovian extensions of Moy's theorem, but I have not been able to convince myself of the validity of the arguments.

The extension provided in Section 2 is quite modest. Our underlying process is required to be nearly Markovian in a precise sense. On the other hand I don't believe the desired conclusion will be true much more generally.

If $(B,\mathcal{B})$ is a measurable space and α and β are two probability measures on it the <u>entropy</u> of β with respect to α is defined by

---

1980 Mathematics Subject Classification 60F15, 60F25.
[1]Supported by NSF Grant MCS 83-01080.

$$h_\alpha(\beta) = \begin{cases} \int (\log \frac{d\beta}{d\alpha}(x))\beta(dx) & , \ \beta << \alpha \\ \infty , \text{ otherwise} \end{cases}$$

where $<<$ denotes the relation of absolute continuity. Donsker and Varadhan [2] have given the following important variational characterisation of entropy:

$$(0.1) \qquad h_\alpha(\beta) = \sup_{\Phi \in \mathcal{B}_b} \left(\int_B \Phi \, d\beta - \log \int_B e^{\Phi} d\alpha \right)$$

where $\mathcal{B}_b$ denotes the class of bounded B-measurable functions.

ACKNOWLEDGEMENT: For the argument involving (3.6) I'm indebted to S.R.S. Varadhan, who set it out in a letter to N. Jain, which he kindly made available to me. The argument is implicit in the proof of Theorem 5.4 of [3].

1. MOY'S THEOREM. For every integer n let $(X_n, \mathcal{S}_n) = (X, \mathcal{S})$ be a measurable space. Let

$$(\Omega, \mathcal{F}) = \prod_{n=-\infty}^{\infty} (X_n, \mathcal{S}_n)$$

Let T be the shift on Ω, defined by $(T\omega)_n = \omega_{n+1}$ for every integer n. Let $\xi_i : \Omega \to X$ be the ith coordinate function $\xi_i(\omega) = \omega_i$, and let $\mathcal{F}_{m,n} = \sigma(\xi_i : m < i \leq n)$. Let $\mathcal{M}$ be the set of probability measures on $(\Omega, \mathcal{F})$. For $\nu \in \mathcal{M}$, $\nu_{m,n}$ is the restriction of ν to $\mathcal{F}_{m,n}$. Let $\mathcal{M}_T$ be the shift invariant elements of $\mathcal{M}$, that is those $\nu \in \mathcal{M}$ satisfying $\nu = \nu \circ T^{-1}$. Let $\mathcal{M}'_T$ denote the class of $\mu \in \mathcal{M}$ satisfying

$$\mu(A | \mathcal{F}_{m,n})(\omega) = \mu(T^{-1}A | \mathcal{F}_{m+1,n+1})(T\omega) \ , \ m < n \ , \ A \in \mathcal{F} \ ,$$

for μ - a.e. ω. (When μ is Markovian, $\mu \in \mathcal{M}'_T$ says that the Markov chain has stationary transition probabilities).

Consider now $\nu \in \mathcal{M}_T$ and $\mu \in \mathcal{M}'_T$ and assume

$$\nu_{m,n} = f_{m,n}\mu_{m,n} \quad , \quad -\infty < m \leq n < \infty$$

by which we mean to indicate that $\nu_{m,n}$ is absolutely continuous with respect to $\mu_{m,n}$ with corresponding Radon-Nikodym derivative $f_{m,n}$.

If η is a random variable an Ω, $T\eta$ is the r.v. $\eta \circ T$. For $-\infty < m < n < \infty$ define a probability measure $(\nu\mu)_{m;n}$ on $\mathcal{F}_{m,n}$ by specifying

$$(\nu\mu)_{m;n}(B \times C) = \int_B \mu(C | \mathcal{F}_{m,n-1}) d\nu \ , B \in \mathcal{F}_{m,n-1} \ , \ C \in \mathcal{F}_{m,n}$$

and extending this set function to a probability on $\mathcal{F}_{m,n}$.

By definition

$$(1.1) \qquad \nu_{m,n} = f_{m,n}\mu_{m,n}$$

and one readily finds

$$(1.2) \qquad \nu_{m,n} = \frac{f_{m,n}}{f_{m,n-1}}(\nu\mu)_{m;n}$$

$$(1.3) \qquad \frac{f_{m,n}}{f_{m,n-1}} = T^{n-1}\left(\frac{f_{m-n+1,1}}{f_{m-n+1,0}}\right)$$

Let

$$r_n = \frac{f_{-n,1}}{f_{-n,0}}, \quad n = 0,1\ldots .$$

Define a probability $\varkappa^n$ on $\mathfrak{F}_{-\infty,1}$ by requiring

$$(1.4) \qquad \varkappa^n(B\times C) = \int_B \mu(C|\mathfrak{F}_{-n,0})d\nu, \quad B\in\mathfrak{F}_{-\infty,0}, \ C\in\mathfrak{F}_{1,1}$$

and extending this set function to a probability on $\mathfrak{F}_{-\infty,1}$, where $n = 0,1,\ldots.$ Restricting $\varkappa^n$ to $\mathfrak{F}_{-n,1}$ one has

$$(1.5) \qquad \varkappa^n_{-n,1} = [\nu\mu]_{-n;1}$$

Then, by (1.1),

$$(1.6) \qquad \nu_{-n,1} = r_n\varkappa^n_{-n,1}$$

observe that

$$(1.7) \qquad \text{If } \mu \text{ is Markovian } \varkappa^n = \varkappa^0 = \varkappa, \quad n = 1,2 \ldots ,$$

where the last equality in (1.7) is the definition of $\varkappa$. Finally note that

$$(1.8) \qquad \frac{1}{n}\log f_{0n} = \frac{1}{n}\log f_{00} + \frac{1}{n}\sum_{j=0}^{n-1} T^j(\log r_j) .$$

The following lemma (whose proof rests on elementary arguments about martingales) is the key fact in Moy's paper; related but weaker results were proved earlier by Perez [6].

LEMMA (Moy [6]). Let $(\Omega',\mathfrak{F}')$ be a measurable space and $(\mathfrak{F}_m)$ a sequence of σ-fields increasing up to $\mathfrak{F}'$. Let ν be a probability measure on $\mathfrak{F}'$, ν_k the restriction of ν to $\mathfrak{F}_k$, and let $\varkappa_k$ be a probability measure on

$\mathfrak{F}_k$, with $\varkappa_k$ being the restriction of $\varkappa_j$ to $\mathfrak{F}_k$ for $1 \leq k \leq j < \infty$. Let

$$\nu_k = g_k \varkappa_k \ , \ k = 1,2\ldots .$$

If for some finite constant M

(i) $$h_{\varkappa_k}(\nu_k) = \int \log g_k d\nu \leq M \ , \ k = 1,2,\ldots$$

then as $k \to \infty$

$$g_k \to g \text{ and } \log g_k \to \log g \text{ in } L_1(\nu) \text{ and } \nu\text{-a.e.}$$

If there is a probability measure $\varkappa$ on $(\Omega', \mathfrak{F}')$ such that

(ii) $\varkappa_k$ is $\varkappa$ restricted to $\mathfrak{F}_k$, $k = 1,2\ldots$

then $\nu = g\varkappa$, $h_k(\nu) < \int \log g \, d\nu$.

REMARK: The $(\varkappa_k)$ in the Lemma form a consistent family of measures, so when Kolmogorov's extension theorem applies (e.g. when X is a complete separable metric space with Borel sets $\mathcal{S}$) the existence of $\varkappa$ follows.

THEOREM (Moy) Let μ be Markovian. The condition

(i) $$\sup_n \int \log r_n d\nu < \infty$$

implies that there exist random variables g and Z, with Z invariant (i.e. $Z = TZ$ ν-a.e) such that as $n \to \infty$

(a) $$\log r_n \to \log g \text{ in } L_1(\nu) \text{ and } \nu\text{-a.e.}$$

(b) $$n^{-1} \log f_{on} \to Z \text{ in } \nu\text{-measure}, \int Z d\nu = \int \log g \, d\nu$$

If both (i) and

(ii) $$\int |\log f_{00}| d\nu < \infty$$

then

(c) $$n^{-1} \log f_{on} \to Z \text{ in } L_1(\nu)$$

Furthermore if (i) holds, then

(d) $$\nu = g\varkappa \ , \ h_{\varkappa}(\nu) = \int \log g \, d\nu$$

where $\varkappa$ is defined in (1.7).

PROOF By (1.6) and (1.7) one may write

$$\nu_{-n,1} = r_n \varkappa_{-n,1} .$$

Using (i) in the Lemma one obtains (a); and (d) is also an immediate consequence of the Lemma. Now rewrite the last term in (1.8) as

$$\text{(1.9)} \qquad \frac{1}{n}\sum_{j=0}^{n-1} T^j \log r_j = \frac{1}{n}\sum_{j=0}^{n-1} T^j \log g + \frac{1}{n}\prod_{j=0}^{n-1} T^j(\log g - \log r_j)$$

As $n \to \infty$ the first term on the right converges to an invariant random variable Z both $L_1(\nu)$ and ν - a.e. by the ergodic theorem, and $\int Z d\nu = \int \log g \, d\nu$. By (a) the last term in (1.9) converges to zero in $L_1(\nu)$. Since the first term on the right in (1.8) certainly converges to zero in ν-measure (b) follows and also it becomes clear that (i) and (ii) imply (c).

2. THE NEARLY MARKOVIAN CASE. The assumption that μ is Markovian made in Moy's theorem can be relaxed; one only needs μ nearly Markovian in a sense made precise in hypothesis (NM) below.

For $\omega = (\ldots\omega_{-1}\omega_0,\omega_1\ldots) \in \Omega$ let $\omega^- = (\ldots\omega_{-1},\omega_0)$ and let $\Omega^- = \{\omega^- : \omega \in \Omega\}$.

(NM) For each positive integer n there exists a regular conditional distribution $\mu_1(\,|\mathfrak{F}_{-n,0})$ of ξ_1 given $\mathfrak{F}_{-n,0}$ and

$$\mu_1(\,|\mathfrak{F}_{-n-1,0})(\omega^-) = (1 + a_n(\,,\omega^-))\mu_1(\,|\mathfrak{F}_{-n,0})(\omega^-)$$

for every $\omega^- \in \Omega^-$, where $1 + a_n(y,\omega^-) > 0$ for all $y \in X$ and $\omega \in \Omega^-$, and

$$a_n^* = \sup_{y \in X,\, \omega^- \in \Omega^-} a_n(y,\omega^-)$$

satisfies

$$\sum_{n=1}^{\infty} a_n^* < \infty .$$

REMARK 1. The Markovian case corresponds to $a_n \equiv 0$.

The required convergence of Σa_n^* is equivalent to the product $\Pi(1 + a_n(y,\omega^-))$ converging uniformly in y and ω^- and this implies

$$\text{(2.1)} \qquad \mu_1(\,|\mathfrak{F}_{-n-m,0})(\omega^-) = \prod_{k=n}^{n+m+1} (1 + a_k(y,\omega^-))\mu_1(\,|\mathfrak{F}_{-n,0})(\omega^-)$$

for all positive integers n and m and all $\omega^- \in \Omega$. We may also let $m = \infty$ in (2.1). From the right side of the equation it is clear that in the limit one does obtain a measure for every fixed ω^- ; and from the left side and the martingale convergence theorem one sees that this must be a regular conditional distribution of ξ_1 given $\mathfrak{F}_{-\infty,0}$. This version of the regular conditional distribution will be used below. Now define $\varkappa^{\infty}$ as in (1.4) with n replaced by ∞.

Since μ is not assumed Markovian the conclusion of (1.7) is no longer valid. However (1.1)-(1.6) and (1.8), (1.9) still holds.

Let us verify

$$(2.2) \qquad \varkappa^{n+m} = \prod_{k=n}^{n+m+1} (1+a_k(\omega_1,\omega^-))\varkappa^n , \quad 0 \leq n < \infty , \quad 0 \leq m \leq \infty$$

Start with $A = B \times C$, $B \in \mathfrak{F}_{-\infty,0}$ and $C \in \mathfrak{F}_{1,1}$. Then

$$\varkappa^{n+m}(A) = \int_B \mu_1(C|\mathfrak{F}_{-(n+m),0})d\nu =$$

$$\int_B [\int_C \mu_1(d\omega_1 \mid \mathfrak{F}_{-n,0})(\omega^-) \prod_{k=n}^{n+m+1} (1+a_k(\omega_1,\omega^-))]\nu(d\omega^-)$$

$$= \int_A \prod_{k=n}^{n+m-1} [1+a_k(\omega_1,\omega^-)]\varkappa^n(d\omega) .$$

Now (2.2) follows by a monotone class argument.

Define

$$e_n(\omega) = E^{\varkappa^n}[\prod_{k=n}^{\infty} (1+a_k) \mid \mathfrak{F}_{-n,1}](\omega)$$

$$g_n(\omega) = \frac{r_n(\omega)}{e_n(\omega)} .$$

We first note that

$$\varkappa^{\infty}_{-n,1} = e_n \varkappa^n_{-n,1}$$

as can be seen from (2.2).

Next

$$(2.3) \qquad \nu_{-n,1} = g_n \varkappa^{\infty}_{-n,1}$$

since our assumptions imply that e_n can be chosen to be strictly positive, so that $\varkappa^n_{-n,1} = e_n^{-1} \varkappa^{\infty}_{-n,1}$ and hence (2.3) follows from (1.6).

Also as n approaches infinity

$$(2.4) \qquad \log g_n - \log r_n = -\log e_n \to 0 \text{ uniformly in } \omega \text{, hence in } L_1(\nu)$$

because

$$|e_n - 1| = |E^{\varkappa^n}[\prod_{k=n}^{\infty}(1+a_k) - 1 \mid \mathfrak{F}_{-n,1}]| \leq \sup_{\omega^-, y} |\prod_{k=n}^{\infty}(1+a_k(y,\omega^-)) - 1|$$

and by (NM) the last term approaches zero.

Now we can prove the claim made in [7]

THEOREM. Let μ satisfy (NM). Then all assertions of Moy's Theorem remain correct, where in (d) one takes $\varkappa = \varkappa^{\infty}$.

PROOF. Assumption (i) of Moy's Theorem together with (2.4) implies that there exists an n_o such that

$$\sup_{n \geq n_o} \int \log g_n \, d\nu < \infty .$$

Now use (2.3) and the Lemma from Section 1 to obtain (a) and (d) of the theorem. The rest of the proof needs no change.

3. ν-a.e. CONVERGENCE. The next theorem deals with convergence ν-a.e.

THEOREM. Let μ satisfy (NM). If condition (i) of Moy's theorem holds, then

(e) $\qquad n^{-1} \log f_{on} \to Z \quad \nu\text{-a.e.}$

PROOF: Consider again (1.8). Evidently the first term on the right converges to zero ν-a.e. For the second term use (1.9). Since the first term on the right of (1.9) approaches 0 ν-a.e. it only remains to examine the last term. By (2.4) then everything reduces to showing that

$$\frac{1}{n} \sum_{j=1}^{n} T^j(\log g - \log g_j) \tag{3.1}$$

converges to zero ν-a.e. Let

$$G_N = \sup_{k \geq N} |\log g_k - \log g|$$

and follow Breiman's argument to note that ν-a.e.

$$\overline{\lim_{n\to\infty}} \left| \frac{1}{n} \sum_{k=0}^{n-1} T^k(\log g_k - \log g) \right| \leq \overline{\lim_{n\to\infty}} \sum_{k=0}^{n-1} |T^k(\log g_k - \log g)| \tag{3.2}$$

$$\leq \overline{\lim_{n\to\infty}} \frac{1}{n} \sum_{k=0}^{n-1} T^k G_N$$

Since

$$(3.3)\qquad G_N \le |\log g| + |\sup_k(\log g_k)|$$

it suffices to show that the right side in (3.3) is integrable in order to identify the last term in (3.2) as

$$(3.4)\qquad \lim_{n\to\infty} \frac{1}{n} \sum_{k=1}^{n} T^k G_N = E[G_N]$$

and since as N tends to infinity G_N approaches zero ν-a.e. one concludes from (3.3) and (3.4) that $E[G_N]$ approaches zero. Hence the first member of (3.2) converges to zero ν-a.e.

It remains to be shown that $E[\sup_k |\log g_k|] < \infty$. Then evidently $E[|\log g|] < \infty$.

Let $\log^+ x = (\log x) \vee 0$, $\log^- x = (-\log x \vee 0)$. It will suffice to show that $\sup_k (\log^+ g_k)$ and $\sup_k (\log^- g)$ have finite expectations. By (d) of the Theorem in Section 2, $\nu = g\mu^\infty$ and by (2.3) $(g_k, \mathfrak{F}_{-k,1}, \mu^\infty, 1 \le k \le \infty)$ is a martingale.

To handle $\sup_k (\log^- g_k)$ we adapt Breiman's argument. Let λ be a positive number and set

$$B = [\sup_k \log^- g_k > \lambda] = [\inf_k g_k < e^{-\lambda}]$$

$$B_n = [g_n < e^{-\lambda},\ g_{n-1} > e^{-\lambda}, \ldots, g_0 > e^{-\lambda}]$$

So $B = \cup_n B_n$ and the B_n are pairwise disjoint.

$$\nu(B_n) = \int_{B_n} g_n d\mu^\infty \le \int_{B_n} e^{-\lambda} d\mu^\infty = e^{-\lambda}\mu^\infty(B_n)$$

Hence $\nu(B_n) \le e^{-\lambda}$ and it follows that

$$(3.5)\qquad \int \sup_k (\log^- g_k) d\nu < \infty .$$

Now we bound $E[\sup_k(\log g_k)^+]$. Let $g^+ = \sup_k g_k$, so that $\sup_k(\log g_k)^+ = \log^+(g^*)$. By (0.1)

$$(3.6)\qquad \int \log^+ g^* d\nu \le \log(\int e^{\log^+ g^*} d\mu^\infty) + h_{\mu^\infty}(\nu)$$

$$= \log \int (g^* \vee 1) d\mu^\infty + h_{\mu^\infty}(\nu)$$

By (d) of the theorem

$$h_{\varkappa^\infty}(\nu) = \int g \log g \, d\varkappa^\infty \leq M < \infty$$

and this condition ensures (see [4] Theorem 3.4 in Chapter VII) that $\int g^* d\varkappa^\infty < \infty$. So the last member of (3.6) is finite, as desired.

BIBLIOGRAPHY

[1] P. Billingsley, Ergodic theory and information, Wiley, New York-London-Sindey, 1965.

[2] M.D. Donsker and S.R.S. Varadhan, Asymptotic evaluation of certain Markov process expectations for large time I, Comm. Pure. Appl. Math. 18 (1975) 1-47.

[3] M.D. Donsker and S.R.S. Varadhan, Asymptotic evaluation of certain Markov process expectations for large time II, Comm. Pure Appl. Math 36 (1983) 183-212.

[4] J. Doob, Stochastic Processes, Wiley, New York, 1953.

[5] J.C. Kieffer, A counter example to Peresz's generalisation of the Shannon-McMillan Theorem, Ann. Prob. (1973), 362-364, 4 (1976), 153-154.

[6] Moy, S.C., Generalisations of the Shannon-McMillan Theorem, Pacific J. Math 11, (1961), 705-714.

[7] Orey, S., Large Deviations in ergodic theory, Symposium on stochastic processes, Evanston, 1984, Ed. Cinlar, forthcoming.

[8] Perez, A. Sur la theorie de l'information daus le cas d'un alphabet abstrait, Trans. first Prague conference on Information Theory, 1956, NCSAV, Prague 1957, 209-243.

[9] Perez, A. Extensions of Shannon-McMillan's limit theorem to more general stochastic processes, Trans third Prague Conference on Information Theory, 1962, NCSAV, Prague 1964, 545-574.

[10] Perez, A., Generalisation of Chernoff's result on the asymptotic discernibility of two random processes, Colloquia Mat. Soc. Janos Bolyai 9, (1972), 619-632.

[11] Perez, A., On Shannon-McMillan's limit theorem for pairs of stationary random processes, Kybernetika 19, (1980), 301-314.

Contemporary Mathematics
Volume 41, 1985

HYDRODYNAMICS OF A ONE-DIMENSIONAL NEAREST NEIGHBOR MODEL

Hermann Rost[1] , Maria Eulália Vares [2]

ABSTRACT. We compute the bulk diffusion coefficient, which appears after renormalizing the fluctuations around equilibrium of an infinite system of one-dimensional diffusions interacting by a nearest neighbor potential. Further in this class of models the asymptotic displacement of a tagged particle can be expressed in terms of the hydrodynamic parameters.

1. INTRODUCTION. This paper forms part of a program dedicated to the study of renormalization of stochastic particle systems, more precisely of fluctuations around equilibrium in such processes, which are always supposed to be stationary in space and time. The goal one wants to achieve is twofold: to find examples of particle systems for which a limit after renormalizations exists and to establish general principles, interesting identities, which hold for those limit objects (see e.g. de Masi et al. 1983, Spohn 1983, Brox and Rost 1984). According to that line of research we proceed here, too: for a certain class of one-dimensional systems, given by an Ito-equation in whose drift term the nearest neighbor interaction between particles enters, we proof existence of the limiting fluctuation field; it is an Ornstein-Uhlenbeck process on some Sobolev space $H^d(R)$ with drift operator $\kappa/_2 \cdot \Delta$, where the bulk diffusion coefficient κ is expressed in a simple way - universally for all interactions in the class considered here - by the first two moments of the distance between two neighboring particles at equilibrium (and the diffusion constant of the driving Brownian motion, of course). On the way we have chosen to follow, almost as a byproduct, we extend a typical one-dimensional result, first established by Harris 1965 for Brownian motions with "elastic collisions" (see also Arratia 1983 for a spatially discrete version), to a large class of models: the displacement of a tagged particle between time 0 and t is asymptotically Gaussian with variance $\beta \cdot t^{1/2}$; again we find a universal identity satisfied by the constant β .

The organization of the paper is the following: most time we look at the process of spacings between consecutive particles, even the construction of the

1980 Mathematics Subject Classification: 60 K 35, 80 A 20
[1]Supported by Deutsche Forschungsgemeinschaft (SFB 123)
[2]Supported by CNPq (Grant Nr. 401299-4)

dynamics goes that way (section 2). One sees easily that the dynamics of the point system can be recovered from the dynamics of the spacings and in addition the displacement of a tagged particle. (Our assumptions guarantee that particles never cross each other). In our case, also the history of a tagged particle is expressable by means of the spacings process. In section 3 the hydrodynamics for the spacings process is established. The proofs are almost identical with the ones in Brox and Rost 1984; the only major difference consists in the consequent use of Dirichlet forms, what seems to be the appropriate tool for this type of problems, in which the underlying process X is reversible and only scalar products of the form $E\, g(X_0)\cdot \underset{x}{S} f(X_t)$, S_x spatial shift by x , intervene. We saw that technique first in Albeverio and Høegh-Krohn 1977. Section 4 contains the result on the displacement of the tagged particle; one sees there, how this displacement, if renormalized, is expressed by the fluctuation field of the spacings process. In section 5, finally, we turn back to the fluctuations of the original particle process; its bulk diffusion coefficient is computed and shown to be in agreement with what one expects, knowing the result of Spohn 1983 on point systems with a Hamiltonian of the form $\sum_{i<j} U(x_i - x_j)$.

2. POINT SYSTEMS AND SPACINGS. The constitutive equation for the point system considered in this article reads as follows (formally):

$$dX_i = \frac{dt}{2}\{\phi'(X_{i+1} - X_i) - \phi'(X_i - X_{i-1})\} + dW_i\ , \qquad i \in \mathbb{Z}\ . \tag{2.1}$$

Here $X_i(t)$ denotes the position of particle i at time t ; the W_i are independent Brownian motions. The labelling of particles is so that the $X_i(0)$ form an increasing doubly infinite sequence. We will see (Th. 1) that the assumptions on ϕ guarantee that the ordering of particles is preserved for all times t .

Assumptions on ϕ :

(A.1) ϕ is a positive, decreasing, convex function on R_+ , twice differentiable;

(A.2) $\phi(0) = \infty$, $\phi(\infty) = 0$;

(A.3) $\int_0^1 (\phi'(x))^2 \cdot \exp(-\phi(x))\cdot dx < \infty$.

Thus ϕ is a repulsive pair potential with a certain minimal strength of repulsion, by (A.3) . The conditions (A) are satisfied e.g. if $\phi(x) = a\cdot x^{-b}$ or $\phi(x) = (1+c)\log(1+x)^{-1}$ with a, b, c > 0 . The convexity of ϕ is useful,because it implies stochastic monotonicity properties; we saw it in an article by Yaguchi (1980) on the construction of classical Hamiltonian dynamics.

The evolution of the random measure $\Sigma\, \delta_{X_i(t)}$ does not depend on the particular labelling we choose at the beginning. If however we take the labelling as important, i.e. if we take a particle by calling it X_0, we can introduce the sequence of spacings:

(2.2) $$U_i(t) := X_{i+1}(t) - X_i(t), \quad i \in \mathbb{Z} ;$$

it evolves according to

(2.3) $$dU_i = \frac{dt}{2}\{\phi'(U_{i+1}) - 2\phi'(U_i) + \phi'(U_{i-1})\} + dW_{i+1} - dW_i .$$

For each $\alpha > 0$ define on R_+ the measure

(2.4) $$f_\alpha(u)du := Z(\alpha)^{-1}\cdot\exp(-\phi_\alpha(u))du$$

where $\phi_\alpha(u) = \phi(u) + \alpha u$, $Z(\alpha) = \int \exp(-\phi_\alpha)du$.

We denote by μ_α the independent product measure of the $f_\alpha\cdot du$ on the sequence space R_+^Z. We fix one μ_α, which we call μ for short; it is a candidate for an equilibrium state of the U-process. The corresponding stationary point process on R, i.e. the renewal process to f_α, will be called $\tilde{\mu}_\alpha$ or shortly $\tilde{\mu}$ and is a candidate for an equilibrium of the X-process.

We proceed now to the construction of the U-process at equilibrium. First we define subdynamics associated to a finite interval $I = \{m, m+1, \ldots n\}$ by

(2.5) $$dU_i = \frac{1}{2}\cdot c_i^I(U)dt + dW_{i+1}^I - dW_i^I$$

where $dW_i^I = dW_i\cdot 1_{\{m<i\leq n\}}$,

$$c_i^I(u) = \begin{cases} \phi'(u_{i+1}) - 2\phi'(u_i) + \phi'(u_{i-1}), & m < i < n \\ \phi'(u_{i+1}) - \phi'(u_i), & i = m \\ \phi'(u_{i-1}) - \phi'(u_i), & i = n \\ 0 & \text{else} \end{cases}.$$

PROPOSITION 2.1. To each finite I there exists a stationary Markov process U on R_+^Z which solves (2.5), with law μ at a fixed time; with probability 1 it never hits the set $\{u_i = 0$ for some $i \in Z\}$.

PROOF. Set $H(u) = \sum_m^n \phi(u_i)$. Fix a $K < \infty$ and take the process given by (2.5) restricted to the set $\{H(u) < K\}$, with reflection at the boundary. It has μ as invariant measure, further, by Ito's formula,

(2.6) $$M_t := H(U_t) - H(U_0) - \int_0^t a(U_s)ds$$

is a continuous martingale up to time

(2.7) $$T_K := \inf\{t : H(U_t) \geq K\},$$

if one sets

(2.8) $$a(u) = -\frac{1}{2}\sum_{m}^{n-1}(\phi'(u_i) - \phi'(u_{i+1}))^2 + \sum_{m+1}^{n-1}\phi''(u_i) + \\ + \frac{1}{2}(\phi''(u_m) + \phi''(u_n)) \; ;$$

its quadratic variation equals

(2.9) $$d\langle M\rangle = dt\cdot\sum_{m}^{n-1}(\phi'(U_i) - \phi'(U_{i+1}))^2 \; .$$

Hence, for fixed t, by stationarity one obtains

(2.10) $$E\sup_{s\leq t\wedge T_K} H(U_s) \leq \int H(u)d\mu + t\cdot\int\sum_{m}^{n}\phi''(u)d\mu \\ + \{4t\int\sum_{m}^{n-1}(\phi'(U_i) - \phi'(U_{i+1})^2 d\mu\}^{\frac{1}{2}} \; .$$

This estimate holds uniformly in K for each t and shows at the same time that $\lim_{K\to\infty} T_K = \infty$ a.s., provided we can show that the right hand side is finite. This follows for the first and third term directly from (A), for the second term from the following proposition.

PROPOSITION 2.2. Under (A) the following two functions are integrable in (0,1):

(2.11) (a) $\phi''(u)\exp(-\phi(u))$, (b) $u^{-2}\cdot\exp(-\phi(u))$

PROOF. We start from the identity

$$-\phi'(x)e^{-\phi(x)} = \int_x^1 \phi''\cdot e^{-\phi}\cdot du - \int_x^1(\phi')^2\cdot e^{-\phi}\cdot du + \text{const.}$$

As $x\to 0$, the second term remains bounded by (A.3), the first one increases; hence

$$L := -\lim_{x\to 0}\phi'(x)e^{-\phi(x)}$$

exists; it is positive, possibly infinite. Since $(\phi')^2e^{-\phi}$ is integrable in $(0,\epsilon)$ whereas $-\phi'$ is not, for any $\varepsilon > 0$, L must be zero. This proves the statement (a) .

To show (b), put $g = e^{-\phi/2}$, $h(x) = g(x)/x$. Then one has

$$\int_0^1 h^2(\cdot x)dx = 2\iint_{0<t<x<1} g(t)\, g'(t)\cdot x^{-2}\cdot dt\, dx \leq \\ \leq 2\int_0^1 g'(t)\cdot g(t)\cdot t^{-1}\cdot dt = 2\int_0^1 g'\cdot h\cdot dt \; .$$

If one applies Schwarz' inequality (by (A.3) g' is in $L^2((0,1)\, , dx)$), one obtains

$$\int h^2\, dx \leq c\cdot(\int h^2\, dx)^{\frac{1}{2}}$$

i.e. h is square integrable in (0,1), what proves (b) .

THEOREM 1. There exists a stationary reversible Markov process U on R_+^Z

with law μ at a fixed time. Its semigroup on $L^2(\mu)$ is the strong limit of the semigroups corresponding to the subdynamics (2.3) . The set $\{u : u(i) = 0\}$ is never hit a.s. for each i . The process U solves (2.3) as far as

$$f(U_t) - f(U_o) - \int_o^t Lf(U_s)ds \tag{2.12}$$

is a continuous martingale for each f of the form $f = \tilde{f} \circ \bar{\mu}_I$, $\tilde{f}$ smooth in R_+^I with compact support in its interior ($\bar{\mu}_I$: projection onto R_+^I), where

$$Lf(u) = \frac{1}{2} \sum_i \frac{\partial f}{\partial u_i} \{\phi'(u_{i+1}) - 2\phi'(u_i) + \phi'(u_{i-1})\} + \sum_i \frac{\partial^2 f}{\partial u_i^2} . \tag{2.13}$$

The process U is invariant under spatial shift (i.e. $i \to i+1$) and reflection (i.e. $i \to -i$) .

SKETCH OF PROOF. The I-subdynamics corresponds to the Dirichlet form D^I on $L^2(\mu)$, defined by

$$D^I(f,f) = \frac{1}{2} \int \sum_m^{n-1} \left(\frac{\partial f}{\partial u_i} - \frac{\partial f}{\partial u_{i+1}} \right)^2 d\mu , \tag{2.14}$$

with domain $\mathcal{D}^I$, the set of all $f \in L^2(\mu)$ for which generalized derivatives $\frac{\partial f}{\partial u_i} - \frac{\partial f}{\partial u_j}$ exists for $i,j \in I$ with finite right hand side in (2.14) .

If $I \subset J$, one has $\mathcal{D}^I \supset \mathcal{D}^J$ and $D^I \leq D^J$ on $\mathcal{D}^J$. Define

$$\mathcal{D} := \{f \in \cap \mathcal{D}^I : \sup_I D^I(f,f) < \infty\} \tag{2.15}$$

$$D(f,f) := \sup_I D^I(f,f) \quad \text{on } \mathcal{D} . \tag{2.16}$$

Then just as in Albeverio and Høegh-Krohn one argues that $(D,\mathcal{D})$ is a closed Markovian Dirichlet form on $L^2(\mu)$; its semigroup is the strong limit of the semigroups to $(D^I,\mathcal{D}^I)$.

The rest is routine: the martingale property of (2.12) follows from convergence of the semigroups, its pathwise continuity from a tightness argument. The invariance of U under shift and reflection comes from the same property for the Dirichlet form D . A last word to the polarity of $\{u_i = 0\}$: one proceeds as in Prop. 2.1, showing that the process $t \to \phi(U_i(t))$ is bounded a.s. on each finite time interval.

Construction of the X-process. A necessary property of a solution of (2.1) is that $U_i = X_{i+1} - X_i$ solves (2.3); so we try to recover the driving noise W from the trajectories of the U-process, if the X-process exists. Denote by $DF = F(t) - F(0)$, t fixed; then

$$-DW_{i+1} + DW_i = -DU_i + \frac{1}{2} \int_o^t [\phi'(U_{i+1}) - 2\phi'(U_i) + \phi'(U_{i-1})]ds \tag{2.17}$$

$$DW_o - DW_k = - \sum_o^{k-1} DU_i + \frac{1}{2} \int_o^t [\phi'(U_{-1}) - \phi'(U_o) + - \phi'(U_{k-1}) + \phi'(U_k)]ds \tag{2.18}$$

DW_o is then the Cesaro limit in L^2 of $DW_o - DW_k$ i.e.

(2.19) $$DW_o = \lim_n \frac{1}{n} \{ \sum_{k=1}^{n} \sum_{i=o}^{n-1} -DU_i + \int(\phi'(U_n) - \phi'(U_o))ds$$
$$+ n \cdot \int(\phi'(U_{-1}) - \phi'(U_o))ds\}$$
$$= \lim_n \cdot \sum_{i=o}^{n} -DU_i(1 - i/_n) + \int_o^t(\phi'(U_{-1}) - \phi'(U_o))ds .$$

Hence any reasonable solution of (2.1) must satisfy, if we add the drift

(2.20) $$DX_k = -\lim_n \sum_i \cdot DU_{k+i}(1 - k/_n)^+ , \quad k \in \mathbb{Z}.$$

We take now (2.20) as possible definition of the X-process, as way to obtain X(t) from X(0), if the U-process is shown to exist. This is summarized in the following.

THEOREM 2. There exists a stationary and reversible Markov process X on the space of increasing point sequences with the renewal measure $\tilde{\mu}$ as its law at fixed time; it is a solution to (2.1) . The set $\{x_{i+1} = x_i$ for some $i \in Z\}$ is never hit a.s.

SKETCH OF PROOF. Realize the initial point configuration X(0) according to the equilibrium $\tilde{\mu}$ and label points so that $X_o(0)$ is the smallest positive point; call $U_i(0) = X_{i+1}(0) - X_i(0)$. Then the law of U(0) is equivalent to μ and the U-process can be constructed as in Th. 1 . Define for each fixed t the displacements $DX_k = X_k(t) - X_k(0)$ by (2.20); then X(t) is defined. The fact that this process solves the martingale problem related to (2.1) follows by approximating it by finite subdynamics, which solve a corresponding martingale problem.

3. RESCALING THE PROCESS OF SPACINGS. In this section we compute the fluctuations around equilibrium for the U-process. By these we mean a family N^ε , $0 < \varepsilon < 1$, of processes with values in a suitable space H^{-d} (d = 5 isa possible choice), which evaluated at a test function $\varphi \in H^d$ are defined as

(3.1) $$N^\varepsilon(\varphi,t) = \sqrt{\varepsilon} \cdot \sum_i \varphi(\varepsilon i)(U_i(t\varepsilon^{-2}) - \rho) , \quad t \geq 0 ;$$

where $\rho = \int u_i \, d\mu = \epsilon U_i(t)$.

(H^d is the completion of $C_0^\infty(R)$ under the norm $f \mapsto \{\int f(x)(-\Delta+x^2)^d f(x)dx\}^{\frac{1}{2}}$).

We are interested in the limit behavior of N^ε as $\epsilon \to 0$. The two main results of this section, the "Boltzmann-Gibbs principle" for the $U_i(t)$ and the Ornstein-Uhlenbeck property for the limit process N are obtained essentially the same way as in Brox and Rost 1984 and Rost 1984. To avoid repetition of arguments, after giving some notations, we just state the main auxiliary results, pointing out where a somehow modified argument has to be used.

NOTATIONS. $\chi := \int (u_i - \rho)^2 \cdot d\mu = \int (u - \rho)^2 \cdot f_\alpha \cdot du$ S_k denotes the shift in R_+^Z, i.e. $(S_k u)_i = u_{i+k}$, $i \in \mathbb{Z}$; if f is a function on R^Z, $S_k f$ denotes the function $f \circ S_k$.

δ_i is the configuration with $(\delta_i)_j = 1$ or 0, if $j = i$ or $j \neq i$.

F_I is the σ-field on R^Z generated by the projection π_I, for I a finite subset of Z.

$L^2(I)$: set of μ-square integrable F_I-measurable functions

$M_+(I)$: set of F_I-measurable functions such that there are positive constants $c(j)$, $j \in Z$ with

(3.2a) $$0 \leq f(\cdot + a\delta_j) - f \leq a \cdot c(j), \quad a \geq 0, \quad j \in Z ;$$

(3.2b) $$\int f \, d\mu = 0$$

(The constants $c(j)$ may be chosen equal to 0 for $j \notin I$).

The semigroup of the U-process is denoted by (T_t) ; L is the operator (2.13), defined on smooth cylindrical functions, which agrees there with the generator of (T_t).

PROPOSITION 3.1. The operators T_t preserve monotonicity. If $f \in M_+(I)$ for some I

(3.3) $$0 \leq T_t f(\cdot + a\delta_i) - T_t f \leq a \cdot \max_j c(j), \quad a \geq 0, \quad i \in Z$$

where $c(j)$ are the constants appearing in (3.2).

PROOF. Follows from convexity of ϕ and a coupling (= comparison) argument.

PROPOSITION 3.2. For $f,g \in M_+(I)$ the expression

(3.4) $$E_t(f,g) := \sum_k \int f \cdot T_t \, S_k g \, d\mu$$

is well defined for all $t \geq 0$; the sum at right hand side converges absolutely. One has the estimate

(3.5) $$E_0(f,f) \leq \chi \cdot (\Sigma \, c(j))^2$$

if $c(j)$ are the constants in (3.2). The quadratic forms E_t are decreasing in t.

We write

(3.6) $$E(f,g) := \lim_{t \to \infty} E_t(f,g)$$

PROPOSITION 3.3. For $f \in M_+(I)$ with constants $c(j)$, $\Gamma \subset Z$ and $t \geq 0$ there exists a function $F = F_{\Gamma,t} \in M_+(\Gamma)$ such that

$$(3.7a) \qquad 0 \leq F(\cdot + a\delta_i) - F \leq a \cdot \Sigma\, c(j) , \quad i \in \Gamma , \quad a \geq 0 ;$$

$$(3.7b) \qquad \int g \cdot F \, d\mu = E_t(g,f) , \quad g \in M_+(\Gamma) .$$

The proof of Th. 1 below proceeds now as in the paper mentioned; the following points of the proof deserve some attention:

PROPOSITION 3.4. For each I the set $M_+(I)$ is contained in the domain $\mathcal{D}$ of the Dirichlet form D (see (2.14) and (2.15)) . One has

$$(3.8) \qquad \frac{d}{dt} E_t(f,g) = - \sum_k D(f, T_t S_k g)$$

for $f,g \in M_+(I)$. (Note that T_t maps $\mathcal{D}$ into itself).

PROPOSITION 3.5. Consider $f \in M_+(I)$ with constants $c(j)$. The functions $F_{\Gamma,t}$, constructed in Prop. 3.3 , converges as $t \to \infty$ weakly in $L^2(\Gamma)$ to a function F_Γ . Moreover

$$(3.9) \qquad D(F_\Gamma, g) = \lim_t D(F_{\Gamma,t} , g)$$

for all g in $M_+(I)$, if I is contained together with all neighboring sites in Γ .

For the family F_Γ one has

$$(3.10) \qquad 0 \leq F_\Gamma(\cdot + a\delta_i) - F_\Gamma \leq a \cdot \Sigma\, c(j) , \quad i \in \Gamma , \quad a \geq 0 ;$$

$$\int g \cdot F_\Gamma \, d\mu = E(f,g) , \quad g \in M_+(\Gamma)$$

$$\int g \cdot F_\Gamma \, d\mu = \int g \cdot F_{\Gamma'} \, d\mu , \quad \Gamma \subset \Gamma' , \quad g \in M_+(\Gamma) .$$

PROOF. We just show (3.9) . It is obvious if $g = \tilde{g} \circ \pi_I$ and $\tilde{g}$ is smooth with compact support in the interior of R_+^I . We want to show that these functions are dense in $M_+(I)$ in the Dirichlet norm defined by $\mathcal{D}$. That is basically the following one-dimensional argument: if h is defined on R_+ , h' bounded and

$$(3.11) \qquad h_\epsilon(u) = \begin{cases} h , & u \geq 2\epsilon \\ 0 & u \leq \epsilon \\ \text{linear between} & \end{cases}$$

then $\lim_{\epsilon \to 0} \int (h' - h'_\epsilon)^2 f_\alpha \, du = 0$.

But this is obvious by Prop. 2.2b .

PROPOSITION 3.6. Let $f \in M_+(I)$ for some I . Define $F_n = F_{\Gamma(u)}$ where

$\Gamma(n) = \{-n, \dots n\}$. Then the following is true

(3.12) $$D(F_n , F_{n+1}) = 0 \qquad \text{for all } n ;$$

(3.13) $$\int \left(\frac{\partial F_n}{\partial u_i} - \frac{\partial F_n}{\partial u_j} \right)^2 d\mu = 0 , \quad i,j \in \Gamma(n) .$$

THEOREM 3. If f_o denotes the particular function $f_o(u) = u_o - \rho$, one has

(3.14) $$E(f_o,g) = E_t(f_o,g) = E_o(f,g)$$

for all $g \in M_+(I)$, $t \geq 0$;

(3.15) $$E(f,g) = \alpha(f) \cdot E_o(f_o,g) \qquad \text{for all } g \in M_+(I)$$

where

(3.16) $$\alpha(f) = \chi^{-1} \cdot E_o(f,f_o) .$$

From that theorem, the asymptotic form of the covariances of $U_o(0)$ and $U_i(t)$ can be deduced; we state it in the final form

THEOREM 4. The processes N^ϵ converge weakly in the Skorokhod space of H^{-d}-valued processes to the Gaussian process N with covariance

(3.17) $$\epsilon N(\varphi,0) \cdot N(\psi,t) = \chi \cdot \int \varphi \cdot R_t \psi \, dx , \quad \varphi,\Psi \in H^d$$

where R_τ is the kernel of Brownian motion, formally written as $\exp(\kappa/_2 \cdot \Delta \cdot t)$ and the constant κ , the bulk diffusion coefficient, is given by

(3.18) $$\kappa = \chi^{-1} \cdot \int \phi'(u) \cdot (u-\rho) \cdot f_\alpha \cdot du .$$

The process N is Ornstein-Uhlenbeck with drift $\kappa/_2 \cdot \Delta$ and covariance operator $-\chi \cdot \kappa \cdot \Delta$ and a stationary measure having covariance χ-times identity.

REMARK 1. Th. 4 can easily extended from Lipschitz continuous functions f to (centered) functions in $L^2(I)$ for some I .

Indeed, differences of $M_+(I)$ are dense in $\{f \in L^2(I) : \int f \, d\mu = 0\}$ in L^2-norm; further, for each fixed I ,

(3.19) $$E_o(f,f) \leq c(I) \cdot \int f^2 \cdot d\mu$$

where $c(I) = |\{k : I \cap (I+k) = \emptyset\}|$.

That means, any in L^2-sense approximating Cauchy sequence in $M_+(I)$ of a general f is also Cauchy in E_o-norm and hence E-norm.

REMARK 2. A simple calculation gives the following identities for κ :

(3.20) $$\kappa = - \left(\frac{d\rho}{d\alpha} \right)^{-1} = \chi^{-1}$$

(α : "chemical potential" determining the equilibrium $\mu = \mu_\alpha$) .

4. MASS CURRENT. For a measure- (or distribution-) valued process N on a one-dimensional space it seems natural to introduce the notion of mass current. We will do it in the special case of the Ornstein-Uhlenbeck process (3.20) . The point there is, as we will see in Th. 5 , that the derivation of the mass current and the limit $N^\varepsilon \to N$ can be exchanged.

Notations: $D_t N := N(t) - N(0)$;

$\varphi_n(x) := 1_{\{x \geq 0\}} \cdot (1 - x/_n)^+$;

for a function φ on R we set $S_x \varphi := \varphi(\cdot - x)$.

PROPOSITION 4.1. Let N be as in (3.17) . For each $t > 0$ the sequence $D_t N(\varphi_n)$ is Cauchy in $L^2(P)$. The family

$$J(t,x) := \lim_n D_t N(S_x \varphi_n) , \quad x \in R , \quad t > 0 \tag{4.1}$$

is a Gaussian one. $J(t,\cdot)$ has an a.s. continuous version and satisfies

$$\int J(t,x)\, \varphi'(x)dx = D_t N(\varphi) \tag{4.2}$$

for all φ of compact support and of bounded variation, if φ' is understood in the distributional sense. Further

$$\epsilon(J(t,x))^2 = \chi \cdot (2\kappa/\pi)^{1/2} \cdot t^{1/2} . \tag{4.3}$$

PROOF. The Cauchy property of $D_t N(\varphi_n)$ will be shown in Prop. 4.2 below. If it holds, (4.1) implies

$$J(t,y) - J(t,x) = D_t N(1_{[y,x]}) , \quad y < x ; \tag{4.4}$$

hence (4.2) is valid for step functions of compact support. But (4.4) also gives the continuity of $x \mapsto J(t,x)$ as difference of two Brownian motions describing the equilibria at time t and 0 . Further, by continuity of $J(t,\cdot)$, (4.2) extends to the largerclass of functions.

The last statement is shown as follows:

$$\epsilon DN_t(\varphi_n) = 2\epsilon\, N(0,\varphi_n) \cdot [N(0,\varphi_n) - N(t,\varphi_n)] =$$

$$= 2\chi \cdot \int \varphi_n(x)\{\varphi_n(x) - (g_{\kappa t} * \varphi_n)(x)\}dx$$

$$(g_\alpha(y) := (2\pi\alpha)^{-1/2} \exp(-y^2/2\alpha))$$

As $n \to \infty$ this tends to

$$2\chi \cdot \int_0^\infty dx \left(\int_x^\infty g_{\kappa t}(y)dy \right) = \chi \cdot (2\kappa/\pi)^{\frac{1}{2}} \cdot t^{\frac{1}{2}} .$$

REMARK. $J(t,\cdot)$ is uniquely determined by (4.2) together with the requirement that

$$\lim_{n} \frac{1}{n} \cdot \int_{0}^{x} J(t,y)dy = 0 \; ; \tag{4.5}$$

as we defined J , (4.5) can be shown to hold using (4.1) and (4.2) .

We call J the <u>mass current</u> process. For fixed x , $J(\cdot,x)$ is a non-Markovian process of stationary increments.

PROPOSITION 4.2. For each t , the family $D_t N^{\varepsilon}(\varphi_n)$ is a Cauchy sequence in n , uniformly in $0 < \varepsilon \leq 1$. If we write

$$J^{\varepsilon}(t,x) = \lim_{n} D_t N^{\varepsilon}(S_x \varphi_n) \; , \tag{4.6}$$

the process $J^{\varepsilon}(t,\cdot)$ converges weakly to $J(t,\cdot)$ as $\varepsilon \to 0$.

PROOF. We only have to show the first assertion, what completes also the proof of Prop. 4.1 . The rest is obvious.

a) Define

$$\begin{aligned} a(k,t) &= \epsilon(U_0(0) - \rho)\cdot U_k(t) \; , \\ b(k,t) &= \epsilon\,(U_0(0) - \rho)\cdot\phi'(U_k(t)) \; ; \end{aligned} \tag{4.7}$$

from the definition of dynamics it follows that

$$\frac{d}{dt} a(k,t) = \frac{1}{2}\,[b(k-1,t) + b(k+1,t) - 2b(k,t)] \tag{4.8}$$

$$\sum_k k^2 a(k,t) = \int_0^t \sum_k b(k,s)ds = t$$

since $\sum_k b(k,s) = b(0,0)$: for all s by (3.14) , and $b(0,0) = 1$.

So we get for

$\hat{a}(\delta,t) := \sum_k a(k,t)e^{i\delta k}$ the estimate

$$\hat{a}(\delta,0) - \hat{a}(\delta,t) \leq \frac{1}{2}\cdot\delta^2\cdot t \; . \tag{4.9}$$

b) Let us assume $n < m$; we have for $\psi = \varphi_n - \varphi_m$

$$\hat{\psi}(\delta) := \sum_k (\psi(k)e^{i\delta k} = \{ \frac{1}{n}(1-e^{in\delta}) - \frac{1}{m}(1-e^{im\delta})\}\cdot|1-e^{i\delta}|^{-2} \; . \tag{4.10}$$

Hence

$$\begin{aligned} \epsilon\,[DN_t^1(\psi)]^2 &= \frac{1}{\pi}\int_0^{2\pi} |\hat{\psi}(\delta)|^2(\hat{a}(\delta,0) - \hat{a}(\delta,t))d\delta \\ &\leq \frac{1}{2\pi}\int |\hat{\psi}(\delta)|^2\cdot t\delta^2\cdot d\delta \; . \end{aligned} \tag{4.11}$$

We split the integral into two parts, from 0 to $1/n$ and from $1/n$ to 2π .

On the first interval the integrand is estimated by

$$Ct\cdot\delta^{-2}\cdot\left[\frac{1}{n}\cdot n\delta+\frac{1}{m}\cdot m\delta\right]^2 = 4Ct ,$$

on the second interval by

$$Ct\cdot\delta^{-2}\left(\frac{2}{n}+\frac{2}{m}\right)^2 \leq 16\cdot Ct\cdot n^{-2}\cdot\delta^{-2} .$$

So the integral (4.11) is estimated by $20\cdot Ctn^{-1}$.

This gives for

$$N_t^{\epsilon}(\varphi_n-\varphi_m) = \sqrt{\varepsilon}\cdot N^1_{t\epsilon^{-2}}(\varphi_{n/\varepsilon}-\varphi_{m/\varepsilon})$$

the final estimate:

$$\epsilon[D_t N^{\epsilon}_{\cdot}(\varphi_n-\varphi_m)]^2 \leq 20\cdot C\cdot t\cdot n^{-1} \tag{4.12}$$

from which everything follows.

THEOREM 5. Consider the dynamical system (2.1) under the Palm measure corresponding to $\tilde{\mu}$. Call x_0 the position of the particle initially at the origin. Then for each $t > 0$ the rescaled displacement

$$\sqrt{\varepsilon}\cdot X_0(t\varepsilon^{-2}) \tag{4.13}$$

is asymptotically normal with mean zero and variance $\chi\cdot(2\kappa/\pi)^{\frac{1}{2}}\cdot t^{\frac{1}{2}}$.

PROOF. It suffices to interprete the quantity (4.13) as equal to $J^{\varepsilon}(t,x)$ for $x \in (-\varepsilon,0)$, which we have already done in (2.20) . Convergence in law of the variables $N^{\varepsilon}(\varphi_n)$, for each n , is due to the fact that φ_n can be approximated by smooth functions w.r.t. the norm $\varphi \to \sup_{\varepsilon}(\sum_k \varepsilon\, \varphi^2(\varepsilon k))^{1/2}$.

5. RESCALING THE PARTICLE PROCESS. Consider at time 0 the point process X with law $\tilde{\mu}$. Label the spacings in such a way that U_0 contains the origin. Then the U_i are independent and all - except U_0 - of law $f_{\alpha}du$. Let N^{ϵ} , as in (3.1), be the fluctuation process to the U's (in the limit $\varepsilon \to 0$ it behaves as if it were in complete equilibrium μ) , and $\tilde{N}^{\varepsilon}$ the corresponding object for the point system, i.e.

$$\tilde{N}^{\varepsilon}(\varphi,t) = \sqrt{\varepsilon}\{\sum \varphi(\varepsilon X_i(t\epsilon^{-2})) - (\varepsilon\rho)^{-1}\cdot\int\varphi\,dx\} . \tag{5.1}$$

We will write $N(A,t)$ shortly for $N(1_A,t)$.

From the invariance principle for $x \to N^{\epsilon}_{\cdot}([0,x],0)$ we get immediately the same result for $y \to \tilde{N}^{\varepsilon}([0,y],0)$; in analogy to an idea used by Bahadur 1966 when comparing empirical and quantile processes we have

PROPOSITION 5.1. Uniformly for x in a bounded set

(5.2) $$\tilde{N}^\varepsilon([0,\rho x],0) = -\rho^{-1}\cdot N^\varepsilon([0,x],0) + o(1)$$

holds, where o(1) goes to 0 in prob. as $\varepsilon \to 0$.

COROLLARY. As $\varepsilon \to 0$ $\tilde{N}^\varepsilon(\cdot,0)$ has a limit $\tilde{N}(\cdot,0)$ which is white noise with variance $\tilde{\chi}$ per unit volume:

(5.3) $$\tilde{\chi} = \chi \cdot \rho^{-3}$$

PROPOSITION 5.2. If $L^\varepsilon \to 0$ in prob. as $\varepsilon \to 0$, then, uniformly for y in a bounded set,

(5.4) $$\tilde{N}^\varepsilon([L^\varepsilon,L^\varepsilon+y],0) = \tilde{N}^\varepsilon([0,y],0) + o(1) .$$

PROOF. Follows from the continuity of the limit process $y \to \tilde{N}([0,y],0)$.

PROPOSITION 5.3. Under the dynamics (2.1) for any fixed t , uniformly for x in a bounded set,

(5.5) $$\tilde{N}^\varepsilon([0,\rho x],t) = -\rho^{-1} N^\varepsilon([0,x],t) + o(1) .$$

PROOF. The point in the statement (5.4) is that $\tilde{N}^\varepsilon(\cdot,t)$ and $N^\varepsilon(\cdot,t)$ are related through the labelling introduced at time 0 . But, if we choose

(5.6) $$L^\varepsilon = \varepsilon X_0(t\varepsilon^{-2})$$

Th. 5 gives us $\epsilon(L^\varepsilon)^2 = O(\varepsilon)$, and application of Prop. 5.2 is possible:

(5.7) $$\tilde{N}^\varepsilon([0,\rho x],t) = \tilde{N}^\varepsilon([L^\epsilon,\rho x+L^\epsilon],t) + o(1)$$

and the right hand side is, by Prop. 5.1 and stationarity, equal to $-\rho^{-1}\cdot N^\varepsilon([0,x],t) + o(1)$.

If we summarize these propositions, in view of (5.5) we arrive at

THEOREM 6. The finite dimensional distributions of $\tilde{N}^\varepsilon(\varphi,t)$, $\varphi \in H^5$, $t \geq 0$, converge to those of an Ornstein-Uhlenbeck process $\tilde{N}$ with covariance

(5.8) $$\epsilon\, \tilde{N}(\varphi,0)\, \tilde{N}(\psi,t) = \tilde{\chi} \cdot \int \varphi\, R_t\psi\, dx$$

where $R_t = \exp(-\tilde{\kappa}\cdot t\cdot\Delta/2)$ and

(5.9) $$\tilde{\kappa} = \kappa\rho^2 .$$

REMARKS. a) By (5.3), the bulk diffusion coefficient $\tilde{\kappa}$ of the X-process is equal to

(5.10) $$\tilde{\kappa} = \tilde{\rho}/\tilde{\chi} ,$$

where $\tilde{\rho} = \rho^{-1}$ is the density of the X-process; this is the form obtained by Spohn 1983 in the case of pairwise interacting diffusions through a Hamiltonian $\sum_{i<j} \phi(X_i - X_j)$.

b) The constant $\chi(2\kappa/\pi)^{\frac{1}{2}}$ appearing in Th. 5 can be expressed by the parameters of the X-process as

(5.11) $$\tilde{\chi}\cdot\tilde{\rho}^{-2}(2\tilde{\kappa}/\pi)^{1/2} \quad ,$$

which is in complete agreement with Harris 1965 and Arratia 1983 ($\tilde{\chi} = \tilde{\rho}$ and $\tilde{\chi} = \tilde{\rho}(1-\tilde{\rho})$ for the Poisson, resp. Bernoulli, measure) .

BIBLIOGRAPHY

1. S. Albeverio, R. Hoegh-Krohn, "Dirichlet forms and diffusion on rigged Hilbert spaces", Z.Wahrscheinlichkeitstheorie, 40 (1977), 1-58.

2. R. Arratia, "The motion of a tagged particle in the simple symmetric exclusion system", Ann.Probability, 11 (1983), 362-373.

3. R.R.A. Bahadur, "A note on quantiles in large samples", Ann.math.Statistics, 37 (1966), 577-580.

4. Th. Brox, H. Rost, "Equilibrium fluctuations of stochastic particle systems", Ann.Probability, 12 (1984), 742-759.

5. Th. Harris, "Diffusions with collisions between particles", J.appl. Prob., 2 (1965), 323-338.

6. H. Rost, "A central limit theorem for a system of interacting particles", In: Stochastic space-time problems. Ed. L. Arnold, P. Kotelenez. Dordrecht: Reidel 1985.

7. H. Spohn, "Equilibrium fluctuations for interacting Brownian particles", Preprint, München 1984.

8. H. Yaguchi, "Construction of a one-dimensional classical dynamical system of infinitely many interacting particles", Hiroshima J.Math., 11 (1981), 255-273.

INSTITUTE FOR APPLIED MATHEMATICS
UNIVERSITY OF HEIDELBERG
D-6900 HEIDELBERG
FRG

IMPA
Estrada Dona Castorina 110
22460, Jardim Botânico
Rio de Janeiro
Brasil

Contemporary Mathematics
Volume 41, 1985

DECAY RATES OF GREEN'S FUNCTIONS FOR THE SCHRÖDINGER OPERATOR

Carolyn L. Schroeder

ABSTRACT. Let $G(x,y)$, $x,y \in \mathbf{R}^d$ denote the Green's function of the operator $-\frac{1}{2}\Delta + V$ where V is a continuous, positive periodic function. Let x be fixed and let $|y| \to \infty$ in a specific direction from the point x. The asymptotic exponential decay rate is obtained using large deviations results of Donsker and Varadhan.

1. **INTRODUCTION.** This paper represents one of a growing number of applications of large deviations results of Donsker and Varadhan to the problems of mathematical physics. In its most general setting, the problem we are concerned with is this: suppose $G(x,y)$, $x,y \in \mathbf{R}^d$ is the Green's function of the Schrödinger operator $-\frac{1}{2}\Delta + V$ where V is a continuous, positive function. Fixing x, for which potentials V will there be an asymptotic exponential decay rate as $|y| \to \infty$, and how can this rate be described?

There is a considerable body of literature concerning the related problem of exponential decay of eigenfunctions of the Schrödinger operator (see, for example, [1],[3],[7], and [9]). If one assumes $V(x) \to \infty$ as $|x| \to \infty$ and $\nabla V(x)/V(x) \to 0$ as $|x| \to \infty$, then it can be shown (Carmona and Simon [3]) that the ground state eigenfunction $\phi(y)$ has an exponential decay rate given by the Agmon metric

$$\rho(y) = \inf\left\{\int_o^1 \sqrt{2V(\gamma(s))}|\dot\gamma(s)|ds; \gamma(0) = 0,\ \gamma(1) = y\right\}.$$

Here the infimum is taken over all γ's that are C^1 paths. This is the exponential decay rate in the sense that

$$\lim_{y\to\infty} \frac{-1}{\rho(y)} \ln\phi(y) = 1.$$

It is not difficult to see that the same result is valid if $G(x,y)$, x fixed, replaces $\phi(y)$ in the equation above.

Another case in which one might hope to have a similar result is when V is periodic. This is the case we will discuss here. Our results, whose details will be given in [8], show for periodic potentials V, the Green's function $G(x,y)$ has an asymptotic exponential decay rate that can be described quite explicitly provided that $|y| \to \infty$ in a fixed direction from the point x.

1980 Mathematics Subject Classification. 60J60, 60F10.

To state the result, let T^d denote the d-fold product $S^1 \times \ldots \times S^1$ and let $C^k(T^d)$ denote the set of continuous functions on $\mathbf{R}^d$ with continuous derivatives of orders $\leq k$ which are periodic of period 1 in each variable. Let $\mathcal{F}$ denote the set of functions $f \in C^2(T^d)$ which are positive and which satisfy

$$\int_{T^d} f(x)dx = 1.$$

Suppose $\vec{\xi}$ is a fixed vector in S^{d-1}, the unit sphere in $\mathbf{R}^d$. Let $\Phi_{\vec{\xi}}$ denote the set of d-dimensional vector valued functions ϕ whose components are in $C^1(T^d)$ and which satisfy

$$\nabla \cdot \phi = 0, \ \int_{T^d} \phi(x)dx = \vec{\xi}.$$

Then we prove, for any $x \in \mathbf{R}^d$ and any $\vec{\xi} \in S^{d-1}$,

$$\lim_{s\to\infty} \frac{1}{s} \ln G(x, x + s\vec{\xi})$$

(1)

$$= -2 \inf_{f \in \mathcal{F}} \left[\left(\int_{T^d} \frac{|\nabla f|^2}{8f} + V f dx \right) \left(\inf_{\phi \in \Phi_{\vec{\xi}}} \int_{T^d} \frac{|\phi|^2}{2f} dx \right) \right]^{1/2}.$$

This is accomplished by proving an upper bound for the lim sup and a lower bound for the lim inf of $1/s(\ln G(x, x + s\vec{\xi}))$ as $s \to \infty$, then showing the bounds are equal.

2. THE LOWER BOUND. To prove estimate (1), let us observe that there is no loss of generality in taking the fixed point x to be the origin. In this section, we show

$$\liminf_{s\to\infty} \frac{1}{s} \ln G(0, s\vec{\xi}) \geq \text{ r.h.s. } (1).$$

The first step that is required is a Harnack-type inequality that will allow us to obtain a pointwise lower bound from an averaged one. In this case, the standard methods (see Aizenman and Simon [2]) will easily show that if α is the upper bound of the potential function V and $x \neq 0$, then

$$G(o, x) \geq d(C(d))^{-1}(\epsilon|x|)^{-1} \exp[-\epsilon\sqrt{2\alpha}|x|] \int_{B(x,\epsilon|x|)} G(0, z)dz$$

where $o < \epsilon < 1$ and $C(d)$ is the surface area of the unit sphere in $\mathbf{R}^d$. Since ϵ is arbitrary, it suffices to obtain the lower bound for

$$\liminf_{s\to\infty} \frac{1}{s} \ln \left(\int_{B(s\vec{\xi},\epsilon s)} G(0. z)dz \right).$$

Let E_o denote expectation with respect to Brownian paths starting at the origin. Using the Feynman-Kac formula,

$$\int_{B(s\vec{\xi},\epsilon s)} G(0, z)dz$$

$$= \int_0^\infty dt \ E_o \left[\exp\left[- \int_o^t V(x(u))du \right] \chi_{B(s\vec{\xi},\epsilon s)}(x(t)) \right]$$

where $\chi_{B(s\vec{\xi},\epsilon s)}$ is the characteristic function of $B(s\vec{\xi},\epsilon s)$. For any time T, this is

$$\geq E_o\left[\exp\left[-\int_o^T V(x(u))du\right]\int_o^T \chi_{B(s\vec{\xi},\epsilon s)}(x(t))dt\right]. \tag{2}$$

Now let Q_o denote Brownian motion with drift b and initial distribution δ_o where

$$b(x) = \frac{\nabla f}{2f}(x) + a\frac{\phi}{f}(x),$$

$f \in \mathfrak{F}$, $\phi \in \Phi_{\vec{\xi}}$, and a is a positive number. An application of the Cameron-Martin formula shows that the right hand side of (2) is

$$(3)\qquad = E^{Q_o}\left[\exp\left[-\frac{1}{2}\int_0^T |b|^2(x(u))du - \int_0^T < b(x(u)), d\beta(u) > - \int_0^T V(x(u))du\right]\cdot\int_0^T \chi_{B(s\vec{\xi},\epsilon s)}(x(t))dt\right]$$

where β is a d-dimensional Brownian motion under Q_o.

Let us pick $\epsilon, \epsilon_1 > 0$ with $\epsilon_1 < \epsilon$ and define $S(T,\epsilon,\epsilon_1)$ to be the set of paths $x(\cdot)$ that satisfy

$$|\frac{1}{T}\int_0^T |b|^2(x(u))du - \int_{T^d} |b|^2(x)f(x)dx| < \epsilon_1 \tag{4a}$$

$$|\frac{1}{T}\int_o^T V(x(u))du - \int_{T^d} V(x)f(x)dx| < \epsilon_1 \tag{4b}$$

$$|\frac{x(t)}{t} - a\vec{\xi}| < a\epsilon_1 \;\; \forall t\epsilon(T-(\epsilon-\epsilon_1)T, \infty). \tag{4c}$$

Then $Q_o[S(T,\epsilon,\epsilon_1)] \to 1$ as $T \to \infty$. To see this, consider the diffusion generated on T^d by

$$L = \frac{1}{2}\Delta + \sum_{j=1}^{d} b^j(x)\frac{\partial}{\partial x_j},$$

$b(x) = (b^1(x),\ldots,b^d(x))$. Recalling that in the definition of $b(x)$, $\nabla\cdot\phi = 0$, we see that f satisfies the equation

$$L^*f = \frac{1}{2}\Delta f - \sum_{j=1}^{d}\frac{\partial}{\partial x_j}(b^j(x)f(x)) = 0$$

so $f(x)dx$ is an invariant measure for this process. Then it follows from the ergodic theorem that equations (4) hold Q_o almost surely as $T \to \infty$.

We now make the choice $T = s/a$. In view of equation (4c), every path in $S(s/a,\epsilon,\epsilon_1)$ spends at least time $(\epsilon-\epsilon_1)s/a$ in $B(s\vec{\xi},\epsilon s)$ before $T = s/a$. The expectation in (3) can only be diminished by restricting the integration to those paths that lie in $S(s/a, \epsilon, \epsilon_1)$, so using equations (4) in (3) gives

$$\int_{B(s\vec{\xi},\epsilon s)} G(0,z)dz \geq$$

(5)

$$\exp\left[-\frac{s}{a}\left(2\epsilon_1 + \frac{1}{2}\int_{T^d} |b|^2(x)f(x)dx + \int_{T^d} V(x)f(x)dx\right)\right]$$

$$\cdot(\epsilon-\epsilon_1)^{\frac{s}{a}} \cdot E^{Q_0}\left[\exp\left[-\int_0^{s/a} < b(x(u)), d\beta(u) >\right]\chi_{S(s/a,\epsilon,\epsilon_1)};\right]$$

where $\chi_{S(s/a,\epsilon,\epsilon_1)}$ is the characteristic function of $S(s/a, \epsilon, \epsilon_1)$. We take a logarithm of both sides, divide by s, and let $s \to \infty$. Jensen's inequality followed by the Schwartz inequality shows that the contribution from the last factor vanishes as $s \to \infty$. Simplifying the remaining expression, again recalling that $\nabla \cdot \phi = 0$, and noticing that ϵ, hence ϵ_1 is arbitrary, shows

$$\liminf_{s\to\infty} \frac{1}{s} \ln\left(\int_{B(s\vec{\xi},\epsilon s)} G(0,z)dz\right)$$

$$\geq \begin{matrix} -\inf \\ a>0, f\in \mathfrak{F}\end{matrix}\left(\frac{1}{a}\int_{T^d}\frac{|\nabla f|^2}{8f} + Vf\ dx + a \begin{matrix}\inf\\ \phi\in\Phi_{\vec{\xi}}\end{matrix}\int_{T^d}\frac{|\phi|^2}{2f}\right).$$

Evaluating at the infimum over $a > 0$ completes the proof of

THEOREM. Let $\mathfrak{F}$, $\Phi_{\vec{\xi}}$ be as defined previously. For any $\vec{\xi} \in S^{d-1}$

$$\liminf_{s\to\infty} \frac{1}{s} \ln G(0, s\vec{\xi})$$

$$\geq -2 \inf_{f\in\mathfrak{F}}\left[\left(\frac{|\nabla f|^2}{8f} + Vf\ dx\right)\left(\inf_{\phi\in\Phi_{\vec{\xi}}}\int_{T^d}\frac{|\phi|^2}{2f}dx\right)\right]^{1/2}.$$

3. THE UPPER BOUND. For each $f \in \mathfrak{F}$ and each direction $\vec{\eta} \in S^{d-1}$, let

$$H(\vec{\eta}, f) = \inf_{W\in C^2(T^d)}\int_{T^d}|\vec{\eta} - \nabla W|^2 f\ dx.$$

Then we have

THEOREM. Let $\vec{\xi}$ be a fixed vector in S^{d-1}. Then

$$\limsup_{s\to\infty} \frac{1}{s} \ln G(0, s.\vec{\xi})$$

$$\leq -2 \sup_{\vec{\eta}\in S^{d-1}} \inf_{f\in\mathfrak{F}}\left[\left(\int_{T^d}\frac{|\nabla f|^2}{8f} + Vf\ dx\right)\frac{\langle\vec{\xi},\vec{\eta}\rangle^2}{2H(\vec{\eta}, f)}\right]^{1/2}.$$

To obtain this estimate, we again begin with a Harnack inequality. Away from the origin, $G(0,\cdot)$ is subharmonic so if $x \neq 0$ and $B(x,r)$ is a ball around x of radius r not containing the origin

$$G(0,x) \leq d(C(d))^{-1} r^{-d} \int_{B(x,r)} G(0,z)dz \tag{6}$$

where again $C(d)$ is the surface area of the unit sphere in $\mathbf{R}^d$. We will set $x = s\vec{\xi}$ and take $r = 1$.

Now let λ be any vector having the direction $\vec{\eta} \in S^{d-1}$. Let us find the supremum of $|\lambda|$ over the set of these λ for which

$$\int_{\mathbf{R}^d} e^{\langle \lambda, z\rangle} G(0,z)dz < \infty.$$

By the Feynman-Kac formula

$$\int_{\mathbf{R}^d} e^{\langle \lambda, z\rangle}\, G(0,z)dz = \int_0^\infty E_o\Big[e^{\langle \lambda, x(t)\rangle}\, \exp\Big[-\int_0^t V(x(u))du\Big]\Big]dt. \tag{7}$$

Let P_o denote the Markov process with initial distribution δ_o and generator

$$L = \frac{1}{2}\Delta + \sum_{j=1}^d \lambda^j \frac{\partial}{\partial x_j}$$

where $\lambda = (\lambda^1, \lambda^2, \ldots, \lambda^d)$. Then it follows from the Cameron-Martin formula that the right hand side of (7) is

$$= \int_0^\infty e^{|\lambda|^2 t/2} E^{P_o}\Big[\exp - \int_0^t V(x(u))du\Big]\, dt.$$

By results of Donsker and Varadhan ([4]-[6])

$$\lim_{t\to\infty} \frac{1}{t} \ln E^{P_o}\Big[\exp - \int_o^t V(x(u))du\Big] = -\sigma(\lambda)$$

where

$$\sigma(\lambda) = \inf_{f \in \mathfrak{F}} \Big(\int_{T^d} \Big(\frac{|\nabla f|^2}{8f} + Vf\Big)dx + \frac{|\lambda|^2}{2} - \frac{|\lambda|^2}{2} H(\vec{\eta}, f)\Big).$$

Then the integral (7) will converge provided that $|\lambda|^2/2 < \sigma(\lambda)$. It is easy to see that the supremum of $|\lambda|$ values for which this inequality is satisfied is that $|\lambda|$ such that $|\lambda|^2/2 = \sigma(\lambda)$. But this is equivalent to

$$|\lambda| = 2 \inf_{f \in \mathfrak{F}} \Big[\Big(\int_{T^d} \frac{|\nabla f|^2}{8f} + Vf dx\Big)/2H(\vec{\eta}, f)\Big]^{1/2}.$$

Since this is true for any $\vec{\eta} \in S^{d-1}$, it follows that

$$\limsup_{s\to\infty} \frac{1}{s} \ln \int_{B(s\vec{\xi},1)} G(0,z)dz$$

$$\leq -2 \sup_{\vec{\eta} \in S^{d-1}} \langle \vec{\xi}, \vec{\eta}\rangle \inf_{f \in \mathfrak{F}} \Big[\Big(\int_{T^d} \frac{|\nabla f|^2}{8f} + Vf dx\Big)/2H(\vec{\eta}, f)\Big]^{1/2},$$

which, in view of (6), gives the theorem.

4. THE EQUIVALENCE OF THE UPPER AND LOWER ESTIMATES.

To complete the proof of (1), we first have

THEOREM. For any $f \in \mathcal{F}$

$$\inf_{\phi \in \Phi_{\vec{\xi}}} \int_{T^d} \frac{|\phi|^2}{f}(x)dx = \sup_{\vec{\eta} \in S^{d-1}} \frac{\langle \vec{\xi}, \vec{\eta}\rangle^2}{H(\vec{\eta}, f)}.$$

The idea of this proof is to introduce Lagrange multipliers to enforce the constraints $\phi \in \Phi_{\vec{\xi}}$. Let $(C^1(T^d))^d$ denote the set of $\phi = (\phi^1, \phi^2, \ldots, \phi^d)$ with $\phi^i \in C^1(T^d), i = 1, \ldots, d$. Then

$$\inf_{\phi \in \Phi_{\vec{\xi}}} \int_{T^d} \frac{|\phi|^2}{f}(x)dx$$

(8)

$$= \inf_{\phi \in (C^2(T^d))^d} \sup_{W \in C^2(T^d),\ c \in \mathbf{R}^d} \int_{T^d} \Big(\frac{|\phi|^2}{f} - 2\langle W, \nabla \cdot \phi\rangle$$

$$-2\langle c, \phi\rangle \Big) dx + 2\langle c, \vec{\xi}\rangle.$$

Here a basic mini-max theorem will justify the interchange of "inf" and "sup". Then an integration by parts shows the left hand side of (8) is

$$= \sup_{W \in C^2(T^d),\ c \in \mathbf{R}^d} \inf_{\phi \in (C^1(T^d))^d} \Big[\int_{T^d} \Big(\frac{|\phi|^2}{f} + 2\langle \nabla W, \phi\rangle$$

$$-2\langle c, \phi\rangle \Big) dx + 2\langle c, \vec{\xi}\rangle \Big].$$

The infimum over $\phi \in (C^1(T^d))^d$ is achieved at $\phi = -(\nabla W - c)f$. Substituting this into the expression above shows that it is

$$= \sup_{W \in C^2(T^d),\ c \in \mathbf{R}^d} \int_{T^d} -|\nabla W - c|^2 f(x)dx + 2\langle c, \vec{\xi}\rangle.$$

We re-write this expression as

$$= \sup_{\vec{\eta} \in S^{d-1}} \Big(\sup_{W \in C^2(T^d),\ r \in \mathbf{R}} \int_{T^d} -|\nabla W - r\vec{\eta}|^2 f(x)dx$$

(9)

$$+2r\langle \vec{\xi}, \vec{\eta}\rangle \Big).$$

Clearly the inner supremum is unchanged in W is replaced by rW. The supremum over $r \in \mathbf{R}$ is achieved when

$$r = \langle \vec{\xi}, \vec{\eta}\rangle / \int_{T^d} |\nabla W - \vec{\eta}|^2 f dx.$$

Then equation (9) is

$$= \sup_{\vec{\eta} \in S^{d-1}} \sup_{W \in C^2(T^d)} \langle \vec{\xi}, \vec{\eta} \rangle^2 / \int_{T^d} |\nabla W - \vec{\eta}|^2 f \, dx$$

$$= \sup_{\vec{\eta} \in S^{d-1}} \langle \vec{\xi}, \vec{\eta} \rangle^2 / H(\vec{\eta}, f).$$

This last theorem shows that the upper and lower bounds are equal up to the interchange of "inf" and "sup". This can be justified by a mini-max theorem proved by Sion (Theorem 4.2 [10]), thus completing the proof of (1).

BIBLIOGRAPHY

[1] Agmon, S., "On exponential decay of solutions of second order elliptic equations in unbounded domains," *Proc. A. Pleijel Conf., Uppsala, September, 1979.*

[2] Aizenman, M., Simon, B., "Brownian motion and Harnack inequality for Schrödinger operators," *Commun. Pure Appl. Math.*, **35** (1982), 209-273.

[3] Carmona, R., Simon, B., "Pointwise bounds on eigenfunctions and wave packets in N-body quantum systems," *Commun. Math. Phys.*, **80** (1981), 59-98.

[4] Donsker, M.D., Varadhan, S.R.S., "Asymptotic evaluation of certain Wiener integrals for large time,"" *Proc. Int. Conf., London, April. 1974, Clarendon Press, Oxford, 1975.*

[5] Donsker, M.D., Varadhan, S.R.S., "Asymptotic evaluation of certain Markov process expectations for large time time, I" *Commun. Pure Appl. Math.*, **28** (1975), 1-47.

[6] Donsker, M.D., Varadhan, S.R.S., "On the principal eigenvalue of second order elliptic differential operators," *Commun. Pure Appl. Math.*, **29** (1976), 595-621.

[7] Lithner, L., "A theorem of the Phragmen-Lindelöf type for second order elliptic operators," *Ark. für Mat.*, **5** (1964), 281-285.

[8] Schroeder, C.L., "Green's functions for the Schrödinger operators with periodic potential," (to appear).

[9] Simon, B., "Schrödinger semigroups," *Bull. Amer. Math. Soc.*, **7** (1982), 447-526.

[10] Sion, M., "On general mini-max theorems," *Pac. J. Math.*, **8** (1958), 171-176.

DEPARTMENT OF MATHEMATICS
BOSTON COLLEGE
CHESTNUT HILL, MA. 02167

Contemporary Mathematics
Volume 41, 1985

DIFFUSIONS WITH RANDOM COEFFICIENTS

Scott Schumacher[1]

Since diffusions with random coefficients are the continuous time analogs of random walks in random environments, we will begin by defining these systems and recalling some of what is known about them.

Let $\{\alpha_i\}$, $i \in \mathbb{Z}$, be a doubly infinite sequence of random variables taking values in $[0,1]$. Let $\mathcal{A}$ be the sigma field generated by the sequence $\{\alpha_i\}$. Define a nearest neighbor random walk on $\mathbb{Z}$ by

$$S_0 = 0,$$
$$P(S_{n+1} = k + 1 | S_n = k, \mathcal{A}) = \alpha_k,$$
$$P(S_{n+1} = k - 1 | S_n = k, \mathcal{A}) = 1 - \alpha_k.$$

Then $\{S_n;\ n \geq 0\}$ is called a random walk in a random environment (RWRE).

The recurrence and transience of the RWRE, in the case where the $\{\alpha_i\}$ are an i.i.d. sequence was studied by Solomon (1975). He discovered the first of a number of results which show that the behavior of the RWRE is much different from that of an ordinary random walk: the condition for recurrence is $E(\log((1-\alpha_i)/\alpha_i))=0$ rather than $E(\alpha_i) = \frac{1}{2}$.

In (1975), Kesten, Koslov, and Spitzer studied the limiting behavior of $(S_n - b_n)/a_n$. They found that in certain cases $a_n=n^\beta$, $\beta<\frac{1}{2}$, and that $\beta\to 0$ as you approach the recurrent case. Ritter (1976) was the first to study the behavior of S_n in the recurrent case. He showed that

$$\sup_{m\leq n}|S_m|/(\log n)^2$$

has a limiting distriubtion as $n\to\infty$, but he could not show that

$S_n/(\log n)^2$ has a limit. This problem was solved by Sinai (1982) in the case where the α_i's are bounded away from zero and one. We will state a result (Theorem 1) that generalizes his result.

Having set the stage, we now consider a diffusion X_t on $\mathbb{R}$ satisfying

$$X_0 = 0, \quad dX_t = dB_t + b(X_t)dt$$

or to be precise

$$X_t = B_t + \int_0^t b(X_s)ds,$$

where B_t is a one dimensional Brownian motion starting at zero, and $b(x,\omega)$ is a random process, independent of B_t. Then $\{X_t;\ t\geq 0\}$ is called a diffusion with random coefficients (DWRC).

The analysis of the DWRC in one dimension is made possible by the fact that after a change of scale, $Y_t = \phi(X_t)$, every diffusion is a time change of Brownian motion. $\phi(x)$ (called the "natural scale") is a function from $\mathbb{R}$ to $\mathbb{R}$ such that $\phi(X_t)$ is a continuous (local) martingale.

In our case, if we let

$$\phi(x) = \int_0^x \exp\left(\int_0^y -2b(z)dz\right)dy$$

and apply Ito's formula (see e.g. Durrett (1984), Chapter 2), we find that

$$\phi(X_t) = \int_0^t \phi'(X_s)dB_s.$$

If we let

$$\sigma(t) = \int_0^t [\phi'(\phi^{-1}(B_s))]^{-2}ds$$

and

$$\gamma(t) = \inf\{s>0:\sigma(s)>t\}$$

then there is a Brownian motion, $\tilde{B}_s$, so that

$$\phi(X_t) = \tilde{B}_{\gamma(t)}.$$

Notice that the random drift appears only through its integral $W_z \equiv \int_0^z -2b(y)dy$. Using the above argument, we can think of our DWRC as being defined by the following ingredients:

(1) a random process, W_z, on $\mathbb{R}$

(2) an independent Brownian motion

(3) the natural scale

$$\phi(x) = \int_0^x \exp(W_z)dz$$

(4) the time change

$$\sigma(t) = \int_0^t [\phi'(\phi^{-1}(B_s))]^{-2} ds.$$

This leads us to look for conditions on W_z that determine the limiting behavior of the DWRC. The first of these is:

Theorem 1. Assume $W_{st}/t^{1/\beta} \Rightarrow U_s$ in $D(-\infty,\infty)$ where U_s satisfies:

(i) U_s is non-degenerate

(ii) if $[a,b]$ is any finite interval, $a<b$, and if $\tau = \inf\{s>a: U_s = \inf_{r\in[a,b]} U_r\}$ then U is continuous at τ.

(iii) if $s \in [a,b]$, $s\neq\tau$ then $U_s > U_\tau$.

Then there exists an explicitly defined function $F:D(-\infty,\infty)\to\mathbb{R}$ such that

$$X_t/(\log t)^\beta - F(W_{s \log^\beta t}/\log t) \overset{p}{\to} 0.$$

Remark: notice that Brownian motion and symmetric stable processes (see e.g. Millar (1973)), satisfy these hypotheses as well as a number of other non-Gaussian self similar processes.

Earlier in the introduction we mentioned that Sinai's theorem was a special case of Theorem 1. This follows from the fact that given a sequence of i.i.d. random variables $\{\alpha_i\}$, $\alpha_i \in [0,1]$, there exists a $b(x,\omega)$ so that if we define $\tau_0=0$, and for $n\geq 0$ $\tau_{n+1} = \inf\{t>\tau_n: X_t \notin (X_{\tau_n}-1, X_{\tau_n}+1)\}$ then

$\{X_{\tau_n}; n\geq 0\} \overset{d}{=} \{S_n; n\geq 0\}$ where $\{X_t; t\geq 0\}$ is the DWRC for

$b(x)$, $\{S_n; n\geq 0\}$ is the RWRE for $\{\alpha_i\}$, and $\overset{d}{=}$ stands for equal in distribution. For this $b(x)$,

$$W_n = \sum_{i=1}^{n} \log((1-\alpha_i)/\alpha_i)$$

so under Sinai's assumptions $(E[\log((1-\alpha_0)/\alpha_0)] = 0$, α_i bounded away from zero and one),

$$W_{st}/t^{\frac{1}{2}} \Rightarrow B_s$$

and Theorem 1 asserts that $X_t/(\log t)^2$ converges in distribution. Further, one can show that if α_i is bounded away from zero and one then there are constants $c_1, c_2>0$ such that $c_1 n \leq \tau_n \leq c_2 n$. So $X_{\tau_n}/(\log n)^2$ converges in distribution.

Theorem 1 shows that for a large class of environments

$X_t \sim (\log t)^\beta$. However, this is not the only possibility. In (1979), Papanicolaou and Varadhan showed that if $X_t \in \mathbb{R}^d$ is the diffusion associated with the generator

$$L = \sum_{i,j=1}^{d} \frac{\partial}{\partial x_i} (a_{ij}(x) \frac{\partial}{\partial x_j})$$

where $a_{ij}(x) = a_{ij}(x,\omega)$ is stationary, ergodic, and uniformly elliptic (i.e. there exists $\varepsilon>0$, independent of x and ω, such that for all $\xi \in \mathbb{R}^d$

$$\varepsilon|\xi|^2 \leq \sum_{i,j} \xi_i a_{ij}(x,\omega)\xi_j \leq \varepsilon^{-1}|\xi|^2)$$

then

$$X_t/\sqrt{t} \Rightarrow N(0,\Sigma)$$

for some non-degenerate Σ.

In one dimension, this says that diffusions governed by generators of the form

$$a(x) \frac{d^2}{dx^2} + a'(x) \frac{d}{dx}$$

will display central limit theorem behavior if $\varepsilon<a<\varepsilon^{-1}$. The next result identifies a class of diffusions of the form Brownian motion plus random drift which exhibit similar behavior.

Theorem 2. Suppose W_s is a stationary process satisfying:

(a) $\lim\limits_{t\to\pm\infty} \frac{1}{t} \int_0^t e^{W_s} ds \xrightarrow{P} c>0$

(b) $E(e^{-W_0}) = k<\infty$

and either

(c) $h(s) = E[(e^{-W_s} - k)(e^{-W_0} - k)] \in L^1(\mathbb{R})$

or

(c$'$) e^{-W_s}, e^{-W_r} are independent if $|s-r|>1$. Then

$$X_t/\sqrt{t} \Rightarrow N(0, \frac{1}{kc}).$$

Note that if W_s is an Orstein-Uhlenbeck process $(dW_t = dB_t - \alpha W_t dt)$, it satisfies a, b, and c. c$'$ is designed to avoid second moment assumptions (see the example below).

Having seen environments where $X_t \sim (\log t)^\beta$ and $X_t \sim t^{\frac{1}{2}}$, it is natural to ask if we can have intermediate behavior. The answer is yes, as illustrated by the following example.

Let $\{Y_i\}_{i\in\mathbb{Z}}$ be an i.i.d. sequence with

$$P(Y_0 \le y) = 1 - e^{-\alpha y}, \quad y \ge 0.$$

Define

$$W_s = -Y_i \quad \text{for} \quad i \le s < i + 1.$$

Then W_s is stationary, and conditions a and c' from Theorem 2 are satisfied. Note that for $y \ge 1$ we have

$$P(e^{-W_0} > y) = y^{-\alpha}.$$

If $\alpha > 1$ then $E(e^{-W_0}) < \infty$ and we can apply Theorem 2. If $\alpha < 1$ then condition b fails and the following results shows that the conclusion of Theorem 2 fails also.

Theorem 3. Let W_z be defined as above with $0 < \alpha < 1$, then

$$X_t / t^{\alpha/\alpha+1} \Rightarrow Z$$

where Z is a non-Gaussian random variable which can be constructed from the local times of a Brownian motion and an independent stable random variable.

Comparing Theorems 1 through 3 shows that the growth rate of the DWRC is controlled by the size of the fluctuations in the environment sequence W_z. In the situation of Theorem 1 ($W_{st} \approx t^{1/\beta}$), the fluctuations are typically large, and this provides stretches of "bad" environments that tend to trap the process, slowing the growth rate. In Theorem 2, or more generally in the example, we have

$$\max_{m \le n} e^{Y_m} \sim n^{1/\alpha}.$$

When $\alpha > 1$, this is $o(n)$, Theorem 2 applies and there is normal central limit theorem behavior. When $\alpha < 1$, these maximums are too big and we have abnormal limiting behavior.

REFERENCES

Durrett, R. (1984). Brownian Motion and Martingales in Analysis. Wadsworth, California.

Kesten, H., Kozlov, M.V. and Spitzer, F. (1975). A limit law for a Random Walk in a Random Environment. Compositio Mathematica 30, 145-168.

Millar, P.W. (1973). Exit Properties of Stochastic Processes with Stationary Independent Incurements. Trans. of the Amer. Math. Soc. 178, 459-478.

Papanicolaou, G.D. and Varadhan, S.R.S. (1979). Boundary Value Problems with Rapidly Oscillating Random Coefficients. Colloquia Mathematica Societatis János Bolyai 27, 835-873.

Ritter, G. (1976). Random Walks in a Random Environment. Ph.D. Thesis, Cornell Univeristy.

Sinai, Y.G. (1982). The Limiting Behavior of a One-Dimensional Random Walk in a Random Media. Theory of Probab. and its Applics. 27, 256-268.

Solomon, F. (1975). Random Walks in a Random Environment. Ann. Probab. 3, 1-31.

Contemporary Mathematics
Volume 41, 1985

DISJOINT OCCURRENCES OF EVENTS:RESULTS AND CONJECTURES

J.van den Berg[1]

ABSTRACT.In a recent joint article with H.Kesten the notion "disjoint occurrences of events" has been intoduced. In the present note we give a brief outline of some results and problems presented in the above article and in a joint article with U.Fiebig.

1. INTUITIVE APPROACH. To explain the notion "disjoint occurrences of events" we start with the following example from percolation theory.

(1.1) EXAMPLE. Consider a finite or countably infinite graph G of which each bond b, independent of the other bonds, is open with probability p_b and closed with probability $1-p_b$. Let Q,R,S and T be sites of G. It follows from a correlation inequality of Harris [9] (which is a special case of the well-known FKG inequality [5]) that the event that there exists an open path from Q to R and the event that there exists an open path from S to T are positively correlated, i.e.

(1.2) $$P\{ Q \to R,\ S \to T\} \geq P\{Q \to R\}\, P\{S \to T\},$$

or, equivalently,

(1.3) $$P\{Q \to R\} \mid S \to T\} \geq P\{Q \to R\}.$$

This is intuitively plausible because, roughly speaking, the information that there exists an open path from S to T makes the bonds of G more likely to be open, which, in turn, makes it more likely that there exists an open path from Q to R. One would also intuitively expect that, on the other hand, the probability that there are open paths from Q to R and from S to T which are disjoint (i.e. have no bonds in common) is smaller than the product of the individual probabilities, i.e.

(1.4) $$P\{\text{ there exist disjoint open paths from Q to R and from S to T }\}$$
$$\leq P\{Q \to R\}\, P\{S \to T\}.$$

1980 Mathematics subject classification 60K35;60C05.
[1]Supported by the Netherlands Foundation for Mathematics SMC with financial aid from the Netherlands Organization for the Advancement of Pure Research ZWO.

This is indeed true and follows from more general results in [3] concerning products of so-called NBU distributions. The following proof, related to that of the clutter theorem (see e.g, [4],[8] and [10]) is more direct. First note that, by obvious limit arguments, we may restrict ourselves to the case that G is finite. Let b be a bond of G.Now replace b by two parallel bonds b' and b" which are both, independent of each other and of all the other bonds, open or closed with the same probabilities as the original bond. Suppose that, in the new graph thus obtained, we allow only those paths from Q to R which do not contain b" and those paths from S to T which do not contain b'. Moreover, we apply this splitting operation step by step to all the bonds of G and at each step we allow only those paths from Q to R which contain only unsplit and '-bonds and those paths from S to T which contain only unsplit and "-bonds. One can check that at each step of this splitting process the l.h.s. of (1.4) increases (i.e. strictly increases or remains unchanged) while both factors in the r.h.s. remain unchanged. The inequality now follows from the above argument and from the following facts: When all bonds have been split, we have obtained two separate copies G' and G" of G so that the event in the l.h.s. of (1.4) is replaced by { there exists an open path from Q to R in G' and an open path from S to T in G"}. The probability of this event is, by independence and symmetry, equal to the r.h.s. of (1.4).

REMARK. It is also easy to check that at each step in the above splitting process the l.h.s. of (1.2) decreases so that this method yields an alternative proof of (1.2).

(1.5) EXAMPLE. Consider again the situation in the first example. However, this time we are not interested in open paths but in alternating paths, i.e. paths which, if followed from one endpoint to the other, exhibit an alternating sequence of open and closed bonds.(These and related models have recently been studied in, e.g. [6]). Simple examples show that the analog of (1.2) fails in this new situation, but we believe that the analog of (1.4) holds. This belief is part of a more general conjecture stated in [3] which will be discussed in the next section.

2. FORMAL STATEMENT OF RESULTS AND CONJECTURES. Let $\Omega = S_1 \times \ldots \times S_n$ and $\mu = \mu_1 \times \ldots \times \mu_n$, where S_i is a finite subset of $\mathbb{N}$ and μ_i a probability measure on S_i, $i=1,\ldots,n$. Elements of Ω are denoted by $\omega = (\omega_1,\ldots,\omega_n)$. A set $A \subset \Omega$ is called increasing (decreasing) if $\omega \in A$, $\omega' \in \Omega$, $\omega_i' \geq \omega_i$, $i=1,\ldots,n$ implies $\omega' \in A$.

Harris' inequality, mentioned in section 1, is the following:

(2.1) THEOREM. If $A,B \subset \Omega$ are increasing, then

(2.2) $\mu(A \cap B) \geq \mu(A)\,\mu(B)$.

If $\omega \in \Omega$ and $K \subset \{1,\dots,n\}$ then the cylinder $[\omega]_K$ is defined as the set $\{\omega' \mid \omega'_i = \omega_i, i \in K\}$. K is called the support of the cylinder.

The set $A \mathbin{\square} B$ of disjoint realisations of A and B is defined as

(2.3) $A \mathbin{\square} B = \{\omega \mid \exists K,L \subset \{1,\dots,n\}$ s.t. $K \cap L = \emptyset, [\omega]_K \subset A, [\omega]_L \subset B\}$.

It has been shown in [3] (as a corollary of results concerning products of NBU distributions) that the probability that two increasing events occur disjointly is smaller than the product of their individual probabilities. In fact [3] only gives a proof for the case that $\Omega = \{0,1\}^n$ but that can easily be extended. So we have:

(2.4) THEOREM. If $A,B \subset \Omega$ are increasing, then

(2.5) $\mu(A \mathbin{\square} B) \leq \mu(A)\,\mu(B)$.

In [3] also the following stronger result has been shown:

(2.6) $\mu(\bigcup_{1\leq i\leq k} A_i \mathbin{\square} B_i) \leq (\mu\times\mu)(\bigcup_{1\leq i\leq k} A_i \times B_i)$,

where A_i, B_i are increasing subsets of Ω, $i = 1,\dots,k$.

Coming back to the example in section 1, if we take $\Omega = \{0,1\}^E$, where E is the set of bonds of G and take $\omega_b = 1$ or 0 according as the bond b is open or closed, then it is clear that (1.2) is a special case of Harris' inequality(2.1) and that (1.4) is a special case of theorem (2.4).

An interesting conjecture in [3] is that the monotonicity condition in theorem 2.4 is immaterial, i.e. that (2.5) holds for all events:

(2.7) CONJECTURE. For all $A,B \subset \Omega$

(2.8) $\mu(A \mathbin{\square} B) \leq \mu(A)\,\mu(B)$.

REMARK. Simple examples show that (2.6) does not hold for all events.

In [2] the above conjecture has been investigated and the following theorem has been proved. First we need some definitions: An event A is called convex if ($\omega,\omega'' \in A$, $\omega' \in \Omega$, $\omega_i \leq \omega'_i \leq \omega''_i$, $i = 1,\dots,n$) implies $\omega' \in A$. It is not difficult to show that an event is convex if and only if it is the intersection of an increasing and a decreasing event. An event A is called permutation invariant if permutation of the coordinates of an element of A always yields again an element of A.

Two cylinders C and C' are said to be perpendicular to each other, denoted by $C \perp C'$, if their supports are disjoint.

(2.9) THEOREM. In the following cases inequality (2.8) holds:

a) A and B are both convex.

b) $\Omega = \{0,1\}^n$ and A and B are both permutation invariant.

c) $A = \bigcup_{i \in I} C_i$, $B = \bigcup_{j \in J} C'_j$, where for each $i \in I$ and $j \in J$, C_i and C'_j are cylinders s.t. $C_i \perp C'_j$ or $C_i \cap C'_j = \emptyset$.

d) $A = \bigcup_{i \in I} C_i$, where each C_i is a maximal cylinder of A and for all $i,j \in I$ $\mathrm{supp}(C_i) = \mathrm{supp}(C_j)$ or $\mathrm{supp}(C_i) \cap \mathrm{supp}(C_j) = \emptyset$.

Note that case a) above yields an extension of theorem 2.4 since each increasing event is convex. The proofs of case a) and b) in [2] are based on the splitting method mentioned in section 1. Ref. [2] also shows corollaries and examples of theorem 2.9.

3. APPLICATIONS. Theorem 2.4 appears to be useful in percolation theory. Several inequalities in [1] and [7] follow easily from our theorem, as has been shown in [3]. Some of these inequalities can be strengthened. Our theorem also yields that, for critical bond percolation on the square lattice,

$$P(B_n) \geq 1/(2\sqrt{n}),$$

where B_n is the event that there exist sites at distance $\geq n$ from (0,0) which can be reached from (0,0) by an open path. Applications of inequality 2.6 to first-passage percolation are shown in [11]. Conjecture 2.7 is interesting in itself, but, if it appears to be true, it will probably have applications, e.g. to alternative percolation models (see example 1.5). Example 1.1 in [2] shows an interpretation outside percolation theory.

BIBLIOGRAPHY

1. Aizenman,M. and Newman,C.M., Tree graph inequalities and critical behavior in percolation models, J.Stat.Phys. 36,107-143(1984).

2. van den Berg,J. and Fiebig,U., On a combinatorial conjecture concerning disjoint occurrences of events, preprint(1984).

3. van den Berg,J. and Kesten,H., Inequalities with applications to percolation and reliability, to appear in J.Appl.Prob.(1985).

4. Campanino,M. and Russo,L., An upper bound on the critical probability for the three-dimensional cubic lattice, to appear in Ann.Probab.(1985).

5. Fortuin,C.M. Kasteleyn,P.W. and Ginibre,J., Correlation inequalities on some partially ordered sets, Comm.Math.Phys. 22,(1971),89 - 103.

6. Halley,J.W., Multicomponent and extended percolation models,preprint(1984).

7. Hammersley,J.M., Percolation processes. Lower bounds for the critical probality, Ann.Math.Statist. 28 (1957), 790-795.

8. Hammersley,J.M., Comparison of atom and bond percolation processes, J.Math.Phys.2 (1961),728-733.

9. Harris,T.E., A lower bound for the critical probability in a certain percolation process, Proc.Cambr.Phil.Soc. 56 (1960),13-20.

10.McDiarmid,C., Clutter percolation and random graphs,Math.Progr.Study 13 (1980),17-25.

11.Kesten,H., Aspects of first-passage percolation,Lectures at Saint-Flour summer school in probability(1984).

DEPARTMENT OF MATHEMATICS AND INFORMATICS
DELFT UNIVERSITY OF TECHNOLOGY
JULIANALAAN 132
2628 BL DELFT
THE NETHERLANDS

Contemporary Mathematics
Volume 41, 1985

DUALITY FOR DIRECTED SITE PERCOLATION

John C. Wierman[1]

ABSTRACT. Dual percolation models have played an important role in the determination of critical probabilities in non-directed percolation models on two-dimensional periodic graphs. A dual model for the directed bond percolation model on the square lattice was recently discovered by Dhar, Barma, and Phani, and used to improve the lower bound for the critical probability in the model.

This paper extends the dual model concept to a class of directed site models on planar periodic graphs. In order to construct the dual model, the directed graph is assumed to have at least one entrance bond and one exit bond at each site, and that at each site there are two disjoint sectors of the plane which contain all exits and all entrances respectively. A site percolation model on such a directed graph has a dual model if it contains no counterclockwise circuits and every bond is contained in a "counter-clockwise" doubly-infinite self-avoiding directed path.

1. INTRODUCTION. Percolation theory provides models for the flow of a fluid through a random medium. The medium is represented by a graph G, from which sites and bonds are deleted by a random mechanism. The fluid is imagined to flow through the resulting network of sites and bonds.

The two classical percolation models are the site and bond models. In a bond percolation model, each bond of G is open with probability p, $0 \leq p \leq 1$, independently of all other edges, while all vertices are considered to be open. In the site percolation model, each vertex is open with probability p independently of all other vertices. An edge is open in the site model if both its endpoint vertices are open. In either model, fluid is permitted to flow only through open bonds and sites. These classical models are non-directed: fluid is permitted to flow through any open edge in either direction.

This paper investigates directed percolation models, in which each edge permits fluid to flow through in only one fixed direction. As in the classical models, the random mechanism may be associated with either the vertices or the edges of the graph.

1980 Mathematics Subject Classification: 60K35
[1]Supported by National Science Foundation grant no. MCS-8303238.

As in the non-directed models, most study has been concentrated on the square lattice. The square lattice has vertex set $\mathbb{Z}^2$, and an edge connects vertices x and y if $\|x-y\| = 1$. The edges parallel to the x- and y-axes will be called horizontal and vertical, respectively. In the square lattice directed percolation model, each horizontal bond is directed to the right, and each vertical bond is directed upward.

A major problem in percolation theory is the determination of the critical probability. Let $P(p)$ denote the probability that there exists an infinite open path beginning at the origin when the parameter of the model has value p. The critical probability is

$$p_{DP} = \inf \{0 \leq p \leq 1: P(p) > 0\}.$$

p_{DP} represents a threshold above which fluid penetrates the medium represented by the lattice.

The exact critical probability is not known for any non-trivial directed percolation model. For any graph, Hammersley (1961) showed that $p_{DP}(b) \leq p_{DP}(s)$, where $p_{DP}(b)$ and $p_{DP}(s)$ denote the bond and site model critical probabilities, respectively. For the square lattice, Hammersley (1957) proved that $.6037 \leq p_{DP}(b) \leq .8496$ by enumeration of self-avoiding paths and circuits. Mauldon (1961) analyzed a sequence of "pseudopercolation" processes which dominated the directed percolation process on the square lattice, to show $p_{DP}(b) \geq .6198$. Bishir (1963) found that $p_{DP}(s) \geq 2/3$ for the square lattice, by applying similar methods. Holley and Liggett (1975), considering an interacting particle system which is equivalent to the square lattice directed site percolation model, proved that $p_{DP}(s) \geq .6774$. A random walk argument of Gray, Smythe, and Wierman (1980) improved the lower bounds for the square lattice to $p_{DP}(b) \geq .6231$ and $p_{DP}(s) \geq .6882$.

While duality has been used extensively with non-directed percolation models, it was not until 1981 that Dhar, Barma, and Phani introduced the first dual model for directed percolation. Their diode-resistor model, described in section 2, is the dual for the square lattice directed bond percolation model. Analysis of the dual model provided a sequence of lower bounds which converge to the critical probability. The fourth bound in the sequence yields $p_{DP}(b) \geq .6298$. Numerical estimates of the critical probability are clustered between .632 and .645. [See Kertesz and Vicsek (1980), Kinzel and Yeomans (1981), Dhar and Barma (1981), and Blease (1977).] Close upper bounds are not available. Durrett (1984) improved Hammersley's bound slightly to $p_{DP}(b) \leq .84$.

This paper generalizes the duality approach of Dhar, Barma, and Phani (1981) to the setting of directed site percolation models on planar periodic graphs.

2. THE DIODE-RESISTOR MODEL. To familiarize the reader with the concept of a dual model for a directed percolation model, we first describe the diode-resistor model introduced by Dhar, Barma, and Phani (1981).

In the diode-resistor model, each edge of the square lattice is a "two-way" bond with probability p, $0 \le p \le 1$, and a "one-way" bond otherwise, independently of all other edges. A one-way bond is directed upward if it is a vertical edge, and is directed to the left if it is a horizontal edge. Dhar, Barma, and Phani describe the model in terms of electrical networks, referring to one-way bonds as "diodes" and to two-way bonds as "resistors".

A path from x to y is an alternating sequence of vertices and edges $x = v_0, e_1, v_1, \ldots, e_n, v_n = y$, where e_i is either a two-way bond or is a one-way bond directed from v_{i-1} to v_i. The wedge filled from v, denoted W(v), is the set of all sites w such that there is a path from v to w. Note that, regardless of the designations of edges as one-way or two-way bonds, for any vertex $v = (v_1, v_2) \in \mathbb{Z}^2$, W(v) contains the entire quadrant $\{y = (y_1, y_2) \in \mathbb{Z}^2 : y_1 \le v_1, y_2 \ge v_2\}$.

For configurations containing sufficiently many two-way bonds, W(v) contains the entire lattice $\mathbb{Z}^2$. The critical probability of the diode-resistor model, denoted p_{DR}, is defined by

$$p_{DR} = \inf \{p \in [0,1]: P_p[W(v) = \mathbb{Z}^2] > 0\},$$

where P_p is the probability measure on configurations under which each edge is a one-way bond with probability p independently. Due to translation invariance of the model, p_{DR} is independent of the choice of source site v, so henceforth we take (0,0) as the source site.

If W(v) is not the entire lattice, then there is an infinite set of one-way bonds which prevents complete filling. This set of edges, which have endpoints in both W(v) and $W(v)^c$, is called the boundary of W(v). When the relationship between the diode-resistor and directed percolation models is established, the boundary of the filled wedge will correspond to a open path in the directed percolation model.

The relationship between the two models makes use of the concept of the dual graph of a planar graph. (See Figure 1.) For any planar graph G, there exists a dual graph G*, constructed by placing a site of G* in each face of G, and connecting two such sites of G* if and only if the two corresponding faces of G share a common edge. This establishes a one-to-one correspondence between edges of G and G*. The square lattice is self-dual: its dual graph is a square lattice with sites in $\mathbb{Z}^2 + (\frac{1}{2}, \frac{1}{2})$.

The specification of a diode-resistor model on the square lattice S induces a directed percolation model on its dual square lattice S* through the one-to-one correspondence between edges. An edge of S* is an open bond if it

FIGURE 1. The diode-resistor model. In (a), solid circles and lines represent the square lattice, and open circles and dashed lines represent the dual graph, which is also a square lattice. The diode-resistor model is shown in (b). Single and double arrows on edges represent one-way and two-way edges respectively. The dashed line indicates the open path in the associated directed percolation model, which corresponds to the boundary of the filled region. s denotes the fluid source site.

(a)

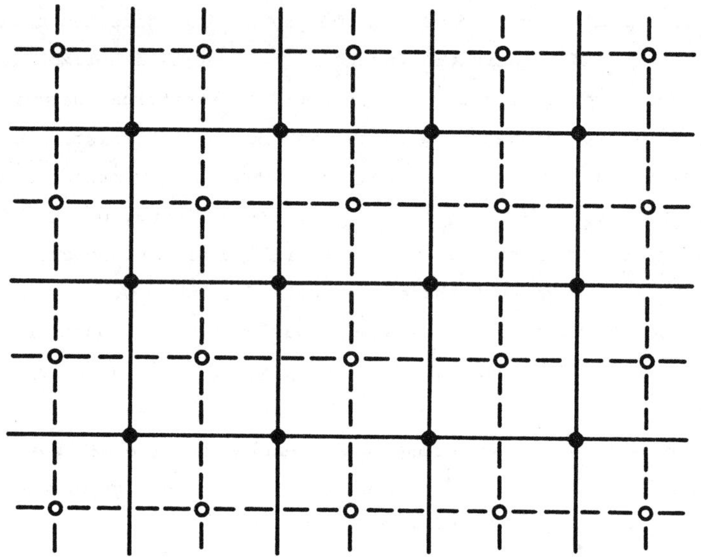

(b)

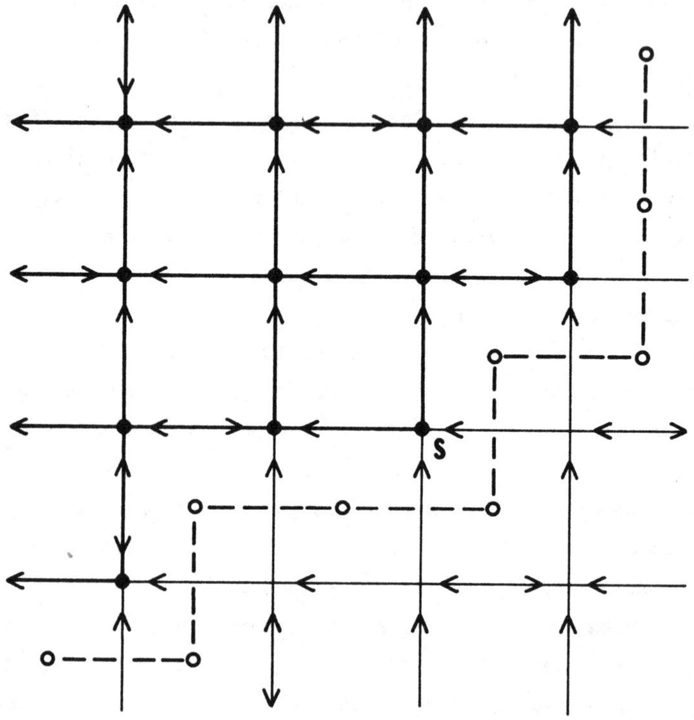

crosses a one-way bond of S, and is a closed bond if it crosses a two-way bond. Since designation of bonds as one-way or two-way are made independently in the diode-resistor model, so are the designations as open and closed in the resulting model. If we consider S* to be a directed graph, with horizontal bonds directed to the right and vertical bonds upwards, then we have a directed percolation model on S*. The boundary of the filled wedge W(v) in the diode-resistor model corresponds to edges which cross a doubly-infinite open path in the directed model. (See Figure 1.) Note that the directions assigned to edges can be reversed to generate a different directed percolation model, in which the boundary of W(v) corresponds to the same doubly-infinite open path with the opposite direction.

With this connection between the diode-resistor and directed models, either the directed model has an infinite open path or the diode-resistor model completely fills $\mathbb{Z}^2$. Thus, either $1 - p \geq p_{DP}$ or $p \geq p_{DR}$, so

$$p_{DP} + p_{DR} = 1.$$

If there is not complete filling in the diode-resistor model, the asymptotic slope of the boundary may be defined as

$$m(p) = \text{a.s.} \lim_{n\to\infty} \frac{a_n}{n}$$

where

$$a_n = \min \{x \in \mathbb{Z} : (n,x) \in W((0,0))\}.$$

Such an asymptotic slope can be shown to exist using results from the theories of sub-additive processes and first-passage percolation. (See Wierman (1983) and Smythe and Wierman (1978).) A similar slope exists in the third quadrant, which by symmetry is $1/m(p)$. If the asymptotic slope m(p) is strictly less than one, then the boundary is "unstable", and the entire plane is filled from the origin with probability one.

Dhar (1982) found upper bounds for p_{DR}, which then provided lower bounds for p_{DP}. The method uses the concept of "backflow paths". For each integer $k \geq 0$, a path r is a k - backflow path if from every vertex $v_i = (x_i,y_i) \in r$, $x_j \geq x_i - k$ for all $j \geq i$. Letting

$$h_n(k) = \{h \in \mathbb{Z} : \exists \text{ a } k \text{ - backflow path from } (0,0) \text{ to } (n,h)\},$$

the limiting slope for the boundary of the wedge filled by k - backflow paths is

$$m_k(p) = \text{a.s.} \lim_{n\to\infty} \frac{h_n(k)}{n}.$$

For all k, $m_k(p) \geq m(p)$, and in fact $m_k(p)$ converges to m(p) as k increases. Upper bounds for p_{DR} are obtained by finding p_k for which

$m_k(p_k) = 1$. Then $m(p_k) \geq 1$ so $p_{DR} \leq p_k$. The sequence $\{p_k\}$ converges to the critical probability p_{DR}. Dhar computed p_k for $k = 0,1,2$ and 3, which led to the lower bound of .6298 for the critical probability of the square lattice directed percolation model.

3. DEFINITIONS AND STATEMENT. A directed graph D consists of a set of vertices V(D) together with a set E(D) of ordered pairs of vertices, which are called edges. An edge $(v,w) \in E(D)$ is said to be directed from v to w, and referred to as an "exit" from v and an "entrance" to w. [In this paper, given a pair of vertices, v and w, at most one of the pairs (v,w) and (w,v) may be in E(D).] $G \equiv G(D)$ denotes the non-directed graph with edges $\{v,w\}$ for which $(v,w) \in E(D)$.

A graph is periodic in $\mathbb{R}^d$ if its edge and vertex sets are invariant under translation by each of a set of d linearly independent vectors (which may always be taken to be the coordinate vectors), is connected, every site is an endpoint of at most z edges for some $z < \infty$, and each compact subset of $\mathbb{R}^d$ contains at most finitely many vertices. For a periodic graph G, the vertex set V(G) may be partioned into N disjoint sets V_i, $i = 1,2,\ldots,N$, such that (with $\xi_1,\ldots,\xi_d$ denoting the coordinate vectors of $\mathbb{R}^d$) for every $i=1,\ldots,N$,

$$v \in V_i \text{ if and only if } v + \sum_{j=1}^{d} k_j \xi_j \in V_i \text{ for any } k_j \in \mathbb{Z},\ 1 \leq j \leq d.$$

$\{V_i, i = 1,2,\ldots,N\}$ is then called a periodic partition. Each V_i is called a periodic class of vertices. For further discussion of periodic graphs, see Kesten (1982), where significant results are obtained for the percolative behavior of site models on periodic two-dimensional (non-directed) graphs.

A path in the directed graph D is an alternating sequence $v_0,e_1,v_1,e_2,\ldots,e_n,v_n$ of vertices $\{v_i\}$ and edges $\{e_i\}$ such that $e_i = (v_{i-1},v_i) \in D$ for all $i = 1,\ldots,n$. The vertices v_0 and v_n are the initial and terminal vertices of the path.

A doubly-infinite path in D is a doubly-infinite alternating sequence $\ldots v_{-1},e_0,v_0,e_1,v_1,\ldots$ of vertices and edges such that every finite segment of the sequence is a path in D. A path is self-avoiding if all its vertices are distinct. A doubly-infinite self-avoiding path r in a planar graph divides the plane into two connected components, one of which, C(r), lies on the left side of the directed path as one proceeds along the path. If C(r) lies entirely in a half-plane, then r is called a "counterclockwise" path.

For each site v in a directed graph D, let P(v,p) denote the probability that there exists an infinite open directed path in D with initial site v. $P(v,\cdot)$ is the percolation probability function at v. The percolation probability function is identical for all sites in each periodic class V_i of a periodic graph, but may differ for sites of different periodic classes.

If for every pair of periodic classes V and W, there exist $v,v' \in V$ and $w,w' \in W$ such that there exists a directed path from v to w and a directed path from v' to w', then $P(s,p) > 0$ if and only if $P(t,p) > 0$ for any sites s and t in D. In this case, we define the critical probability for D by

$$p_{DP}(D) = \inf \{p \in [0,1] = P(v,p) > 0\}$$

in terms of a specific site $v \in D$, noting that p_{DP} is in fact independent of the choice of v. $p_{DP}(D)$ represents a threshold above which fluid penetrates the medium represented by D.

We will require the graph D to be "directed at each site", that is, there is a preferred direction at each site, but not necessarily a common preferred direction. For each site $v \in D$, there exist two disjoint sectors $O(v)$ and $I(v)$, defined by

$$O(v) = \{x \in \mathbb{R}^2: x = v + (r\cos\theta, r\sin\theta), r \geq 0, \theta_1 \leq \theta \leq \theta_2\}$$
$$I(v) = \{x \in \mathbb{R}^2: x = v + (r\cos\theta, r\sin\theta), r \geq 0, \gamma_1 \leq \theta \leq \gamma_2\}$$

such that $O(v)$ is the minimal sector containing all exits from v, and $I(v)$ is the minimal sector containing all entrances to v. (See Figure 2.) As a convention, we choose $0 \leq \theta_1$ and $\theta_2 < \gamma_1 < \theta_2 + 2\pi$. Note that the angles θ_i and γ_i may depend on the vertex v, but are identical for all v in a periodic class of vertices. For paths p, p_1 and p_2 from v, if p must be crossed in moving counterclockwise from p_1 to p_2, we say that p lies between p_1 and p_2.

In the general setting, as in the diode-resistor model, a filling model will be defined, and called the dual of a directed percolation model if there is incomplete filling if and only if there is an infinite open path in the directed model. If this is the case, the critical probabilities of the two models sum to one, so upper bounds for one critical probability provide lower bounds for the other.

In the diode-resistor model discussion, the filled wedge lay to the left of the directed path as it is travelled. By reversing the direction of all the one-way bonds in the diode resistor model, one obtains another diode-resistor model which lays to the right of the directed path. In the general setting, when the directed graph is not symmetric, there may be distinct "right-dual" and "left-dual" filling models. It is also possible that a left-dual, but no right-dual, exists, or vice versa. The following theorem states conditions insuring existence of a left-dual filling model. The right-dual for a directed model is the left-dual of the directed percolation model obtained by reversing the direction of every bond.

THEOREM. Let D be a periodic planar directed graph. The directed site percolation model on D has a left-dual filling model if the following hold:

(a) For every pair of periodic classes, V and W, there exist $v,v' \in V$ and

FIGURE 2. A portion of a directed graph D is shown in (a), containing a vertex v. (a) indicates the sectors O(v), R(v), and I(v), and the face F(v). Arrows in (b) are the exits from v when v is a restricted site. Arrows in (c) are the exits from v when v is an unrestricted site.

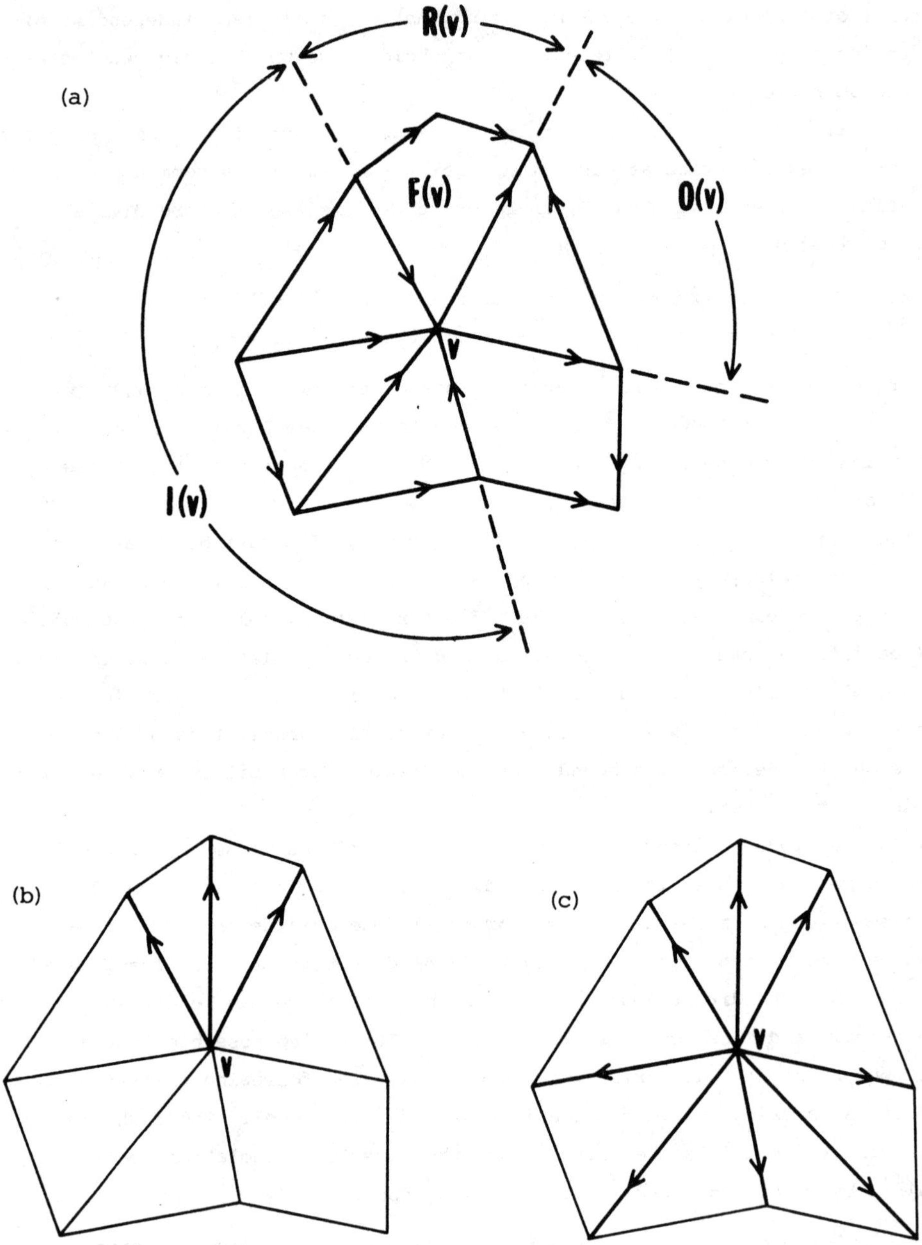

$w, w' \in W$ such that there exists a directed path from v to w and a directed path from w' to v'.

(b) Each site of D has at least one exit and one entrance.

(c) For each site of D, there exists two disjoint sectors, one containing all entrances, the other all exits.

(d) D contains no counterclockwise circuits.

(e) Every edge in D is in a counter-clockwise doubly-infinite self-avoiding directed path.

Condition (b) is effectively no restriction on the graph D, since any directed graph may be modified to satisfy (b) without changing its percolative behavior. If a vertex $v \in D$ has no exit, there cannot be any infinite path originating at v. Thus, deleting v (and any incident edges) does not affect the probability of existence of an infinite open path. If a site v has no entrance, it may be in an infinite open path only if it is the initial vertex of the path. In this case, deletion of v leaves an infinite open path from an endpoint of an exit from v. Note that in a periodic graph, one periodic class may be deleted in such a way that the resulting graph is periodic.

Condition (a) insures the existence of a critical probability for the directed percolation model. Conditions (b) and (c) are used to define a filling process similar to the diode-resistor model, producing a candidate for the role of dual model. Condition (d) ensures that the filled wedge has an infinite boundary, which will correspond to a doubly-infinite path, and (e) ensures that this path is a directed path.

4. FORMULATION OF THE DUAL MODEL. Recall that in the diode-resistor model, there were two choices for the directions of the edges. Each choice provided a square lattice directed percolation model containing a doubly-infinite directed path corresponding to the boundary of the filled wedge. In either of the directed percolation models, the set of vertices that can be reached from a given vertex v is a quadrant. Two quadrants remain that cannot be reached from v in either model. The diode-resistor model was defined to fill one of these prohibited quadrants automatically. The construction of the general filling model is based on automatic filling of a region prohibited to a directed percolation model and its reverse.

The filling model is defined on the matching graph of the underlying non-directed graph G. To define the concept of matching graph, consider a planar graph M. To "close-pack" a face F of M, we insert a diagonal bond $\{v,w\}$ for each pair of vertices v and w of F which are not adjacent in M. Suppose that a graph G is constructed by close-packing all faces in some subset $\mathcal{F}$ of faces of M. The matching graph of G, denoted G*, is constructed by close-packing all faces in $\mathcal{F}^c$. Note that G and G* have identical vertex sets. The matching graph concept, introduced by Sykes and Essam (1964), plays an important role in the determination of critical probabilities for non-directed site models.

Two types of sites, restricted and unrestricted, will be present in the filling model. If a vertex v is an unrestricted site, each bond of G* incident to v is an exit from v. To describe the restricted site, for each vertex v we define the sector

$$R(v) = \{x \in \mathbb{R}^2 : x = v + (r \cos \theta, r \sin \theta), r \geq 0, \theta_2 \leq \theta \leq \gamma_1\}.$$

The boundary rays of R(v) coincide with two edges incident to v in a neighborhood of v. These edges border on a common face of G, denoted F(v). If v is a restricted site, then each edge of G* which is incident to v and another site w on F(v) is an exit from v. Note that it is possible for an edge $\{v,w\} \in G^*$ to be both an exit from v and exit from w. In this case, the edge $\{v,w\}$ is called undirected.

A filling path is an alternating sequence of vertices and edges $v_0, e_1, v_1, e_2, \ldots, e_n, v_n$, where e_i is either undirected or directed from v_{i-1} to v_i for each $i=1,\ldots,n$. An automatic filling path is a filling path in which e_i is an exit from v_{i-1} which lies in $R(v_{i-1})$ for each $i=1,\ldots,n$. In this case, e_i would be an exit from v_{i-1} even if v_{i-1} were a restricted site.

The wedge filled from v, denoted W(v), is the set of all sites w such that there is a filling path from v to w. The wedge automatically filled from v, denoted A(v), is the set of all sites w such that there is an automatic filling path from v to w.

It is convenient at times to visualize the filling process and automatic filling process in terms of faces. In this formulation, a site v that is filled either fills the face F(v) (if v is restricted) or fills all adjacent faces of G (if v is unrestricted). (See Figure 3.) Any site of a filled face is considered to be filled, from which the process may continue. With this formulation, one need not deal with matching graphs. In the following discussions, the formulation that is most convenient for each step will be adopted for that step.

A directed percolation model on D induces a filling model on G*. If v is open in the directed percolation model, it is a restricted site in the filling model. If v is closed in the directed model, it is an unrestricted site in the filling model.

5. PROOF OF THEOREM. The first step is to show that a doubly-infinite open path in the directed percolation model prevents the filling process from completely filling the graph.

Suppose that p is a doubly-infinite self-avoiding open directed path in

FIGURE 3. The two formulations of the filling process on the square lattice. Solid circles and open circles represent restricted and unrestricted vertices respectively. Arrows in (a) represent exit edges. Arrows in (b) indicate faces of the graph that are filled from the corresponding vertex. Filled faces are shaded in (b). s is the source site for the filling process.

(a)

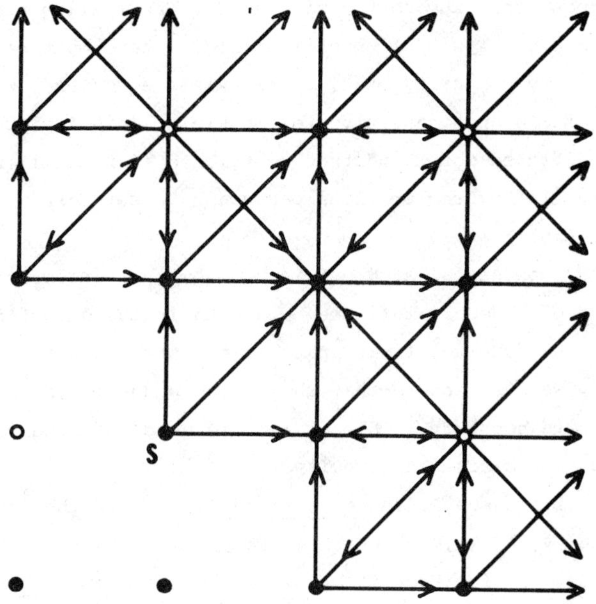

(b)

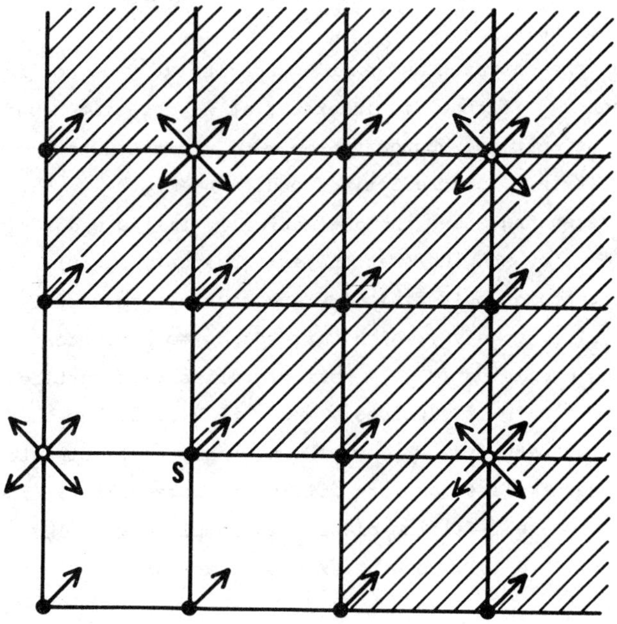

the directed percolation model. Note that p divides the plane into two connected components, L(p) and R(p), where L(p) lies on the left side and R(p) lies on the right side of p as p is travelled. Each site $v \in p$ has an entrance and an exit in p. The entrance lies in I(v) and the exit lies in O(v), so R(v) and hence F(v) lies between them in L(p).

Now suppose that the source site v_0 for the filling process is in L(p). If there is a site z in R(p) which is filled, there must be a filling path r from v_0 to z. By planarity of G, there exists a vertex y which lies on both r and p, which fills an adjacent vertex in R(p). This is impossible, since y is open in the directed model and thus restricted in the filling model, filling only sites in L(p) by the previous paragraph. Therefore, the wedge $W(v_0)$ is contained in $L(p) \cup p$.

The more substantial argument is to show that incomplete filling implies the existence of a doubly-infinite open path in the directed percolation model on D. Since the doubly-infinite path is to correspond to the boundary of the filled wedge, we must insure that the filled wedge is infinite in order to have an infinite boundary. This is a consequence of condition (d), that D contain no counterclockwise circuits.

In the filling model, the <u>exit boundary</u> at v is the infinite sequence path $v = v_0, e_1, v_1, e_2, v_2, \ldots$, where e_i is the edge incident to v_{i-1} in both $R(v_{i-1})$ and $O(v_{i-1})$ for each $i \geq 1$. The <u>entrance boundary</u> at v is the infinite sequence $v = v_0, e_1, v_1, e_2, v_2, \ldots$, where e_i is the edge incident to v_{i-1} in both $R(v_{i-1})$ and $I(v_{i-1})$ for each $i \geq 1$. The exit and entrance boundaries together bound the automatically filled wedge A(v). Note that in D the exit and entrance boundaries at v are directed paths, directed from v and toward v respectively.

Note that the entrance and exit boundaries at v do not intersect at any site other than v. If they did intersect at w, the exit boundary from v to w followed by the entrance boundary from w to v constitute a counterclockwise directed circuit in D, contradicting condition (d).

Similar reasoning shows the automatically filled wedge A(v) is infinite for every site v, by showing that the exit boundary is infinite. If the exit boundary at v contains only finitely many distinct vertices, it must contain a segment $v_k, v_{k+1}, \ldots, v_n = v_k$, which corresponds to a counterclockwise circuit in D, violating condition (d). A periodic class V of vertices is <u>self-filling</u> if there exists a pair of vertices v and w in V such that there is an automatic filling path from v to w. We now show that complete filling of any class of vertices implies complete filling of some self-filling class. Suppose that a non-self-filling class U is completely filled. Let $u \in U$, and let $u = v_0, e_1, v_1, e_1, e_2, \ldots$, be the exit boundary of u. Since there is a finite number N of periodic classes, the vertices $v_0,\ldots,v_N$ cannot be from distinct periodic classes. Thus, there exist $i < j \leq N$ with v_i and v_j in the same periodic class, which is then a self-filling class V. Using translates of the exit boundary from u to v_i, each site of V is automatically filled from a site of U, regardless of the configuration of edges. Thus, V is completely filled also.

Suppose that a self-filling class V is not completely filled. Then there exists a vertex $v \in V$ which is not filled. Since V is self-filling, there are vertices x and y in V such that there is an automatic filling path from x to y. Thus, there is an infinite automatic filling path ending at v, consisting of translates of the path from x to y, which passes through sites of the form $v - n(y-x)$, $n = 1,2,\ldots$. No vertex on this path is filled, since otherwise v would be filled also. In the alternate formulation, no face with an edge on the path is filled, so there is an infinite connected component of the plane which is not filled. Therefore, W(s), where s denotes the source site, has an infinite boundary which is doubly-infinite path in G. The fact that this path is the boundary of W(s) implies that all its vertices are restricted sites in the filling model, and thus open sites in the directed percolation model. It remains to show that it is a directed path in D.

Consider two adjacent vertices v and w on the boundary of W(s). The edge $\{v,w\} \in G$ borders two faces of G, one which is filled and one which is not filled. In terms of the automatic filling process, there are three possible relationships between v and w. Either v automatically fills w, w automatically fills v, or neither automatically fills the other.

If v automatically fills w, then since w is adjacent to v on G, it is on the boundary of A(v), the wedge automatically filled from v. Thus, the edge $\{v,w\}$ is directed so that A(v) is on the left, as desired. The same reasoning applies to the case when w automatically fills v.

Suppose that neither v nor w automatically fills the other. Then there exists another site z, unrestricted in the filling model, which fills both v and w. Viewing the process in terms of faces, z fills the face incident to v, w, and z. We need to show that that edge $\{v,w\}$ is directed so that the filled wedge is on the left.

The automatically filled wedges A(v) and A(w) may or may not intersect. If A(v) and A(w) do not intersect, the entrance and exit boundaries of v and w are all "parallel" in the following sense: Two exit or entrance boundary paths q and r are parallel if there exist vertices $q_1,q_2 \in q$ from a self-filling class and $r_1,r_2 \in r$ from a self-filling class for which the line segments from q_1 to q_2 and r_1 to r_2 are parallel. Since it is easily seen that exit boundaries of all sites have a common direction, as do all entrance boundaries, and the entrance and exit boundaries do not intersect, those of v and w as above must be parallel. Thus, if A(v) and A(w) do not intersect, there is an infinite strip of finite width between them, bounded by {v,w}, one exit boundary and one entrance boundary. In this case, since z fills v and w, the probability of each vertex being unrestricted is $p > 0$. By considering translates of a path from a vertex of A(v) to a vertex of A(w), each of which has the same positive probability of all sites being unrestricted, we see that the infinite strip is crossed by infinitely many filling paths. Hence, it does not contribute to the infinite boundary path of W(s) in G.

The union of A(v), A(w), and the edge {v,w} has an infinite boundary (disjoint from the boundary of the infinite strip if A(v) and A(w) do not intersect). Without loss of generality, suppose that this boundary consists of the entrance boundary of v, the exit boundary of w, and the edge {v,w}. If a dual model is to exist, {v,w} must be directed from v to w, so the filled region is on the left of {v,w}. To prove this by contradiction, we assume that {v,w} is directed from w to v. (See Figure 4.)

Under this assumption, by condition (c) there are no entrances to w in the sector between (w,v) to the exit boundary of w. Therefore, all entrances to w lie in the sector between the entrance boundary of w and (w,v), since there are no bonds strictly in the face F(w). By condition (e), there is a counter-clockwise doubly-infinite self-avoiding path r through (w,v), so it must contain an entrance to w through this sector. Similarly, r must exit v through an edge in the sector between the edge (w,v) and the exit boundary to v.

Since r_1, the portion of r preceding (w,v), is directed to w, the entrance boundary to w must lie between the exit boundary of w and r_1. Similarly, r_2, the portion of r following (w,v), is directed from v, so the exit boundary of v lies between r_2 and the entrance boundary of v. Since the path r is self-avoiding, r_2 lies between r_1 and the exit boundary of v.

If A(v) and A(w) intersect, the exit boundary of v and the entrance boundary of w intersect, so r is finite, contradicting condition (e).

If A(v) and A(w) do not intersect, the entrance and exit boundaries of v and w are parallel. The path r is contained in the half-infinite strip of finite width between A(v) and A(w). Thus r divides the plane into two components, but bounds the only component contained in a half-plane clockwise rather than counterclockwise, contradicting condition (e).

Therefore, {v,w} must be directed from v to w, so the filled region is on its left. Since this reasoning applies to every edge on the boundary of W(s), the filled region is bounded by a counterclockwise doubly-infinite self-avoiding open path in the directed percolation model on D.

FIGURE 4. (a) shows vertices v and w with the edge between directed to v. The automatically filled wedges A(v) and A(w) are shaded. The doubly-infinite path r passes through (w,v). Arrows indicate the directions of edges in D. (b) provides a directed graph in which the automatically filled wedges A(v) and A(w) do not intersect.

(a)

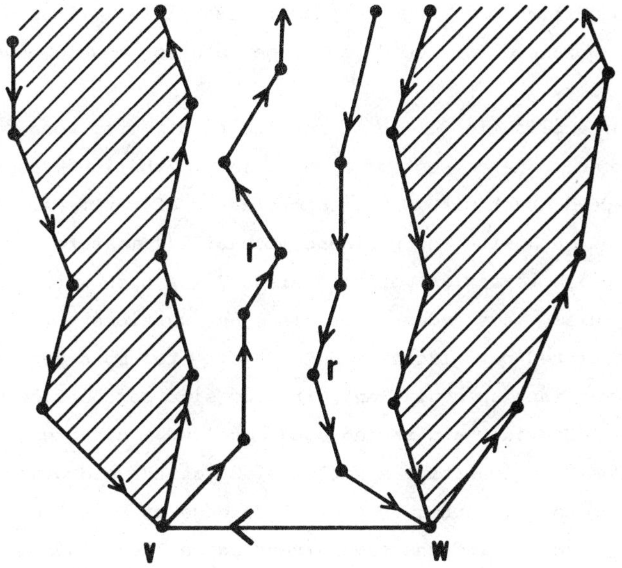

(b)

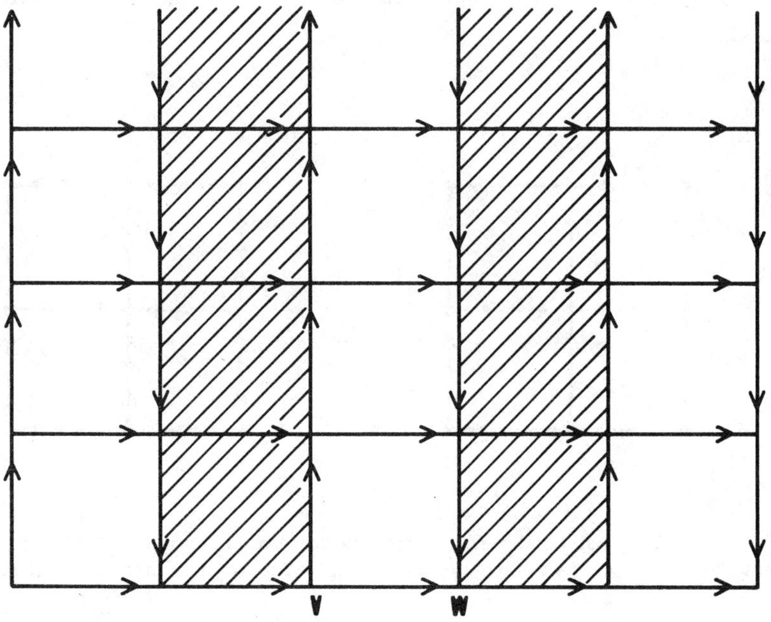

6. REMARKS. As in the diode-resistor model, one may define k-backflow paths, and find upper bounds for the critical probability of the filling model as the asymptotic slope of the wedge filled from the source site through k-backflow paths. Using the methods of first-passage percolation and subadditive process theory, it can be shown that the bounds converge to the critical probability as $k \to \infty$. The techniques of proof are illustrated in Wierman (1983) and Smythe and Wierman (1978), but may be simplified with the use of the ergodic theorem for subadditive processes due to Liggett (1984). Thus, the sequence of lower bounds obtained for the directed percolation model converge to its critical probability.

To consider application of these results to directed bond percolation models, we introduce the concept of a periodic probability measure. Let D be a periodic directed graph with N periodic classes of vertices. In the percolation model on D with the periodic probability measure with parameters $(p_1, p_2, \ldots, p_N)$, each vertex in periodic class V_i is open with probability p_i and closed otherwise, independently of all other vertices.

As for undirected percolation models, a directed bond percolation model may be transformed into an equivalent directed site percolation model on a different graph. Consider a directed bond model on a directed graph D. Construct D_p from D by inserting a vertex of D_p at the midpoint of each edge of D, in addition to the vertices of D. Each edge of D is thus divided into two edges of D_p, each having the same direction as the original edge of D. (See Figure 5.). Let each vertex of D_p that was inserted be open if and only if the corresponding edge of D is open. Let the vertices of D_p that are also vertices of D be open with probability one. Then an open directed path on D is transformed into an open directed path on D_p, yielding the equivalence of the models.

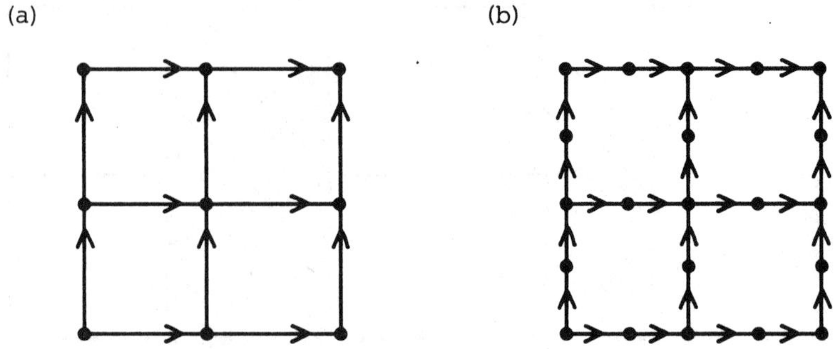

FIGURE 5. The directed bond percolation model on the square lattice in (a) is equivalent to the directed site percolation model on the graph in (b).

Note that if the original graph D is planar and periodic, the resulting graph D_p is also planar and periodic. A bond model on D_p with a periodic probability measure will transform into a site model on D_p with a periodic probability measure.

In the terminology of the classical non-directed percolation models, D_p is obtained by first constructing the "covering graph" of D, introduced by Fisher and Essam (1961), directing or deleting edges to be consistent with D, then taking its planar modification (see Kesten (1982), Chapter 2).

Extensions of the duality result of this paper are being investigated. It is reasonably clear that in the presence of conditions (a) - (d), condition (e) is necessary and sufficient for the existence of a dual filling model. There are examples of percolation models on graphs with both directed and non-directed bonds for which a dual filling model exists. See, for example, the "stiff" model in Smythe and Wierman (1978), Chapter 9. There are also two-dimensional directed graphs which are nonplanar for which duals exist. The results may also be extendable to irregular graphs based on points in a two-dimensional point process.

BIBLIOGRAPHY

1. Bishir, J., "A lower bound for the critical probability in the one-quadrant oriented-atom percolation process", J. Royal Stat. Soc. B, 45 (1963), 401-404.

2. Blease, J., "Directed bond percolation on hypercubic lattices", J. Phys. C, 10 (1977), 925-936.

3. Dhar, D., "Percolation in two and three dimensions I", J. Phys. A, 15 (1982), 1849-1858.

4. Dhar, D., and M. Barma, "Monte Carlo simulation of directed percolation on a square lattice", J. Phys. C, 14 (1981), L1-L6.

5. Dhar, D., M. Barma, and M. Phani, "Duality transformations for two-dimensional directed percolation and resistance problems", Phys. Rev. Letters, 47 (1981), 1238-1241.

6. Durrett, R., "Oriented percolation in two dimensions", Ann. Probability 12 (1984), 999-1040.

7. Fisher, M. E., and J. W. Essam, "Some cluster size and percolation problems", J. Math. Phys., 2 (1961), 609-620.

8. Gray, L., R. T. Smythe, and J. C. Wierman, "Lower bounds for the critical probability in percolation models with oriented bonds", J. Appl. Prob., 17 (1980), 979-986.

9. Hammersley, J. M., "Bornes superieurs de la probabilité critique dans un processus de filtration", Le calcul des probabilités et ses applications. CRNS Vol. 87 (1957), 17-37.

10. Hammersley, J. M., "Comparison of atom and bond percolation processes", J. Math. Phys., 2 (1961), 728-733.

11. Holley, R. A., and T. M. Liggett, "Ergodic theorems for weakly interacting infinite systems and the voter model", Ann. Probability, 3 (1975), 643-663.

12. Kertesz, J., and T. Vicsek, "Oriented bond percolation", J. Phys. C, 13 (1980), L343-L348.

13. Kesten, H., Percolation theory for mathematicians, Birkhauser, Boston, 1982.

14. Kinzel, W., and J. Yeomans, "Directed percolation: a finite size renormalization approach", J. Phys. A, 14 (1981), L163-L168.

15. Liggett, T., "An improved subadditive ergodic theorem", 1984, preprint.

16. Mauldon, J., "Asymmetric oriented percolation in the plane", Proc. Fourth Berkeley Symposium, I (1961), 337-345.

17. Smythe, R. T., and J. C. Wierman, First-passage percolation on the square lattice, Lecture Notes in Mathematics Vol. 671, Springer-Verlag, Berlin, 1978.

18. Sykes, M. F., and J. W. Essam, "Exact critical percolation probabilities for site and bond problems in two dimensions", J. Math. Phys., 5 (1964), 1117-1127.

19. Wierman, J. C., "On square lattice directed percolation and resistance models", J. Phys. A, 16 (1983), 3345-3351.

MATHEMATICAL SCIENCES DEPARTMENT
JOHNS HOPKINS UNIVERSITY
BALTIMORE, MARYLAND 21218

ABCDEFGHIJ – AMS – 898765